Differences between means:

Large samples: $$z = \frac{\bar{x}_1 - \bar{x}_2}{\sqrt{\dfrac{\sigma_1^2}{n_1} + \dfrac{\sigma_2^2}{n_2}}} \approx \frac{\bar{x}_1 - \bar{x}_2}{\sqrt{\dfrac{s_1^2}{n_1} + \dfrac{s_2^2}{n_2}}}$$

Small samples: $$t = \frac{\bar{x}_1 - \bar{x}_2}{\sqrt{\dfrac{(n_1-1)s_1^2 + (n_2-1)s_2^2}{n_1+n_2-2}}\sqrt{\dfrac{1}{n_1}+\dfrac{1}{n_2}}}$$ (and use d.f. $= n_1 + n_2 - 2$)

Hypotheses concerning variances:

$F = \dfrac{s_1^2}{s_2^2}$ (if $s_1^2 > s_2^2$) or $F = \dfrac{s_2^2}{s_1^2}$ (if $s_2^2 > s_1^2$), with d.f. $= n_1 - 1,\ n_2 - 1$

One-way ANOVA: $$SS_{\text{total}} = \sum(x^2) - \frac{(\sum x)^2}{n}$$

$$SS_{\text{factor}} = \frac{\sum(T^2)}{c} - \frac{(\sum x)^2}{n}$$

$$SS_{\text{error}} = SS_{\text{total}} - SS_{\text{factor}}$$

$$\text{d.f.}_{\text{total}} = n - 1,\ \text{d.f.}_{\text{factor}} = r - 1$$

$$\text{d.f.}_{\text{error}} = \text{d.f.}_{\text{total}} - \text{d.f.}_{\text{factor}} = r(c-1)$$

$$MS_{\text{factor}} = \frac{SS_{\text{factor}}}{\text{d.f.}_{\text{factor}}},\ MS_{\text{error}} = \frac{SS_{\text{error}}}{\text{d.f.}_{\text{error}}}$$

$$F = \frac{MS_{\text{factor}}}{MS_{\text{error}}}$$

Chi-square: $\chi^2 = \sum \dfrac{(A-E)^2}{E}$ For contingency table application, d.f. $=$ (# rows $-$ 1)(# columns $-$ 1).

Coefficient of correlation: $r = \dfrac{n\sum(xy) - (\sum x)(\sum y)}{\sqrt{n\sum(x^2) - (\sum x)^2}\sqrt{n\sum(y^2) - (\sum y)^2}}$ with d.f. $= n - 2$

Regression line: $y = bx + a$, where $b = \dfrac{n\sum(xy) - (\sum x)(\sum y)}{n\sum(x^2) - (\sum x)^2}$

$$a = \frac{\sum y \sum(x^2) - \sum x \sum(xy)}{n\sum(x^2) - (\sum x)^2}$$

Sign test: $z = \dfrac{2x - n}{\sqrt{n}}$ with continuity correction

Mann-Whitney *U*-test: $U = n_1 n_2 + \dfrac{n_1(n_1+1)}{2} - R_1$ or $U' = n_1 n_2 + \dfrac{n_2(n_2+1)}{2} - R_2$

$$\mu_U = \frac{n_1 n_2}{2},\quad \sigma_U = \sqrt{\frac{n_1 n_2(n_1+n_2+1)}{12}},\quad z = \frac{U - \mu_U}{\sigma_U}$$

Kruskal-Wallis *H*-test: $H = \dfrac{12}{n(n+1)}\left(\dfrac{R_1^2}{n_1} + \dfrac{R_2^2}{n_2} + \ldots + \dfrac{R_k^2}{n_k}\right) - 3(n+1)$

and compare with χ^2 (d.f. $= k - 1$)

Runs test: $\mu_R = \dfrac{2n_1 n_2}{n_1+n_2} + 1,\quad \sigma_R = \sqrt{\dfrac{2n_1 n_2(2n_1 n_2 - n_1 - n_2)}{(n_1+n_2)^2(n_1+n_2-1)}},\quad z = \dfrac{R - \mu_R}{\sigma_R}$

Spearman's rho (rank correlation coefficient): $r_s = 1 - \dfrac{6\sum(d^2)}{n(n^2-1)},\quad z = r_s\sqrt{n-1}$

Elementary Statistics

Winthrop Publishers, Inc. Cambridge, Massachusetts

Donald R. Burleson

Middlesex Community College
Bedford, Massachusetts

Elementary Statistics

Library of Congress Cataloging in Publication Data

Burleson, Donald R
Elementary statistics.

Bibliographies.
Includes index.
1. Statistics. I. Title.
QA276.12.B87 519.5 79-19885
ISBN 0-87626-213-2

Design by David Ford

17 Dunster Street, Cambridge, Massachusetts 02138

 Printed in the United States of America.

10 9 8 7 6 5 4 3 2 1

This book is dedicated to

HOWARD PHILLIPS LOVECRAFT

who is the central intellectual force in my life.

Contents

Exercises and Examples in Particular Areas of Application (P="Ponderables" Problems; E=Examples)

Chapter, Section		Natural Sciences	Social Sciences	Business-Economics	Education	General Applications
1	A	P8		P2	9	6
	B	3, 4				E1, E2
	C					1, 4
2	A	17		10	9, 18, 20, E6	1, 2, 3, 4, 5, 16, E7, E13
	B		P10		6, 9	E7, E13
	C	E5, E8	3, 9, E7	P1		4, 12, P2
3	A	5	7, 10, P2	8, P7		9, 11, 12, P1 P3, E8, E9, E10
	B	5	3, E1	1		3, E3
4	A	8, 11, P4	1, 7, 14, P6, P8	3, 9, 13, E6		5, P9, E4, E8 E9
	B	1, 5, 6	P5, E1, E3	3, 4, E2, E4		2, 7, 8, P2
	C	5	8, 13, E3, E4	1, P1	10, 11	E1, E2, E5, E6
5	A	5, 7, 9, P1, P3	14, 15, P2	1, 11, 13, 16, 17, 19, 20, 21, 22, E2, E3, E7, E8		3, E1, E4, E5
	B	P4, P7	11, 12, 13	1, 2, 7, 8, 9, 10, 15, 17, P1, P9		3, 4, 5, 6, 19, 20, P3, P5, E1, E2, E3, E4, E5, E6, E7
	C	14, E3, E4	7, 9, 12, 13, P1, P4	1, 2, 16, 18, P2, P6	10, 11	3, 4, 5, 6, 20, E1, E2, E5, E6
	D		3, 8	1, 2, 10		4, 5, 9, E1, E2, E3
6	A	2, 11	1, 12, 14, P1, P2, P3	5, 6, 7, 9, 16	15, E1	3, 8, 10, E2
	B	12, 14	10, 13, 15, 16, 17, P3, E3	6, 7, 8, 18, E4	3, 11	1, 2, 4, P4, E1
7	A	5, 9	6, 7, 8, 11, E1	4, E3	1, 3	2, 10, E4
	B	1, 3, 4, 5, 6, 7, 10, 11, P1, P2	8, 9, E1		2	E2
8	A		8		3, 6, 7, E1, E2	1, 2, 5
	B	10	E1	1, 2	5	3, 6, 7, 9, E2
	C	6	4	3, E1		1, 2, 5, 7, 9
	D		6, 10	1, 2		3, 4, 5, 7, 8, 9, E2
	E	6		8	1, 2, 7	3, 4, 5, 9, 10, E1, E2

Topics Introduced in "Ponderables"

1	**A**	Geometric mean, harmonic mean, midrange, quartiles, midquartile
	B	Square root interpolation, mean deviation, semi-interquartile range (Q)
	C	Frequency distribution standard deviation, frequency distribution mode
2	**B**	Odds, conditional probability, independent events
	C	Hypergeometric distribution
3	**A**	Tchebycheff's Theorem
5	**A**	Small-sample t-test
	B	Two-sample t-test
	C	Differences between means
	D	Double interpolation in F tables
6	**A**	Factorial design, interaction, two-way ANOVA
	B	Yates' correction
8	**A**	Sign test for the median

Preface

This book is designed to serve a standard one-semester general statistics course at the college freshman level, for students with perhaps only a very modest background in mathematics.

Having written a number of mathematics textbooks and other books, and having taught statistics and a great variety of other courses for many years, I have again employed here an approach that I have long found effective both in writing and in teaching: an informal and conversational exposition, with examples worked out in detail at every turn to illustrate what has just been introduced and explained. I have made every effort to avoid "skipped steps" and "gaping holes" in the discussions, so that the student has the best chance of following the examples smoothly. With this discussion-and-examples format the student should be encouraged to go through all the examples carefully, because they have been designed to bring out various important aspects of the topics in question, as have the exercises. Many examples and exercises are object lessons in such matters as the effect of sample size on the computations.

I have sought to make the book flexible, both in terms of course content and level. The number of topics available should make it easy for the instructor to select whatever is desired, and the exercise sets are plentiful enough and the discussion inclusive enough to make it possible to stress any particular topic as needed. In fact, the exercises themselves provide considerable flexibility in course level, because each section includes a regular exercise set and special problems called "Ponderables." These addititional problems offer either more challenging questions about the section or an extension of the topics covered, or both. The instructor may assign these or not as preferred, or may use them for classroom discussion.

For their kind assistance I wish to thank all those at Winthrop Publishers who have worked with me on the book's production, as well as the reviewers who have made useful suggestions in manuscript. I want to express appreciation to my colleagues and students at Middlesex Community College for their words of encouragement. Finally, but especially, I want to thank my wife Sue and my sons Bruce and Brian for the patience and understanding that they have shown once again during a long and difficult time; I well know that when a community college instructor writes a textbook and keeps up with a heavy teaching load, the demands on the family are inordinate.

Donald R. Burleson

Elementary Statistics

1 Statistics: A Beginning

Well begun is half done.—Horace

A MEASURING CENTRAL TENDENCY

Probably it would surprise no one to say that we live in an increasingly complex world—and a world increasingly *numerical*. We are surrounded, in our daily lives, by numerical information. Advertisers hurl conflicting figures and claims at us from all sides, parading data before us to sell us their products. Our banking facilities and our insurance policies are founded on stunning amounts of numerical analysis. Our jobs often call for struggling through mazes of numerical information. Our physical health depends to no small degree upon the applications of numerical methods to medical science, to the production and distribution of food, to the preservation of the environment. Every day, weighty decisions are made in business and government—from the village store to the giant corporation, from the town level to the international level—by considerations of numerical data. Our cities flourish because of intelligent numerical planning and decision making, or crumble because of the lack of it. Our industries, our governments, all facets of our lives are shaped by the manipulation of numerical information.

It is therefore natural that the science of *statistics* has evolved as one of the most important of all human endeavors. In this book we will examine some of the concepts and procedures belonging to this far-reaching subject, to gain an acquaintance with the fundamentals of its applications to human life.

Statistics has so many aspects and so many uses that a brief definition of it, although necessary, is by itself unlikely to convey any true picture of its real nature. However, we may start by defining **statistics** as

the science of organizing and analyzing numerical data for the purposes of description and decision making.

That is, statistics concerns itself with certain numerical *descriptions* of the world, or parts of the world, and with the making of *decisions* on numerical grounds—intelligent decisions that often must be made earlier than we would have liked, and on the basis of only partial information.

Let us look at a few preliminary terms and ideas.

First, the total body of information that we can be concerned about in any given situation—the "sea of information" on whose shores we stand looking and wondering—is called a **population.**

In statistics a population may or may not have anything to do with people. For example, a population *may* be all the people in Los Angeles, or all the people in California, or in the United States, or in the world. Or it may be all the catsup bottles that come off a conveyor belt in a given year, or the heights of all the trees in a certain forest, or all the members of a certain rare family of butterflies, or all the possible readings on a meter for a chemical experiment. In short, a population may be a lot of things—it is, in any given situation, the *total* body of information that one would theoretically have to deal with.

The trouble with populations is that they tend to be *large*. Sometimes they are enormous, in fact. If the population in question, for example, is composed of all the actual weights of the schoolchildren in a given country, that population may contain millions of numbers. (A population, for that matter, may even be *infinite*. For example, if a population consists of all the *possible* weights of schoolchildren, then it literally contains infinitely many numbers.)

For this reason, statisticians concern themselves with *sampling*.

A **sample,** as one would expect, is a selection of items taken from a population. For example, a sample could consist of 100 schoolchildren selected from among *all* the schoolchildren in Massachusetts (a population consisting of many thousands). Or a sample could consist of 250 observations of an experiment that can be performed any number of times: a sample of size 250 taken from an infinite population.

The reason for sampling is clear: a population, typically, is very large, so that the statistician cannot possibly hope to handle *all* the information about it, but must be content with looking only at *some* of the information (a sample).

A statistician hopes that the sample studied will paint some reasonably accurate and enlightening picture of the population from which it was taken. This is one of the ways in which statistics makes it possible to make informed decisions based upon partial information. But if this approach is to work sufficiently well, the sample taken should (if possible) be of good size, and should be *random*.

A sample is **random** if it is drawn from the population in such a way that no one could have predicted which items would be selected and which would not, each item having the same chance of being selected. This selection, needless to say, must be done with some care. For example, if out of a population of accounts in file folders we wish to select one folder at random (to audit that account, say), we may assign numbers to the folders and then take a number from a table of *random digits*, and select that corresponding folder (or keep taking random digits until one set of digits matches a folder number). This way, no one could say, "You picked my folder because you don't like me!" because no one could have predicted which folder we would select. On the other hand, it would *not* be a random procedure simply to close our eyes and reach into the drawer containing the alphabetized folders and pick one, because (for example) we would be more likely to choose a folder starting with M in the middle of the drawer than to pick one starting with A or Z.

If a sample is selected in such a way that it is *not* random, then we may say that it is **biased.** For a very obvious example, suppose we are sampling people and measuring their heights to gain an idea of the public's heights in general. If we were to take our sample by stopping basketball players on their way to and from the locker room, our sample would be so thoroughly biased that it would provide no real idea of most people's heights.

Now, suppose that from some population, a random sample of numbers is taken.

One aspect of the sample which we may wish to measure is **central tendency:** We may wish to determine, in some sense, where the "middle" of the data is. And to do so, we should compute some sort of *statistic* that measures central tendency. In popular terminology, we want to compute some sort of "average" for the sample numbers. It turns out that there are a number of different ways to do this. (In computing an "average" of some sort for a sample of numbers drawn from a population, we become involved with **descriptive statistics:** computations which simply *describe* some aspect of a sample or a population. This descriptive aspect of statistics we will look at in various forms before pursuing **inferential statistics:** the science of drawing probable inferences or conclusions about populations by analyzing samples. More about this later.)

The most common kind of "average"—the better term being "measure of central tendency"—is called the **mean** (also sometimes called the **arithmetic mean** to distinguish it from certain other means).

Given a set of sample numbers

$$\{x_1, x_2, x_3, \ldots, x_n\}$$

(where n is the number of items in the sample), the *mean* of the sample is denoted $\bar{x}$ and is defined to be

$$\bar{x} = \frac{x_1 + x_2 + x_3 + \ldots + x_n}{n}.$$

That is, we simply add the numbers and divide by the number of things added. (This is what most people naturally think of when the term "average" is used.) Because it is a little awkward to have to write the sum of the x's "strung out" as shown, we may more compactly write

$$x_1 + x_2 + x_3 + \ldots + x_n = \sum x$$

where the $\sum$ (Greek capital letter *sigma*) simply means "the sum of . . . ." Thus

$\sum x$ (read "sigma x") means "the sum of all the x's." Thus we can write the formula for the sample mean $\bar{x}$ more simply as

$$\bar{x} = \frac{\sum x}{n}.$$

EXAMPLES

1. Given the eleven sample numbers

$$\{52, 57, 47, 50, 40, 41, 52, 42, 40, 59, 47\}$$

the sample mean is computed as

$$\bar{x} = \frac{\sum x}{n} = \frac{\sum x}{11}$$

$$= \frac{52+57+47+50+40+41+52+42+40+59+47}{11}$$

$$= \frac{527}{11} = 47.909090\ldots \approx 47.9$$

(where $\approx$ means "is approximately equal to"). We have retained the mean to the nearest tenth here in keeping with a general principle:

Always compute a sample statistic correct to one *more* decimal place than the given data.

(Thus, having given data in integer form calls for a mean to the nearest tenth.)

2. Given the eight sample numbers

$$\{23.09, 38.20, 29.71, 24.58, 31.34, 28.12, 34.50, 22.76\}$$

the sample mean is

$$\bar{x} = \frac{\sum x}{n} = \frac{232.30}{8} = 29.0375,$$

which must be rounded off. Because the given data are in hundredths, we should round this result off to the nearest thousandth; again, to one more decimal place than the given numbers. The question here is whether to round off to 29.037 or 29.038 (because 29.0375 is exactly halfway between them). The general practice, when a number occurs halfway like this, is to round the number off to an *even* digit. Thus we would conclude

$\bar{x} \approx 29.038.$

(But, for instance, 29.0765 would have been rounded to 29.076.)

There are other ways, besides the mean, to measure central tendency. One such way is called the **median,** denoted $\tilde{x}$. To compute the median $\tilde{x}$ for a set of numbers, we first arrange the numbers *in order,* say smallest to largest, then count to the physical center of the list and take the number encountered there. (If *two* numbers are encountered there, we take the mean of those two.) Thus, in the ordered list, just as many numbers will precede the median value as will follow it. Formally, the median stands in position $\frac{n+1}{2}$ in the list, where n as usual is the number of numbers involved. Thus in a list of 25 numbers the median is the number standing in position $\frac{25+1}{2} = 13$ in the list, that is, the thirteenth number. In a list of 30 numbers, the median stands in position $\frac{30+1}{2} = 15.5$, that is, halfway between the fifteenth number and the sixteenth.

$\tilde{x}$ = the score which divides the ordered list of data into two equal parts.

EXAMPLES

3. Given the 13 sample numbers

{81, 70, 67, 76, 88, 84, 67, 83, 78, 81, 76, 73, 74}

to find the median $\tilde{x}$ we first arrange the numbers in order:

{67, 67, 70, 73, 74, 76, 76, 78, 81, 81, 83, 84, 88}

and, counting to the middle and finding the seventh number $\left(\text{the number in position } \frac{n+1}{2} = \frac{13+1}{2} = 7\right)$ to be 76, we conclude that

$\tilde{x} = 76.0.$

Again, we have expressed a sample statistic to one more decimal place than the given data.

4. Given the 14 sample numbers

{35.6, 38.1, 31.7, 25.4, 38.5, 25.9, 28.1, 34.7, 35.8, 37.2, 30.5, 28.1, 35.3, 28.1} to find the median $\tilde{x}$ we order the numbers:

{25.4, 25.9, 28.1, 28.1, 30.5, 31.7, 33.6, 34.7, 35.3, 35.6, 35.8, 37.2, 38.1, 38.5}

and count to the middle, finding the *two* numbers

33.6, 34.7

so that we are obliged to take their mean. The desired median is

$$\tilde{x} = \frac{33.6+34.7}{2} = \frac{68.3}{2} = 34.15.$$

We express this statistic in hundredths, one more decimal place than the given data, which were expressed in tenths.

Thus the not very informative word "average" can signify at least two different things: mean and median. There are others. Unfortunately, we are sometimes not told, when the word "average" is used, *which* meaning was intended, and there is even a possibility of deliberate manipulation (e.g., reporting the median instead of the mean, or vice-versa, just because it makes the speaker's case sound better). An honest commentator will specify *what* measure of central tendency is intended if the word "average" is used. After all, $\bar{x}$ and $\tilde{x}$ do not have to turn out equal to each other. For example, for the numbers

{75, 91, 84, 60, 48, 82, 78, 71, 77, 71}

the mean is

$$\bar{x} = \frac{737}{10} = 73.7$$

but the median is

$$\tilde{x} = \frac{75+77}{2} = 76.0.$$

In some cases, the difference between the two may be quite considerable. For example, suppose we measure the diameters of the tree trunks in a certain little grove and find them to be (in centimeters)

{32, 36, 34, 29, 39, 38, 40, 31, 27, 37}.

The mean is $\bar{x} = \frac{343}{10} = 34.3$, and the median is $\tilde{x} = \frac{34+36}{2} = 35.0$, very nearly the same. *But:* Suppose, standing at the edge of the grove, there is an enormous tree whose trunk diameter is 215 centimeters (or 2.15 meters). If this new tree were added to the sample as an eleventh item, then the new mean would be

$$\bar{x} = \frac{558}{11} \approx 50.7 \text{ centimeters}$$

(a considerable change from 34.3), but because the numbers would then be arranged

$$\{27, 29, 31, 32, 34, 36, 37, 38, 39, 40, 215\}$$

the median $\tilde{x}$ would only be shifted from 35.0 to 36.0. The introduction of the "atypical" number 215 has distorted the original mean far more than it has distorted the original median, and in this situation it may well be that the median paints a more informative picture for us (concerning what *most* of the trees look like) than the mean.

However, the mean is often preferred over the median in statistical analysis because of its stability—its *reliability.* That is, in certain circumstances we can rely on it to have certain kinds of values. If we repeatedly take different random samples (at a fixed sample size) from the same population and compute their sample means, these means will tend to fluctuate by chance much less than the medians would. For that reason, sample means, together with other information, may be put to good use for estimating certain characteristics of the population itself. More of this later.

One further measure of central tendency which we may examine here is the statistic called the **mode,** which will be denoted $\hat{x}$. The mode of a set of numbers is simply that number (if any) which occurs more often than any of the others. The mode may or may not exist, and when it exists may or may not be unique.

EXAMPLES

5. For the set of sample numbers

$$\{38, 53, 49, 38, 47, 38, 50, 47, 38, 45\}$$

the mode is

$$\hat{x} = 38.0$$

because 38 occurs four times and none of the other numbers occurs as many as four times.

6. For the set of sample numbers

{29, 31, 35, 34, 38, 30, 33, 37, 39, 32, 36, 40}

there is *no* mode, because no number occurs even twice.

7. For the set of sample numbers

{81, 88, 90, 88, 79, 82, 88, 79, 74, 95, 79}

either 88 or 79 would be an equally compelling candidate to be called the mode, because each one occurs three times (and nothing else occurs that many times). There is, then, no unique mode. Because there are two candidates for it, this set of data may be said to be **bimodal.**

A mode may be discerned among nonnumerical items, of course. If you ask 30 people what their favorite color is, and if 5 of them say red, 13 say green, 4 say yellow, and 8 say blue, then for that sample, green is the "modal" (most popular) color.

Occasionally the mode of a set of numbers may have a particularly appealing role as a measure of central tendency. For example, if in a company of 20 employees, 19 of them are paid $10,000 a year and the twentieth person (the owner) is paid $50,000 a year, then the *mode* of the 20 salaries is clearly $\hat{x} = \$10{,}000$. And that "average" salary certainly conveys a more realistic picture to a prospective new employee than the mean would: $\bar{x} = \$12{,}000$. However, in most situations the mean would be the preferred statistic, because, taking all the data into account in its computation, it tends to be more informative.

NEW TERMS IN THIS SECTION

(Starred items occur only in the special "Ponderables" problems.)

statistics
population
sample
random
bias
central tendency
descriptive statistics
inferential statistics
(arithmetic) mean
median
mode
bimodal data
***geometric mean**
***harmonic mean**
***midrange**
***quartiles**
***midquartile**

EXERCISES 1-A

1. Find the mean of each of these sets of sample numbers, expressing your results correct to the appropriate numbers of decimal places:
(a) {15, 19, 11, 26, 14, 29, 20, 19, 26, 25, 18, 16}

(b) {53, 47, 56, 60, 49, 57, 45, 61, 59, 57, 53}
(c) {35.2, 29.9, 31.8, 37.3, 27.4, 24.7, 33.8}
(d) {107, 112, 103, 98, 100, 116, 94, 101, 103}
(e) {99.7, 96.4, 103.2, 90.1, 88.6, 92.4, 105.8, 93.2}
(f) {435, 406, 471, 422, 417, 448, 459, 401, 397, 466, 428, 451, 433}
(g) {21.59, 30.41, 25.77, 27.90, 31.23, 20.78, 36.60, 28.22}
(h) {0.006, 0.017, 0.019, 0.012, 0.008, 0.007, 0.028, 0.013, 0.026}

2. Find the median $\tilde{x}$ for each of the sets of sample numbers in Exercise 1-A.1.

3. For each of these sets of numbers, compute and compare the mean and the median; comment on the observed differences.
(a) {58, 51, 55, 47, 264, 43, 46, 50}
(b) {739, 750, 732, 728, 744, 15, 735}
(c) {0.081, 0.586, 0.077, 0.073, 0.092, 0.084, 0.071, 0.088}

4. For each set of numbers in Exercise 1-A.3, remove the "atypical" or unusual number and recompute the mean and the median for the remaining numbers.

5. If a set of numbers consists of the heights of the members of a fourth-grade school basketball team, together with the height of their school coach, what measure of central tendency do you suspect one should use to express the "average" height?

6. In a certain company, 6 employees have had no college education, 9 have had one year of college, 13 have had two years, 11 have had three years, and 8 have had four years; this count includes all the company's employees. What is the "average" number of years of college for these employees in terms of
(a) the mean?
(b) the median?
(c) the mode?

7. Find $\bar{x}$, $\tilde{x}$, and $\hat{x}$ (if any) for these numbers:

22	31	41	33	27	36	29
42	29	17	41	35	18	33
32	43	29	38	32	29	21
21	35	34	38	17	47	31
29	50	26	48	29	37	30
47	20	43	33	45	23	29
46	29	27	46	32	36	24

8. In terms of random sampling, what is the major strength of the *mean* as a measure of central tendency?

9. Three students (Jones, Smith, and Brown) have taken 10 quizzes, with the resulting scores as follows.

Jones: 86, 74, 75, 80, 84, 80, 70, 78, 80, 79.

Smith: 96, 77, 81, 75, 77, 94, 76, 77, 98, 97.
Brown: 75, 86, 88, 96, 75, 83, 72, 75, 89, 91.

How would the three students be ranked (highest to lowest) in terms of
(a) the means?
(b) the medians?
(c) the modes?

PONDERABLES 1-A

1. Can a *median* ever be "computed" for nonnumerical data? How could it be done, and under what circumstances?

2. Why do we hear (for example) that the "average family" in a certain city has 2.4 children (presuming that there are in fact no fractional children running about)?

3. Suppose we know all of the values in a set of 11 sample numbers except one (which we will denote x), as follows:

$\{48.3, 51.7, 43.9, 40.8, 55.1, 52.6, 44.3, 46.4, 47.2, 54.8, x\}$

and suppose we know that the mean of the 11 numbers is $\bar{x} = 49.2$. Can the eleventh number x be determined? If so, how, and what is it?

4. Can a similar thing be done (as in the previous question) with the median? The mode? Why or why not?

5. Given a set of numbers, how is the mean $\bar{x}$ affected if each of the numbers is increased by a constant k? If each of the numbers is multiplied by a constant k? Can you prove your answers?

6. What can be said for the mean and the median of a set consisting of two numbers? What about the mode?

7. Another (rather infrequently seen) measure of central tendency is the **geometric mean,** which we may denote x_g. For a given set of numbers

$\{x_1, x_2, x_3, \ldots, x_n\}$

the geometric mean is

$$x_g = \sqrt[n]{x_1 \cdot x_2 \cdot x_3 \cdot \ldots \cdot x_n},$$

that is, the nth root of the product of the numbers. For two numbers ($n = 2$) we would take *square* root, for three numbers ($n = 3$) we would take *cube* root, and so on. Find the geometric mean and the regular (arithmetic) mean for:
(a) $\{8, 18\}$
(b) $\{5, 8, 25\}$

8. Still another measure of central tendency is the **harmonic mean,** which we will denote x_h. For a given set of nonzero numbers

$$\{x_1, x_2, x_3, \ldots, x_n\}$$

the harmonic mean is defined to be

$$x_h = \frac{n}{\Sigma \frac{1}{x}},$$

that is, the number of numbers divided by the sum of their reciprocals. The harmonic mean is employed only in very specialized circumstances. For example, it can be shown that for a set of numbers representing the amounts of time (all in the same units) required for a sequence of pipes to fill a tank with water (each pipe working alone), the amount of time required for *all* the pipes (working together) to fill the tank is

$$\frac{1}{n} \cdot x_h = \frac{1}{\Sigma \frac{1}{x}}.$$

If there are five pipes, and if the times required are:

Pipe	Minutes
A	20
B	40
C	25
D	50
E	10

compute x_h for the five numbers and determine how many minutes a tank would require to be filled if all five pipes were working.

9. One very simple measure of central tendency is the **midrange:** the mean of the largest number and the smallest number present. Clearly the midrange is of limited significance because it ignores all but the largest number and the smallest number. Construct an example to show that two samples may have the same midrange but different *medians.*

10. Related to the concept of *median* as a measure of central tendency (where the median is found by "counting to the middle" of the ordered data and taking the value which stands in position $\frac{n+1}{2}$), we have also a concept called **quartile,** a measure of location. Quartiles are similar to the median, but divide the ordered data up into fourths instead of halves. Given a set of data ordered from smallest to largest, the *first quartile,* denoted Q_1, is the value lying at position $\frac{1}{4}(n+1)$ in the list, so that one-fourth of the data precede Q_1 in the list. The *second quartile* Q_2 lies in position $\frac{2}{4}(n+1) = \frac{n+1}{2}$ and is identical to the median $\tilde{x}$. The *third quartile* Q_3 is

the value lying at position $\frac{3}{4}(n+1)$ in the list, so that three-fourths of the data lie previous to Q_3 in the list. Thus, for example, given the thirteen numbers

$\{35, 36, 36, 39, 41, 43, 43, 45, 47, 47, 48, 48, 51\}$

the first quartile Q_1 lies in position $\frac{1}{4}(13+1)=3.5$, halfway between the third and fourth numbers:

$Q_1=37.5$

and the third quartile Q_3 lies in position $\frac{3}{4}(13+1)=10.5$, halfway between the tenth and eleventh numbers:

$Q_3=47.5.$

Find Q_1 and Q_3 for each set of data in Exercise 1-A.1. Keep one more decimal place than in the given data, and remember if necessary to round off numbers ending in 5 so that the results are even.

11. Another measure of central tendency is the **midquartile:**

$\frac{1}{2}(Q_1+Q_3).$

Find an example to show that the midquartile is *not* necessarily the same as the median.

B MEASURING DISPERSION

In the previous section we examined the question of measuring central tendency for a sample of numerical data, finding that the *mean* is one common and reliable statistic for doing so, but also that there are other statistics (the median, the mode, and even some others) that measure central tendency.

But it turns out that central tendency is definitely not the only aspect of a sample of numbers that we may need to measure. Consider, for example, the following.

Suppose you and I each have a car and we keep track of our gasoline mileage during a certain period of time, each buying gas 10 times. Suppose your mileage figures (in miles per gallon) are as follows:

$\{14.2, 13.9, 15.2, 14.5, 13.1, 15.0, 13.4, 14.7, 15.3, 13.7\}.$

Note that these figures constitute a *sample*—a sample out of a population, the population being *all* of the possible mileage figures that you *might* compute over the lifetime of your car. We may immediately compute the sample *mean:*

$$\bar{x} = \frac{\sum x}{n} = \frac{143.0}{10} = 14.30,$$

your "average" (in the sense of mean) mileage over the given time period.

Now, suppose that when I measure *my* mileage over the same time period I obtain these figures, again in miles per gallon:

{13.6, 10.1, 17.8, 9.4, 15.5, 16.8, 12.7, 17.2, 19.0, 10.9}.

Again, this is a sample drawn from a large population of possible mileage figures. Computing my sample mean:

$$\bar{x} = \frac{\sum x}{n} = \frac{143.0}{10} = 14.30.$$

Lo and behold! My mean mileage turns out to be exactly the same as yours.

But: Is it not true that, in some sense, my mileage results are different in nature from yours? Look at those sets again. Your mileage never seems to *vary* much from the mean 14.30 miles per gallon. You go as low as 13.1 (1.2 miles per gallon below the mean) and as high as 15.3 (1.0 mile per gallon above the mean)—not very much variation! You can probably *predict* your mileage on any given occasion reasonably well; you can probably even estimate the cost of an extended trip within a few cents.

But *my* scores—they seem to be considerably scattered. I go as low as 9.4 (4.9 miles per gallon below the mean) and as high as 19.0 (4.7 miles per gallon above the mean). I would probably have difficulty predicting my mileage on any particular occasion, or estimating the cost of a trip.

The difference between these two samples is a matter of **dispersion,** or "scattering." My results are considerably more dispersed (scattered, dissimilar) than yours are, even though our sample means are equal. After all, if we graph your results and mine on separate axes and compare them, the difference in their distributions becomes quite evident, as shown in Figure 1.1.

One of the important jobs to be done in statistics is to *measure* such dispersion: How scattered are your scores, and how scattered are mine? We

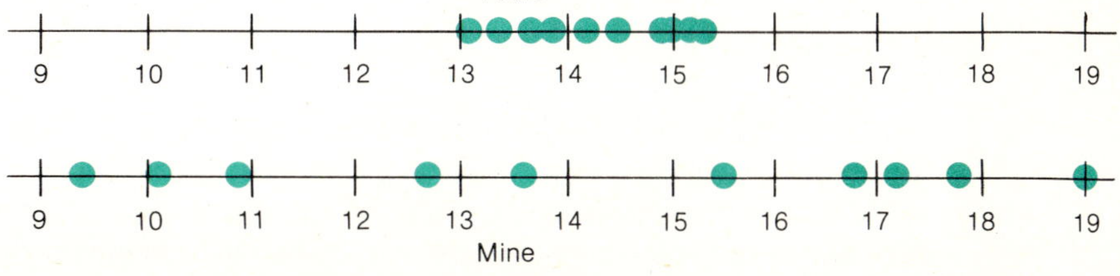

Figure 1.1

must have some sort of sample statistic that will measure dispersion. (Again, as when we were measuring central tendency, we are involved in *descriptive* statistics; this time, the aspect of the sample that we are seeking to describe is dispersion.)

The simplest way to measure dispersion for a sample of numbers would be to compute its

range: the largest number in the sample minus the smallest number.

Thus, for the given data, *your* range would be 15.3 − 13.1 = 2.2, and *my* range would be 19.0 − 9.4 = 9.6. However, the drawback is that the range *ignores* all but the lowest and highest scores; two very differently dispersed sets of numbers could have the same range. Like most things that are easy to compute, the range is of little value.

A more revealing approach should be to compare *all* of the numbers in a sample *with each other.* But this is awkward to do. (Even in a sample containing only 10 numbers, it would require our making 45 different comparisons, as you may verify if you think it through.)

Thus we will appeal to a more reasonable method: we will use the *mean* of the numbers as a central reference point and compare each of the numbers with the mean.

This procedure requires some elaboration and caution. If we subtract the mean from each number, for example, in the sample consisting of *your* gasoline mileage figures, we get

$$
\begin{aligned}
14.2 - 14.3 &= -0.1\\
13.9 - 14.3 &= -0.4\\
15.2 - 14.3 &= 0.9\\
14.5 - 14.3 &= 0.2\\
13.1 - 14.3 &= -1.2\\
15.0 - 14.3 &= 0.7\\
13.4 - 14.3 &= -0.9\\
14.7 - 14.3 &= 0.4\\
15.3 - 14.3 &= 1.0\\
13.7 - 14.3 &= -0.6.
\end{aligned}
$$

But if we add all these resulting differences, some being positive and some negative, we get zero!

$$3.2 + (-3.2) = 0.$$

Now *that* can be no way to measure dispersion: the very quality we have tried to measure seems to have vanished, because the negative differences cancelled out the positive ones. This would not happen if we *squared* the differences—and it turns out to be statistically desirable to follow this procedure. Thus:

$$(14.2-14.3)^2=(-0.1)^2=0.01$$
$$(13.9-14.3)^2=(-0.4)^2=0.16$$
$$(15.2-14.3)^2=\quad(0.9)^2=0.81$$
$$(14.5-14.3)^2=\quad(0.2)^2=0.04$$
$$(13.1-14.3)^2=(-1.2)^2=1.44$$
$$(15.0-14.3)^2=\quad(0.7)^2=0.49$$
$$(13.4-14.3)^2=(-0.9)^2=0.81$$
$$(14.7-14.3)^2=\quad(0.4)^2=0.16$$
$$(15.3-14.3)^2=\quad(1.0)^2=1.00$$
$$(13.7-14.3)^2=(-0.6)^2=0.36.$$

And *now* if we add the results we get

5.28.

Next, it is desirable to "average" (in the sense of taking the mean) the squared differences. This would customarily be done by dividing by 10, the number of terms added, but for reasons which will be mentioned later, it is the practice in this situation to *divide by one less than the number of terms:*

$$\frac{5.28}{10-1}=\frac{5.28}{9}\approx 0.587.$$

What we have at this point is called the sample **variance** and is denoted s^2.

$$s^2=\frac{\sum(x-\bar{x})^2}{n-1}$$

We have one final consideration: what **units** are we talking about? That is, 0.587 *what?* We started by talking about *miles per gallon*, but we have *squared* the numbers, so that instead of miles we are now dealing in *square* miles, in fact in "square miles per square gallon," whatever that means. (If the original units had been feet, or seconds, or grams, then the resulting 0.587 would have meant square feet, or square seconds, or square grams.)

To get back to the original units, then, we should take the *square root* of the number obtained:

$$\sqrt{0.587}$$

and to do so requires using the table "Digit Sequences for Square Roots" in the back of the book.

The table is used as follows. *Mark off the digits of the number in pairs, going both left and right away from the decimal point* (filling in 0's, if necessary, to make pairs):

$$\sqrt{\overleftarrow{00}\,.\,\overrightarrow{58}\,\overrightarrow{70}}$$

(The desired answer will be written above the square root sign, with the decimal point placed as shown.) Now, each *pair* of digits marked off in this manner will give rise to a *single* digit in the answer desired. Thus the pair of 0's on the front gives rise to *one* 0 before the decimal point, the digits 58 give rise to *one* digit following the decimal point, and so on. In fact, we consider the pair $\overrightarrow{58}$ and ask: what is the largest one-digit number that we can square and *not exceed* 58? The answer is 7 (because $7^2 = 49$ but $8^2 = 64$, too large). Thus the desired square root must start out 0.7 . . . :

$$\begin{array}{l} \quad 0.\quad 7\quad ?\quad ?\quad ? \\ \sqrt{\overleftarrow{00}.\;\;58\;\;70\qquad} \end{array}$$

Knowing that the answer "starts with a 7," so to speak, we look up the digits 587 in the square root table, and we see:

587 2423 7662

and the idea is that *one* of the digit sequences, 2423 or 7662, must be the desired one. It is 7662 in this case, because we know that the square root that we seek starts with the digit 7:

$$\begin{array}{l} \quad 0.\quad 7\quad 6\quad 6\quad 2 \\ \sqrt{\overleftarrow{00}.\;\;\overrightarrow{58}\;\;\overrightarrow{70}\;\;\overrightarrow{00}\;\;\overrightarrow{00}}. \end{array}$$

That is,

$$\sqrt{0.587} \approx 0.7662,$$

an answer that is, as is almost any square root, an approximation. We should round this result to one decimal place more than the original data. The data were given in tenths, so that for the computed statistic we should accept hundredths:

$$\sqrt{0.587} \approx 0.77.$$

This important result is called the **sample standard deviation** for the given sample numbers, and is denoted s. Because to obtain it we

1. subtract the mean $\bar{x}$ from each given number x,
2. square the differences,
3. add the squared differences,

4. divide by *one less than the sample size* (arriving at the sample variance), and
5. take the square root,

the formula for the sample standard deviation of a set of numbers

$\{x_1, x_2, x_3, \ldots, x_n\}$

is

$$s = \sqrt{\frac{\sum (x - \bar{x})^2}{n - 1}}.$$

The quantity s is measured in the same units as the given data, so that (for example) our 0.77 obtained means 0.77 miles per gallon. (Remove the square root sign and you have a formula for the sample **variance s^2**.)

It turns out that the concept of standard deviation, as a measure of dispersion, is one of the most important concepts in applied statistics.

EXAMPLES

1. To compute the sample standard deviation (s) of the *other* set of mileage figures given previously (*my* mileages):

$\{13.6, 10.1, 17.8, 9.4, 15.5, 17.8, 12.7, 16.2, 19.0, 10.9\}$

we note that the mean is $\bar{x} = 14.30$. Subtracting the mean from each number (each *datum*) and squaring the differences, we obtain:

$(13.6 - 14.3)^2 = (-0.7)^2 = 0.49$

$(10.1 - 14.3)^2 = (-4.2)^2 = 17.64$

$(17.8 - 14.3)^2 = (3.5)^2 = 12.25$

$(9.4 - 14.3)^2 = (-4.9)^2 = 24.01$

$(15.5 - 14.3)^2 = (1.2)^2 = 1.44$

$(16.8 - 14.3)^2 = (2.5)^2 = 6.25$

$(12.7 - 14.3)^2 = (-1.6)^2 = 2.56$

$(17.2 - 14.3)^2 = (2.9)^2 = 8.41$

$(19.0 - 14.3)^2 = (4.7)^2 = 22.09$

$(10.9 - 14.3)^2 = (-3.4)^2 = 11.56$

and the sum of the squares is $\sum (x - \bar{x})^2 = 106.70$. We must now divide this sum by one less than the sample size:

$$\frac{\sum (x-\bar{x})^2}{n-1} = \frac{106.70}{10-1} = \frac{106.7}{9} = 11.85555\ldots,$$

which we must round to three significant digits to use the square root table: $11.855555\ldots \approx 11.9$. (This is the sample *variance* s^2.) Marking off the digits of 11.9 in pairs:

$$s = \sqrt{\frac{\sum (x-\bar{x})^2}{n-1}} = \sqrt{\overleftarrow{11}.\overrightarrow{90}}$$

we ask, for the first block of digits ($\overleftarrow{11}$): What is the largest integer that we can square and not exceed 11? The answer is 3, so that the desired square root starts with 3:

$$\begin{array}{l} \;\;3. \\ \sqrt{\overleftarrow{11}.\;\;\overrightarrow{90}} \end{array}$$

Looking up the digit sequence 119 in the square root table, we find

119 1091 3450

and conclude that the desired digit sequence is the one starting with 3:

$$\begin{array}{llll} \;\;3. & 4 & 5 & 0 \\ \sqrt{11.} & 90 & 00 & 00 \end{array}$$

Thus $s = 3.45$ (miles per gallon), a standard deviation noticeably larger than the 0.77 (miles per gallon) obtained for the first set of mileage figures. This difference is to be expected—standard deviation measures *dispersion*, and this example consists of sample numbers much more dispersed, intuitively, than those of the previous set.

2. The heights of 15 people taken at random, as a sample, are recorded in centimeters as follows:

$$\{162, 174, 157, 178, 168, 151, 165, 176, 153, 148, 181, 170, 160, 173, 146\}.$$

To find the sample standard deviation s we may first compute the mean by writing

$$\bar{x} = \frac{\sum x}{n} = \frac{2462}{15} \approx 164.1,$$

and then compute the squared differences:

$(162.0-164.1)^2=(-2.1)^2 = 4.41$

$(174.0-164.1)^2=(9.9)^2 = 98.01$

$(157.0-164.1)^2=(-7.1)^2 = 50.41$

$(178.0-164.1)^2=(13.9)^2 = 193.21$

$(168.0-164.1)^2=(3.9)^2 = 15.21$

$(151.0-164.1)^2=(-13.1)^2=171.61$

$(165.0-164.1)^2=(0.9)^2 = 0.81$

$(176.0-164.1)^2=(11.9)^2 = 141.61$

$(153.0-164.1)^2=(-11.1)^2=123.21$

$(148.0-164.1)^2=(-16.1)^2=259.21$

$(181.0-164.1)^2=(16.9)^2 = 285.61$

$(170.0-164.1)^2=(5.9)^2 = 34.81$

$(160.0-164.1)^2=(-4.1)^2 = 16.81$

$(173.0-164.1)^2=(8.9)^2 = 79.21$

$(146.0-164.1)^2=(-18.1)^2=327.61$

And the sum is $\sum (x-\bar{x})^2=1801.75$, so that the sample *variance* is

$$s^2=\frac{\sum (x-\bar{x})^2}{n-1}=\frac{1801.75}{15-1}\approx 128.70$$

and the sample standard deviation is

$$s=\sqrt{128.7}\approx\sqrt{129}\approx 11.36\approx 11.4 \text{ (centimeters)}.$$

We have looked at the process of taking a *sample* of numbers

$$\{x_1, x_2, x_3, \ldots, x_n\}$$

(chosen from a larger population) and computing the sample variance and the sample standard deviation (square root of the sample variance), both computed on the basis of the sample mean $\bar{x}=\sum x/n$. If we were capable of computing the actual **population mean** taking into account all the numbers of the entire population in question—and this might well be a task requiring us to work with thousands or millions of numbers—the true population mean would be denoted μ (the Greek letter *mu*) and we would have

$$\mu = \frac{\sum x}{N}$$

where N is the number of items in the entire population and $\sum x$ is the sum of *all* the population's members. The true mean μ is an example of a **population parameter:** something that measures some aspect (in this case, central tendency) of a whole population. A parameter is not to be confused with a sample *statistic* such as $\bar{x}$, which measures some aspect of a *sample* from the population.

Another important population parameter, similarly, is the **standard deviation** for the whole population. It is denoted σ (lower-case Greek *sigma*) and is defined as

$$\sigma = \sqrt{\frac{\sum (x-\mu)^2}{N}}$$

where N is the population size and the summation is taken for all the population's members x; note that the true population mean μ is used in the subtractions. (Without the square root sign we would have σ^2, the **population variance,** another parameter.)

Notice that in computing the *population* standard deviation σ, we divide by N (the population size), but in computing a *sample* standard deviation s, we divide by $n-1$ (one *less* than the sample size). One reason is that it can be shown (though a proof of the fact is beyond the scope of this book) that when the parameter σ is approximated by computing the statistic s (from a sample), s would generally *under*estimate σ if we divided by n. Dividing by $n-1$ slightly increases the quotient and is a hedge against this underestimating tendency.

Standard deviation is one of the most important and useful concepts in all of statistics, and we will be meeting it again in various settings.

NEW TERMS IN THIS SECTION

dispersion
range
(sample) standard deviation (s)
(sample) variance (s^2)
population parameter
population mean (μ: *mu*)
population standard deviation (σ: *sigma*)
population variance (σ^2)
***linear interpolation**
***mean deviation**
***semi-interquartile range**

EXERCISES 1-B

1. Use the square root table to approximate each of the following square roots, rounding the given number under the radical sign to three digits first if necessary; express your answers using

all available digits from the table.

(a) $\sqrt{381}$ (b) $\sqrt{38.1}$ (c) $\sqrt{3.81}$ (d) $\sqrt{0.381}$
(e) $\sqrt{0.0381}$ (f) $\sqrt{57.2}$ (g) $\sqrt{5720}$ (h) $\sqrt{572}$
(i) $\sqrt{5.72}$ (j) $\sqrt{0.0000572}$ (k) $\sqrt{7.39}$ (l) $\sqrt{73.9}$
(m) $\sqrt{739}$ (n) $\sqrt{73{,}900}$ (o) $\sqrt{0.0739}$ (p) $\sqrt{90.71}$
(q) $\sqrt{755.4}$ (r) $\sqrt{9.227}$ (s) $\sqrt{0.0007853}$ (t) $\sqrt{178{,}600}$

2. For each of these sets of sample numbers, compute the sample mean $\bar{x}$, sample variance s^2, and (its square root) the sample standard deviation s:

(a) {39, 46, 32, 31, 45, 43, 49, 35}
(b) {83, 74, 89, 85, 70, 69, 81, 77, 78, 84}
(c) {19, 24, 27, 21, 13, 11, 29, 20, 15, 10, 31}
(d) {105, 98, 101, 111, 91, 107, 114, 94, 88}
(e) {60, 63, 68, 50, 59, 66, 53, 52, 64, 70, 60, 61}
(f) {61, 48, 67, 80, 45, 72, 77, 65, 63, 54, 70, 73, 57, 51}
(g) {61, 74, 53, 68, 80, 59, 83}
(h) {21, 20, 22, 19, 16, 23, 20, 24, 17, 23, 18, 16, 23}
(i) {0.019, 0.011, 0.027, 0.008, 0.014, 0.036, 0.005, 0.023, 0.031, 0.025, 0.016, 0.029, 0.007, 0.040, 0.022, 0.015}
(j) {207.3, 209.9, 204.1, 201.8, 202.5, 210.6, 205.7, 208.4, 199.3, 203.2}

3. In a science lab one student reads a meter eight times in succession and gets readings (in volts) of:

{32.3, 35.7, 34.1, 37.0, 34.9, 36.4, 33.6, 35.2}.

Another student reads the meter eight times and gets readings of

{35.3, 33.0, 36.8, 33.9, 34.3, 34.8, 36.2, 36.5}.

In terms of sample standard deviation, whose results are more *consistent*?

4. Two experimental drugs are designed to lower a patient's blood pressure. Drug A is tried on 10 patients with the result that it lowers their blood pressure 12.0, 11.0, 13.0, 10.0, 11.0, 10.0, 13.0, 12.0, 14.0, and 12.0 millimeters, respectively. Drug B is tried on 10 patients with the result that it lowers their blood pressure 10.0, 13.0, 12.0, 11.0, 12.0, 14.0, 11.0, 13.0, 13.0, and 9.0 millimeters, respectively. Compute the sample means and sample standard deviations. What (if any) conclusions might be drawn?

5. Recall that the *range* of a set of numbers is found by subtracting the smallest value from the largest one. For each of the following sets of sample data, compute the mean, the range, and the sample standard deviation. (In each case, do separate computations for set A and set B.)

(a) A = {34, 25, 39, 47, 38, 40, 24, 44, 25, 42, 40, 27, 39, 41, 37, 26, 31, 42, 50, 29}
B = {95, 80, 94, 89, 81, 95, 99, 92, 84, 80, 79, 105, 96, 94, 97, 97, 86, 102, 82, 93}

(b) A = {30, 23, 15, 24, 21, 17, 19, 13, 10, 28}
B = {29, 30, 40, 29, 27, 31, 20, 24, 28, 32}
(c) A = {20, 26, 24, 24, 22, 26, 24, 22, 24, 28}
B = {26, 22, 26, 22}
(d) A = {37, 42, 45, 47, 47, 47, 48, 48, 49, 49, 49, 49, 50, 50, 50, 51, 52, 53, 55, 62}
B = {40, 41, 41, 42, 43, 44, 44, 45, 45, 45, 47, 47, 52, 56, 57, 57, 58, 58, 58, 60}

6. Observe, from your results in Exercise 1-B.5, that for two sets of sample data A and B, it is possible to have

(a) the same range and same sample standard deviation but different means.
(b) the same range but different sample standard deviations (because much of the dispersion is "internal," and range ignores this).
(c) different ranges but the same sample standard deviations.
(d) the range of A *larger* than the range of B, but the sample standard deviation of A *smaller* than the sample standard deviation of B.

7. What is the sample variance and what is the sample standard deviation for a sample consisting of 50 identical numbers?

PONDERABLES 1-B

1. What happens to the mean $\bar{x}$ of a set of numbers if a constant K is added or subtracted (for each of the numbers)? Can you prove your answer?

2. For a set of sample numbers

$\{x_1, x_2, x_3, \ldots, x_n\}$

for which the sample standard deviation is some value s, what can be said about the sample standard deviation for the set of numbers

$\{x_1 - K, x_2 - K, x_3 - K, \ldots, x_n - K\}$

where the numbers have been derived from the given ones by uniformly subtracting a constant K? Can you prove your answer?

3. Do the results of the first two problems suggest a way of simplifying the calculation of s for some sets of numbers? For example, in Exercise 1-B.2(j), what could be done?

4. By squaring $x - \bar{x}$ and then replacing $\bar{x}$ with $\sum x/n$ in the formula

$$s = \sqrt{\frac{\sum (x - \bar{x})^2}{n - 1}}$$

and simplifying algebraically, try to produce an alternative formula for s which does not involve computing $\bar{x}$ first. What may be said about the use of such a formula? Is it necessarily shorter or easier than the usual formula? (*Remark*: For some statistical procedures it is necessary to know the sample mean $\bar{x}$ in any event, so that we are not necessarily able to avoid computing it even when there is a formula for s that avoids it.)

5. You will notice that when using the square root table, we are only able to look up a number containing three significant digits. Thus we may find

$\sqrt{71.8}$ or $\sqrt{7180}$ or $\sqrt{0.000718}$

by looking up 718 and placing the decimal point appropriately in the resulting digit sequence. But if we want to look up the square root of 7184, we "round down" to 7180 first; or if we want to look up the square root of 7187 we first "round up" to 7190. In such a situation, where there is a "fourth digit" to deal with, we may also appeal to an approximation method called **linear interpolation:** a method of proportionally estimating values that lie between (hence *inter*polation) the table values, thus obtaining a little additional accuracy. For example, to compute

$\sqrt{7187}$,

we look up

$$\sqrt{7180} = 84.73$$

and $\sqrt{7190} = 84.79$

and take the desired square root $\sqrt{7187}$ to be (approximately) *seven-tenths* of the way between 84.73 and 84.79. The difference between the table values is

$$84.79 - 84.73 = 0.06$$

and seven-tenths of that difference is

$$(0.06)(0.7) = 0.042$$

which we *add* to the smaller value $\sqrt{7180}$:

$$84.73 + 0.042 = 84.772 \approx 84.77$$

and this is our approximation to $\sqrt{7187}$. (The assumption here is that the square roots are proportional to the numbers themselves: not an entirely true assumption, but very nearly descriptive of the truth, so that only tiny amounts of error are likely to be generated by making the assumption.) By this method, compute the following square roots, rounding your answers to the same number of decimal places as would have been obtained directly from the table:

(a) $\sqrt{2174}$ (b) $\sqrt{38.62}$ (c) $\sqrt{514.8}$ (d) $\sqrt{7.065}$
(e) $\sqrt{4.916}$ (f) $\sqrt{83.33}$ (g) $\sqrt{12.51}$ (h) $\sqrt{6447}$
(i) $\sqrt{2.028}$ (j) $\sqrt{47.19}$

6. Besides standard deviation, another measure of dispersion is the **mean deviation,** given by

$$\text{M.D.} = \frac{\sum |x - \bar{x}|}{n},$$

where the vertical bars (absolute value bars) around each difference $x - \bar{x}$ simply indicate that the subtraction is done in the nonnegative order. (For example, $|20-18| = 2$ but also $|18-20| = 2$ and not -2.) Although the mean deviation is easier to compute than the standard deviation because there is no squaring or square root involved, it turns out that the mean deviation does not lend itself well to further analysis or fruitful application, hence the standard deviation is generally much preferred. Still, so that you will be well aware of both measures of dispersion and the similarities and differences in computing them, find both M.D. and s for the following sets of data and compare your results each time:

(a) {28, 35, 21, 30, 33, 26, 24, 23, 31, 37}
(b) {73, 78, 82, 68, 78, 63, 75, 66, 77, 84, 64, 80}
(c) {54, 46, 59, 40, 43, 56, 50, 52, 49, 45, 57, 60, 54, 45, 53, 39, 58, 47, 51, 62}

7. Still another measure of dispersion is the **semi-interquartile range,** denoted Q: half the difference between the third and the first quartiles (see Ponderable 1-A.10):

$$Q = \frac{1}{2}(Q_3 - Q_1).$$

Find Q_1, Q_3, and Q for each set of data in Exercise 1-B.1.

C FREQUENCY DISTRIBUTIONS

Suppose the following data were given:

30	56	21	49	34	58	22	38	27	31
35	41	53	25	34	48	33	58	20	34
30	50	26	52	32	63	25	50	36	29
34	21	61	33	51	20	41	30	57	26
28	45	36	59	26	60	42	21	63	56
36	54	24	43	30	27	56	26	35	32

We will not, for present purposes, assign any particular meaning to these numbers; they could be various kinds of collected data. (For example, these numbers might stand for the daily numbers of traffic tickets handed out by a certain town's police force; such data would be called **count data,** because each number would have been obtained simply by counting. Or the numbers might stand for the lengths, in centimeters, of 60 cats chosen at random; such data would be called **measure data,** because each number would have been obtained by measuring—and *rounding off* to the nearest centimeter.)

As things stand, this array of numerical data is not very informative. The numbers are simply dropped into our laps and are not arranged in any order or organized in any fashion. If we wished to gain a quick idea as to the nature of the numbers—for example, if we wished to know the median $\tilde{x}$—unfortunately no quick glance at the numbers would tell the tale.

For this reason, such sets of numbers are often organized (grouped) into a format called a **frequency distribution.** Let us see what this means and how it is done.

First, we observe (and even this observation requires some cautious searching) that the smallest number is 20 and the largest is 63. Thus the *range* of the data is

$$63-20=43.$$

Now, 43 is not a very convenient range, because it is not divisible by anything (except 43 and 1, of course)—we will want to divide the range into equal segments. Suppose, then, we adjust this range and think of it as running (say) from 20 to 65: a range or "spread" of 45 this time. Because 45 may, for example, be factored as $45=9 \cdot 5$, we may think of the range of the data as being segmented into nine **classes** as follows:

20 to 24
25 to 29
30 to 34
35 to 39
40 to 44
45 to 49
50 to 54
55 to 59
60 to 64

(and the next class, if there were one, would start with 65). Because we expected each of the nine classes to represent a little "spread" of 5 units (because $45=9 \cdot 5$), each class starts 5 units farther along than the last one—hence the endpoints "20 to 24," "25 to 29," and so on are determined.

Now we simply go back to the **raw data**—the original unorganized, ungrouped 60 numbers—and simply *count* how many numbers occur from 20 to 24 inclusive (that is, how many times we see 20, 21, 22, 23, or 24 in the listing), how many numbers occur from 25 to 29 inclusive, and so on. The results would be:

20 to 24	7
25 to 29	10
30 to 34	13
35 to 39	6
40 to 44	4
45 to 49	3
50 to 54	6
55 to 59	7
60 to 64	4

and as a check we may add these counts and get

$$7+10+13+6+4+3+6+7+4=60.$$

Note, then, that the nine classes were chosen in such a way that *no data are omitted.*

These numbers (7, 10, 13, . . .) are called the **frequencies** of the given classes. The frequency of a class is the number of items present in that class. Frequencies are denoted *f*, so that in this case we may write

$$\sum f=60$$

(the sum of all the frequencies is 60).

The grouping of the numbers into classes with the frequency shown for each class thus produces a sort of summary table called a *frequency distribution.*

There is some more terminology involved.

The *endpoints* of each class (for instance, 20 and 24 for the first class) are called class **limits:** for the first class, 20 is the *lower class limit*, and 24 is the *upper class limit.*

However, we sometimes need to think of each class as being theoretically "extended half a unit" on each side, so that the first class would be thought of as running from 19.5 to 24.5, the next class running from 24.5 to 29.5, and so forth. This extension has the effect of filling in the gaps between the class limits (between 24 and 25, for example). These extended numbers (24.5, 29.5, and so on) are called **class boundaries.** Thus we may show all the class boundaries for this particular frequency distribution in parentheses as follows:

Class	*f*
(19.5) 20 to 24 (24.5)	7
(24.5) 25 to 29 (29.5)	10
(29.5) 30 to 34 (34.5)	13
(34.5) 35 to 39 (39.5)	6
(39.5) 40 to 44 (44.5)	4
(44.5) 45 to 49 (49.5)	3
(49.5) 50 to 54 (54.5)	6
(54.5) 55 to 59 (59.5)	7
(59.5) 60 to 64 (64.5)	4

We are concerned with class boundaries for several reasons; certain computations are based on them. For one thing,

the **width** or (as it is often called) **interval** of each class is defined to be the difference between the *upper boundary* and the *lower boundary.*

Thus, for example, the interval (or width) of the class 30 to 34 is

$$34.5 - 29.5 = 5$$

as we would expect: *All* classes are 5 units wide, because there are nine classes and we started by factoring the range 45 as $9 \cdot 5$. Note that the interval of a class is *not* the difference of its class *limits*: *not* $34 - 30$, for example, which would only be 4.

Another term to consider is the **class mark** for each class: the midpoint value of the class, found either by taking the mean of the class's limits or the mean of its boundaries. Thus the class mark for the class 30 to 34 is given either by

$$\frac{30+34}{2} = 32 \quad \text{or} \quad \frac{29.5+34.5}{2} = 32.$$

You may verify that the class marks for this particular frequency distribution are: 22, 27, 32, 37, 42, 47, 52, 57, and 62. (Generally, they do not have to be integer values, however.)

We may draw a diagram of our frequency distribution by placing the class *boundaries* on a horizontal scale, then drawing for each class a rectangle whose height (measured against a vertical scale) represents the class *frequency*, as shown in Figure 1.2.

Such a bar-type graphical representation is called a **histogram.** We note that now—unlike the raw, ungrouped 60 given numbers—the **grouped data** take on a visual aspect that makes it possible to gather some quick impressions. (For example, the data seem to "bunch up" around 24.5 to 34.5, but seem to "thin out" in other places.)

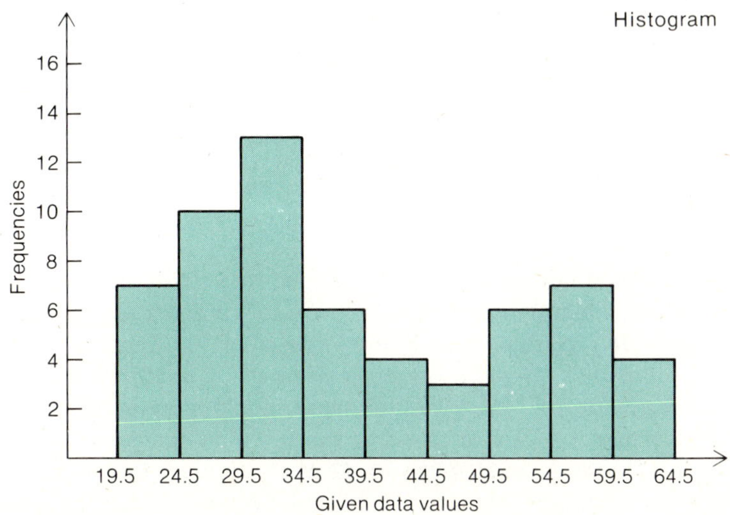

Figure 1.2

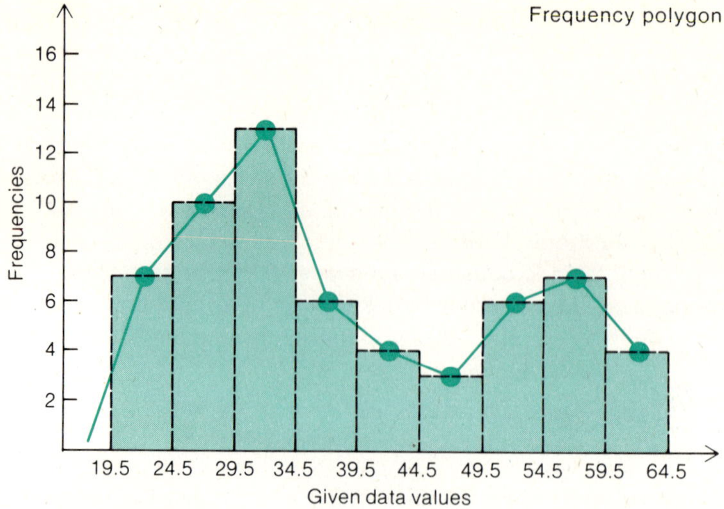

Figure 1.3

Another visual way to display the grouped data in a frequency distribution is to draw line segments connecting the midpoints of the rectangle tops in a histogram, producing a jagged line diagram (as shown in Figure 1.3) called a **frequency polygon.**

Some general guidelines for constructing a frequency distribution are:

1. Make all classes of equal interval (width).
2. Do not allow the first or the last class to be empty.
3. Do not allow classes to overlap—they should intersect only at the boundaries (for example, at 24.5, 29.5, and so on in the example given).
4. Do not use too many or too few classes: usually from 6 to 15. (The use of too few classes may tend to oversimplify the data, placing too many dissimilar items into one class. And the use of too many classes may tend to make the frequency distribution not much more informative or convenient than the original ungrouped data.)

EXAMPLE

1. The same 60 numbers may be arranged into a frequency distribution in other ways. For example, if we think of the total spread 20-to-63 (range 43) as being adjusted to run from 18 to 66 for a spread of 48 units, we may take advantage of the fact that $48 = 6 \cdot 8$. Suppose we sort the data into eight classes, each of interval (width) 6. Then the first class would start at 18, the second at 24 (6 units farther along), and so on, so that we may determine the classes to be: 18 to 23, 24 to 29, 30 to 35, 36 to 41, 42 to 47, 48 to 53, 54 to 59, and 60 to 65. (And the next class, if there were one, would start at 66.) Then going through the raw data and tallying up how may items fall into each class, we have:

Class	f
18 to 23	6
24 to 29	11
30 to 35	15
36 to 41	6
42 to 47	3
48 to 53	7
54 to 59	8
60 to 65	4
Check:	$\sum f = 60.$

Note again that the interval of each class can be verified to be 6 by subtracting class *boundaries:* For the "36 to 41" class, for example, $41.5 - 35.5 = 6$. You may draw a histogram and frequency polygon for this frequency distribution and compare them with the previous results.

It should be clear that taking a set of raw sample data and making a frequency distribution is an act of simplification and *summary.* Like any summary, a frequency distribution involves the loss of some information: we no longer see all the original numbers, but rather only how many of them lie within each class. If we do not have the original ungrouped data, then some information has been lost—for example, in the first frequency distribution examined in this section, in which the first class runs from 20 to 24 with a frequency of 7, we do not know how many (if any) of the seven numbers are 21's. Also, we do not know how many (if any) of the seven numbers are less than 24 (because some of them may or may not actually be equal to 24). Nevertheless, this slight loss of information is generally offset by the convenience gained when we have the data grouped for easier comprehension.

Even though a frequency distribution is only a summary of the original raw data, we may—even if we have *only* the summary and not the original numbers—compute some statistics from the frequency distribution itself, statistics that parallel and closely approximate those which would have been computed from the raw data.

First, let us see how a *mean* is computed from a frequency distribution. We will denote such a mean $\bar{X}$ (using the capital letter to indicate that the statistic is computed from a frequency distribution, as distinct from $\bar{x}$, which would be computed from the ungrouped raw data). It is important to understand that we will compute this mean $\bar{X}$ from a frequency distribution alone, without recourse to the original data.

Let us take, for example, the frequency distribution initially constructed in this section, and determine the class *mark* (midpoint, denoted x') for each class:

Class	f	Class mark x'
20 to 24	7	22
25 to 29	10	27
30 to 34	13	32
35 to 39	6	37
40 to 44	4	42
45 to 49	3	47
50 to 54	6	52
55 to 59	7	57
60 to 64	4	62

Next, to compute the frequency distribution mean $\bar{X}$, we make a working assumption about each class in the distribution: we assume (though we know that it may not literally be true) *that each number in a given class lies at the class mark*. Thus, for example, in the first class we proceed as if all seven of the member numbers there are 22's. (Although this assumption is usually false, in general we are sometimes *under*estimating and sometimes *over*estimating the actual numbers by assuming them to be at the mark, so that the error generated tends to be self-correcting.)

Now, under the assumption that all the data lie at the class marks, we add seven 22's, ten 27's, thirteen 32's, six 37's, four 42's, three 47's, six 52's, seven 57's, and four 62's (so that we can sum everything before dividing to compute the mean):

$$\begin{aligned} 7 \cdot 22 &= 154 \\ 10 \cdot 27 &= 270 \\ 13 \cdot 32 &= 416 \\ 6 \cdot 37 &= 222 \\ 4 \cdot 42 &= 168 \\ 3 \cdot 47 &= 141 \\ 6 \cdot 52 &= 312 \\ 7 \cdot 57 &= 399 \\ 4 \cdot 62 &= 248 \end{aligned}$$

Adding these frequency-times-class-mark products, we get

$$\sum fx' = 2330$$

and dividing by the total number of items involved (which in this case is $\sum f = 60$), we get the **frequency distribution mean**

$$\bar{X} = \frac{\sum fx'}{\sum f} = \frac{2330}{60} \approx 38.8.$$

The formula, then, for a frequency distribution mean is

$$\bar{X} = \frac{\sum fx'}{\sum f}.$$

The product fx' (frequency times class mark) is computed separately for each class, the products added, and the resulting sum divided by the sum of the frequencies (which is the total number of data involved).

This $\bar{X} = 38.8$, just computed, is regarded as an approximation to the original "raw mean" $\bar{x}$, because $\bar{X}$ can be computed even if the ungrouped original numbers are no longer available to compute $\bar{x}$ with. But because in this example we *do* still have the original 60 ungrouped numbers, it may be interesting to return to these numbers and compute the raw mean $\bar{x}$ (for comparison with $\bar{X}$), obtaining, as you may verify:

$$\bar{x} = \frac{\sum x}{n} = \frac{2363}{60} \approx 39.4.$$

The comparison shows that $\bar{X} = 38.8$ (from the frequency distribution) misses $\bar{x} = 39.4$ (from the raw data) by only about six-tenths of a unit—not bad, considering that $\bar{X}$ is computed without knowledge of the 60 individual numbers!

EXAMPLE

2. Let us compute a mean $\bar{X}$ from the second frequency distribution in this section, the eight-class distribution shown in Example 1:

Class	f	Class mark x'	fx'
18 to 23	6	20.5	$6(20.5) = 123.0$
24 to 29	11	26.5	$11(26.5) = 291.5$
30 to 35	15	32.5	$15(32.5) = 487.5$
36 to 41	6	38.5	$6(38.5) = 231.0$
42 to 47	3	44.5	$3(44.5) = 133.5$
48 to 53	7	50.5	$7(50.5) = 353.5$
54 to 59	8	56.5	$8(56.5) = 452.0$
60 to 65	4	62.5	$4(62.5) = 250.0$
	$\sum f = 60$		$\sum fx' = 2322.0$

and the desired frequency distribution mean is

$$\bar{X} = \frac{\sum fx'}{\sum f} = \frac{2322.0}{60} = 38.7.$$

This is almost the same as the result (38.8) obtained from the first frequency distribution, and again we see that it misses the true mean ($\bar{x} = 39.4$ from the raw data) by very little—this time by about seven-tenths of a unit, a little more error than before. (The present frequency distribution, with only eight classes instead of nine, is a little cruder than the first one. In approximating $\bar{x}$ with $\bar{X}$, a greater number of classes in the distribution *tends* to improve the approximation, though this effect is not strictly guaranteed in every case.)

Just as a mean $\bar{X}$ can be computed from a frequency distribution, even in the absence of the original raw data, a **frequency distribution median $\tilde{X}$** can also be computed from the distribution alone. Let us see how this is done.

As illustration of the procedure, we will return to the eight-class frequency distribution of this section:

Class	f
18 to 23	6
24 to 29	11
30 to 35	15
36 to 41	6
42 to 47	3
48 to 53	7
54 to 59	8
60 to 65	4

Recall that finding an ordinary "raw data" median $\tilde{x}$ requires counting to the middle of the numbers when they are arranged in order. The same idea applies here, essentially. Because 60 items are involved (i.e., $\sum f = 60$), we need to count half of those (i.e., 30 items). We do so by starting at the top of the distribution and adding the frequencies as we go along, until we reach or exceed 30:

First frequency: 6
First two frequencies: $6+11=17$
First three frequencies: $6+11+15=32$

and we have for the first time exceeded 30. Thus we expect the median to lie within the range 30-to-35 (and this class will be called the *medial class*).

Now we look at the fact that when adding up the frequencies *prior* to the medial class we get only $6+11=17$, *leaving 13 items still to be counted to reach 30.* This difference (13 in this example) will be denoted d.

Next, what we must do is this: "Go into" the medial class 30-to-35 the appropriate distance *by adding the appropriate fraction of the class interval* (w) *to the lower class boundary* (B). And the "appropriate fraction" is d/f (the difference d previously described, over the class frequency f). Thus in these terms the formula for the frequency distribution median $\tilde{X}$ is

$$\tilde{X} = B + \frac{d}{f}(w).$$

Remember, d is the number of items still necessary to count (to reach half the total number of items) at the time when the frequencies *prior* to the medial class have been added.

In this case, because $B=29.5$, $d=13$, $f=15$, and $w=6$ (all classes in this distribution being of width 6), we have

$$\tilde{X}=29.5+\frac{13}{15}(6)$$

$$=29.5+\frac{13}{5}(2)=29.5+\frac{26}{5}=29.5+5.25$$

$$=34.75\approx 34.8.$$

Because in this case we do still have the original 60 numbers, we may (going to some trouble) place them in order and determine the actual "raw" median $\tilde{x}$:

20	20	21	21	21	22	24	25	25	26
26	26	26	27	27	28	29	30	30	30
30	31	32	32	33	33	34	34	34	(34)
(35)	35	36	36	36	38	41	41	42	43
45	48	49	50	50	51	52	53	54	56
56	56	57	58	58	59	60	61	63	63

the raw median, then, being

$$\tilde{x}=\frac{34+35}{2}=34.5.$$

Thus our *estimate* $\tilde{X}=34.8$, obtained from the frequency distribution alone, has missed the true value by only about three-tenths of a unit—again, not bad, considering that $\tilde{X}$ was computed *without* the raw data!

EXAMPLE

3. Suppose the same 60 raw data had been arranged in a *six*-class frequency distribution as follows:

Class	f
18 to 25	9
26 to 33	17
34 to 41	12
42 to 49	5
50 to 57	10
58 to 65	7
	$\sum f=60$

To compute the frequency distribution median $\tilde{X}$, we first note that *half* the data would be $\frac{1}{2}(60)=30$. (*Remark:* In some situations, this need not even turn out to be an integer; for instance, with 65 numbers, half of 65 is 32.5, and this is the value actually used.) Now, to count 30 items, we start adding the frequencies:

9

$9+17=26$

$9+17+12=38$

until the sum exceeds 30. We find that the median must lie in the 34 to 41 range. That is, the third class is the medial class. Now, in the formula

$$\tilde{X}=B+\frac{d}{f}(w)$$

B is the medial class lower boundary ($B=33.5$), f is the medial class frequency ($f=12$), w is the medial class interval or width ($w=41.5-33.5=8$), and the numerator d is the amount remaining to be counted (in trying to reach 30) when only the previous frequencies have been counted ($9+17=26$, falling 4 short of 30, so that $d=4$). Thus we have

$$\tilde{X}=33.5+\frac{4}{12}(8)=33.5+\frac{1}{3}(8)$$
$$\approx 33.5+2.67=36.17\approx 36.2.$$

This time we have missed the true median $\tilde{x}=34.5$ by about 1.7 units, obtaining a respectable but somewhat less accurate estimate—but then after all, the frequency distribution employed here consists of only six classes. A distribution with very few classes *may* produce a close approximation to $\tilde{x}$, but there is no guarantee. (On the other hand, a frequency distribution with a greater number of classes does not always guarantee a closer approximation to $\tilde{x}$ either. Having a greater number of classes in a frequency distribution is rather more likely to improve the approximation of $\bar{x}$ than the approximation of $\tilde{x}$.)

In a frequency distribution, instead of listing the frequency of each class individually, we may sometimes prefer to *add* the frequencies as we go along, in a manner (for example) such as the following:

Class	(Cumulative) frequency
<26	9
<34	26
<42	38
<50	43
<58	53
<66	60

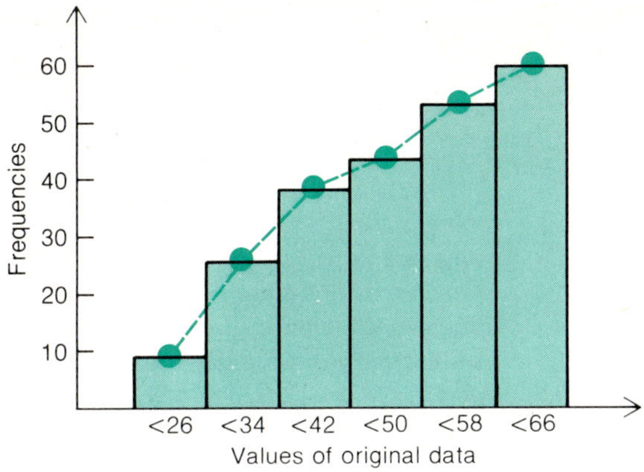

Figure 1.4

where < means "less than." This is a **cumulative frequency distribution:** we have listed that there are (among the original data) nine items less than 26, twenty-six items less than 34, thirty-eight items less than 42, and so on. This cumulative distribution is derived from the regular frequency distribution in Example 3 by using the class limits on the left. The cumulative distribution shows how the items "accumulate" as we go along, and it may be represented by a histogram or a frequency polygon as shown in Figure 1.4. (A frequency polygon, like the dotted line in Figure 1.4, is called an **ogive,** pronounced "OH-jive," when it relates to a *cumulative* distribution.)

EXAMPLE

4. We may construct a cumulative frequency distribution from the eight-class frequency distribution in Example 2 using the symbol ≤ ("less than or equal to") with the *righthand* class limits:

Class	Cumulative frequencies
≤23	6
≤29	17
≤35	32
≤41	38
≤47	41
≤53	48
≤59	56
≤65	60

showing that there are six items less than *or equal to* 23: seventeen items less than or equal to 29; and so on. Note that there are some questions that we cannot resolve if all we have is this cumulative frequency distribution; for example, if we wish to know how many items are strictly *less than* 29, we cannot tell, because the seventeen items (that are known to be less than *or equal to* 29) may or may not, some or all of them, actually be equal to 29.

NEW TERMS IN THIS SECTION

count data	grouped data
measure data	frequency polygon
frequency distribution	frequency distribution mean ($\bar{X}$)
class	frequency distribution median ($\tilde{X}$)
raw (ungrouped) data	cumulative frequency distribution
frequency	ogive
class limits	*frequency distribution standard deviation
class boundaries	*modal class
class interval (width)	*frequency distribution mode ($\hat{X}$)
class mark	
histogram	

EXERCISES 1-C

1. The heights of 72 kindergarten children are measured (in centimeters) with the following results:

95	85	93	98	90	96	94	87	96
100	92	104	93	92	103	85	94	89
91	101	89	104	94	103	100	89	94
94	86	104	92	109	90	96	92	90
104	94	94	107	87	89	107	91	105
91	97	100	84	102	96	92	90	98
102	95	108	95	104	101	103	86	95
86	97	91	101	90	109	94	90	93

Construct a frequency distribution (drawing a histogram and frequency polygon in each case)

(a) with seven classes, the first class starting at a lower limit of 82 and the last class running to an upper limit of 109.

(b) with six classes, the first class starting at a lower limit of 82 and the last class running to an upper limit of 111.

(c) with nine classes, the first class starting at a lower limit of 83 and the last class running to an upper limit of 109.

2. For each frequency distribution obtained in Exercise 1-C.1, compute the frequency distribution mean $\bar{X}$ and median $\tilde{X}$.

3. Compute the "raw" mean $\bar{x}$ and median $\tilde{x}$ of the given 72 numbers in Exercise 1, and compare them with your results in Exercise 1-C.2. Which required more time and trouble, the $\bar{x}$ computations or the $\bar{X}$ computations?

4. A commuter records his driving time to work (in minutes) for each of 81 days, with the following results:

29	26	30	27	23	32	30	25	28
28	31	23	33	39	26	22	33	27
36	31	25	43	24	24	34	27	28
25	42	27	30	28	26	29	31	23
25	40	28	46	22	39	30	30	26
34	22	35	36	27	29	24	48	28
26	35	28	30	26	31	36	29	25
28	29	32	27	28	39	22	28	27
27	23	37	25	38	25	33	23	32

(a) Construct a frequency distribution with six classes, the first class having lower limit 20 and the last class having upper limit 49. Draw a histogram and frequency polygon.

(b) Find $\bar{X}$ and $\tilde{X}$ for the frequency distribution constructed in part (a).

(c) Construct a frequency distribution with seven classes, the first class having lower limit 21 and the last class having upper limit 48. Draw a histogram and frequency polygon.

(d) Find $\bar{X}$ and $\tilde{X}$ for the frequency distribution constructed in part (c).

(e) Construct a frequency distribution with nine classes, the first class having lower limit 22 and the last class having upper limit 48. Draw a histogram and frequency polygon.

(f) Find $\bar{X}$ and $\tilde{X}$ for the frequency distribution constructed in part (e).

5. For the ungrouped raw data given in Exercise 4, compute the raw mean $\bar{x}$ and the raw median $\tilde{x}$ and compare the results of Exercises 1-C.4(b), (d), and (f) with the true values of $\bar{x}$ and $\tilde{x}$. Again, which is more trouble to compute, $\bar{x}$ or $\bar{X}$?

6. For the frequency distribution

Class	f
70 to 76	11
77 to 83	13
84 to 90	17
91 to 97	20
98 to 104	14
105 to 111	10

specify

(a) the class limits of each class

(b) the class boundaries of each class

(c) the class mark of each class

(d) the interval of each class

(e) $\sum f$

(f) $\bar{X}$ (can it be checked to see how close an approximation to $\bar{x}$ it is?)

(g) $\tilde{X}$ (can it be checked to see how close an approximation to $\tilde{x}$ it is?)

7. Compute $\bar{X}$ and $\tilde{X}$ for each of the following frequency distributions:

(a)

Class	f
35 to 41	24
42 to 48	29
49 to 55	35
56 to 62	43
63 to 69	33
70 to 76	21
77 to 83	15

(b)

Class	f
15 to 19	6
20 to 24	10
25 to 29	14
30 to 34	24
35 to 39	15
40 to 44	13
45 to 49	11
50 to 54	7

(c)

Class	f
54 to 59	18
60 to 65	24
66 to 71	38
72 to 77	33
78 to 83	20
84 to 89	17

8. Compute $\tilde{X}$ for the first frequency distribution given in this section (with nine classes, beginning with 20 to 24). What is d in the formula for $\tilde{X}$ in this case?

9. Change the frequency distribution

Class	f
35 to 39	21
40 to 44	33
45 to 49	46
50 to 54	82
55 to 59	97
60 to 64	75
65 to 69	51
70 to 74	30

into

(a) a $<$ type cumulative frequency distribution (draw the histogram and frequency polygon or ogive)

(b) a ≤ type cumulative frequency distribution (draw the histogram and frequency polygon or ogive).

10. Given the cumulative frequency distribution

	f
<20	32
<30	68
<40	105
<50	177
<60	201
<70	226
<80	243

tell which of the following are possible to determine from the distribution, and give the information when possible:

(a) How many items are less than 40?
(b) How many items are greater than 30?
(c) How many items are greater than or equal to 40?
(d) How many items are greater than 60 but less than 70?
(e) How many items are greater than or equal to 50 but less than 60?

11. Given the cumulative frequency distribution

	f
≤ 25	41
≤ 30	62
≤ 35	90
≤ 40	110
≤ 45	126
≤ 50	135

tell which of the following are possible to determine from the distribution, and give the information when possible:

(a) How many items are less than 40?
(b) How many items are greater than 30?
(c) How many items are greater than or equal to 25?
(d) How many items are greater than 45 but less than or equal to 50?
(e) How many items are greater than 45 but less than 50?

PONDERABLES 1-C

1. Recall that in constructing a frequency distribution we should not allow the first or the last class to be empty. Can we avoid having "internal" classes (classes other than the first or last class) empty, necessarily?

2. In terms of increasing the number of classes for a frequency distribution representing a given set of raw data, employing more and more classes, what is the "ultimate" frequency distribution? (Remember, however, we do not usually *want* too many classes.)

3. Recall that we develop the formula

$$\bar{X} = \frac{\sum fx'}{\sum f}$$

for the mean of a frequency distribution (x' being the class mark of each class) by computing as if each datum in a class lies *at* the class mark. Using the same idea, convince yourself that a formula for the sample **standard deviation** of a frequency distribution would be

$$S = \sqrt{\frac{\sum (x' - \bar{X})^2 \cdot f}{(\sum f) - 1}}$$

as an approximation to the "raw" sample standard deviation s. (The idea is that in each class, each number is assumed to lie at the class mark x', so that subtracting the mean $\bar{X}$ from x' is like subtracting the mean from each number. The difference is squared, and multiplying by the class frequency f assures that the squared difference will be "counted" the appropriate number of times. Dividing by $\sum f - 1$ is of course equivalent to dividing by $n-1$.) Compute $\bar{X}$ and S for this frequency distribution:

Class	f
40 to 44	11
45 to 49	14
50 to 54	19
55 to 59	22
60 to 64	20
65 to 69	14.

4. Given the fifty sample numbers

24	35	28	33	37	25	41	37	25	38
36	39	28	35	39	37	38	35	31	27
38	26	41	36	28	30	41	24	44	38
39	40	26	41	29	36	27	26	30	32
40	43	37	40	25	45	29	38	27	36

that are organized into the frequency distribution

Class	f
22 to 25	5
26 to 29	11
30 to 33	5
34 to 37	11
38 to 41	15
42 to 45	3

(a) Compute $\bar{X}$ and S from the frequency distribution alone.

(b) Compute $\bar{x}$ and s from the ungrouped data and compare $\bar{X}$ and S with the true values of $\bar{x}$ and s.

5. In a frequency distribution, the class (if any) with the highest frequency is called the **modal class.** How would one reasonably define, then, the *mode* ($\hat{X}$) of a frequency distribution? What, for example, would $\hat{X}$ be for the first frequency distribution examined in this section? Does $\hat{X}$ necessarily exist (as a unique statistic) in general?

6. Under what conditions, without raw data, is it possible to construct a frequency distribution from another frequency distribution?

7. What would be the class boundaries for the following frequency distribution, in which the unit of measure is *tenths:*

Class	f
20.5 to 30.4	20
30.5 to 40.4	31
40.5 to 50.4	51
50.5 to 60.4	40
60.5 to 70.4	23
70.5 to 80.4	15

8. What, then, is the median $\tilde{X}$ for the frequency distribution in the preceding problem? To how many decimal places should it be expressed?

SUGGESTED READINGS

Flaws and Fallacies in Statistical Thinking, by Stephen K. Campbell (Prentice-Hall, Inc., 1974). This book provides an amusingly eye-opening and forthright discussion of the various possible abuses of statistical terminology and procedures, with specific numerical examples.

How to Lie with Statistics, by Darrell Huff (W. W. Norton & Company, Inc., 1954). This oft-reprinted little volume provides an entertaining and revealing general discussion of the misuses of statistics, with emphasis on such things as misleading graphs.

Graphs, Models, and Finite Mathematics, by Joseph Malkevitch and Walter Mayer (Prentice-Hall, Inc., 1974). This book offers, in Chapter 7, a general discussion of statistical thinking, including comparisons between different measures of central tendency.

2 Fundamentals of Probability

Uncertainty and expectation are the joys of life.—Congreve

A COUNTING WITHOUT COUNTING

The science of statistics frequently concerns itself with decision making in the face of the unknown. As shown in later chapters, we often apply statistical methods of analysis to situations in which decisions must be made, *conclusions* must be drawn and acted on, on the basis of only partial information—typically, *sample* information from a population largely unseen and unknown. Such conclusions often take the form of highly educated guesses, and are entertained with the understanding that we know with high *probability*—but not with certainty—that we are drawing the proper conclusions.

Thus, in the study of statistical methods, some basic acquaintance with the fundamentals of *probability* will be useful.

First, some terminology.

In the subject of probability theory, we often use the term **experiment.** Conversationally, the word conjures up mental pictures of laboratories with beakers bubbling and instrument panels humming and white-frocked scientists giving injections to white mice. Such situations do indeed imply experimenting, but in probability theory we use the word "experiment" in a much broader sense.

For our purposes, an *experiment* is anything which can be done and observed with the intention of seeing which of a number of possible results is obtained.

Thus, to use a perhaps overworked but pertinent example, flipping a coin (to see whether heads or tails turns up) is an experiment—admittedly, a very simple experiment. Throwing a pair of dice (to see whether the total number coming up is 2, 3, 4, 5, 6, 7, 8, 9, 10, 11, or 12) is an experiment.

The selection of one card out of a deck of 52 playing cards (to see, simply, which card is drawn) is an experiment. The selection of one card out of a deck of 52 playing cards (to see what *suit* is drawn: spades, clubs, hearts, or diamonds) is another experiment, with a different purpose. The selection of one playing card for the purpose of seeing only what *color* is drawn (black or red) is yet another experiment.

The seating of six people around a picnic table—to see which of a number of possible seating arrangements is obtained—is an experiment. Observing a diner enter a café and order one of a number of possible meal combinations is an experiment.

The recording of the high temperatures on each of 12 consecutive summer days in a given city (to see what sequence of numbers is obtained, among an infinitude of possible sequences) is an experiment.

And, indeed, such a thing as giving a laboratory mouse an injection to see what reaction it has (among a number of possible reactions) is an experiment.

Given any experiment, we call the set of all possible *outcomes* of the experiment, the **sample space.**

EXAMPLES

1. Given the experiment of flipping an ordinary coin two times to see whether heads (denoted H) or tails (denoted T) will turn up, the sample space is

   $\{H, T\}$

   because these are the possible outcomes. (We ignore other possible outcomes if they are bizarre—for example, we do not list the possibility that when the coin is in the air it will vaporize or be carried away by a seagull. We list, in a sample space, only those possible outcomes among which we reasonably expect to have to choose.)

2. Given the experiment of flipping a coin twice in succession, to see what heads-or-tails results turn up in sequence, the sample space is

   $\{HH, HT, TH, TT\}$.

3. Given the experiment of selecting one card out of a deck of 52 playing cards, for the purpose of seeing *which* card is chosen, the sample space is

   {A♤, A♧, A♡, A♢, K♤, . . . , 2♢},

   that is, a list of the whole deck: all 52 cards.

4. Given the experiment of choosing one playing card, for the purpose of seeing what *suit* (spades, clubs, hearts, or diamonds) is obtained, the sample space is

   {♤, ♧, ♡, ♢}.

5. Given the experiment of choosing one playing card, for the purpose of seeing only what *color* of card is obtained, the sample space is

   {black, red}.

It is often important, in the study of probability, to determine *how many* things there are in a sample space—and it may be necessary to determine how many *without actually counting* the elements of the set, because they may be quite numerous. Thus one may have to have ways of "counting without counting." Like Elizabeth Barrett Browning, one must sometimes say, "Let me count the ways," and there are a number of mathematical principles that will help.

First, we consider the **multiplication rule.** This is a rule by which we can tell how many elements are in a sample space when the experiment consists of doing several things in sequence.

We begin with an example. Suppose the experiment is to order a meal in a cafeteria which offers *three* meat dishes, *five* vegetables, *two* kinds of bread, *four* drinks, and *three* desserts. It is further understood that the meal must consist of exactly one choice out of each category. In how many ways can such a meal be ordered? (That is, how many things—how many different such meals—are there in the sample space?)

We find the answer by *multiplying* the numbers of ways the successive choices can be made:

$$3 \cdot 5 \cdot 2 \cdot 4 \cdot 3 = 360.$$

Believe it or not, 360 different meals can be ordered! (See Figure 2.1.) Such are the large sample spaces that can be obtained, and we certainly need to be able to determine *how* numerous the possible outcomes are without actually trying to list and count them.

This example illustrates a general *rule of multiplication*:

If Task 1 can be done in n_1 ways, and if Task 2 can be done in n_2 ways, and if Task 3 can be done in n_3 ways, . . . and if Task k can be done in n_k ways, then it is possible to do Task 1 *and* Task 2 *and* Task 3 *and* . . . *and* Task k in

$$n_1 \cdot n_2 \cdot n_3 \cdot \ldots \cdot n_k$$

different ways.

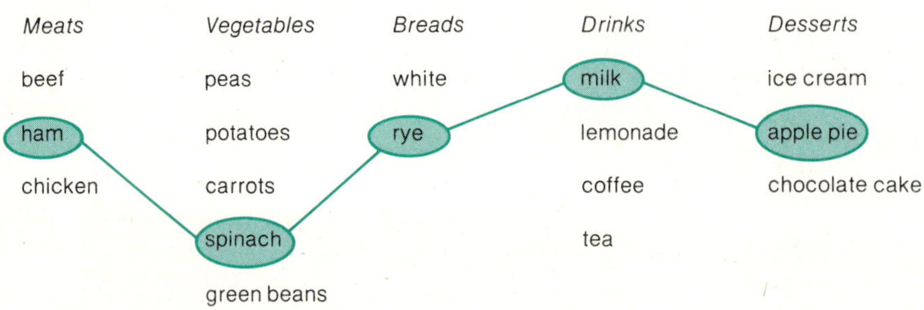

Figure 2.1 *One* possible meal out of 360 such possibilities.

EXAMPLES

6. Suppose an experiment consists of a student's choosing one mathematics course (from among 6 choices), one English course (from among 5 choices), one science course (from among 4 choices), one social science course (from among 8 choices), and one additional elective course (from among 10 choices). Then (assuming all possible combinations are permissible) the number of elements in the sample space of the experiment—the total number of ways to choose such a slate of courses—is found by multiplying:

$$6 \cdot 5 \cdot 4 \cdot 8 \cdot 10 = 9600.$$

Needless to say, we can be happy to arrive at this conclusion without having to *list* the 9,600 possibilities.

7. Suppose an experiment consists of getting dressed in the morning when one has a choice of 5 pairs of pants, 8 shirts, 11 ties, 2 pairs of shoes, 4 jackets, and 3 hats. Assuming that exactly one item in each category will be worn, the number of elements in the sample space of the experiment (the number of possible ways of dressing) would be

$$5 \cdot 8 \cdot 11 \cdot 2 \cdot 4 \cdot 3 = 10{,}560.$$

(This is not to say, of course, that all of the possible clothing combinations would necessarily be desirable!)

As a special case of the multiplication rule, let us look now at this question: Given n objects, in how many different ways can the objects be arranged in order? The multiplication rule will help.

For example, suppose we have a family of 10 people who have to be seated in some order every night around the dinner table, and we wish to know how many such seating arrangements there are to choose from.

First numbering the seats around the table from 1 to 10, we would reason as follows (see Figure 2.2). How many ways are there of filling Seat 1? *Ten*, of course, because there are 10 people. Then for *each* way that we may fill Seat 1, how many ways are there to fill Seat 2? There are 9 (since for each person who may sit in Seat 1, there will be some 9 people still waiting to be seated). Thus (by the multiplication rule) there are already $10 \cdot 9 = 90$ ways to fill Seats 1 and 2.

Then for each of these 90 ways, there are 8 ways to fill Seat 3 (since whoever sits in Seats 1 and 2 would leave some 8 people still waiting to be seated). Therefore Seats 1, 2, and 3 can be filled in $10 \cdot 9 \cdot 8 = 720$ ways.

For each of these 720 ways there are, likewise, 7 ways to fill the next seat, and so on, so that continuing around the table in this manner and applying the

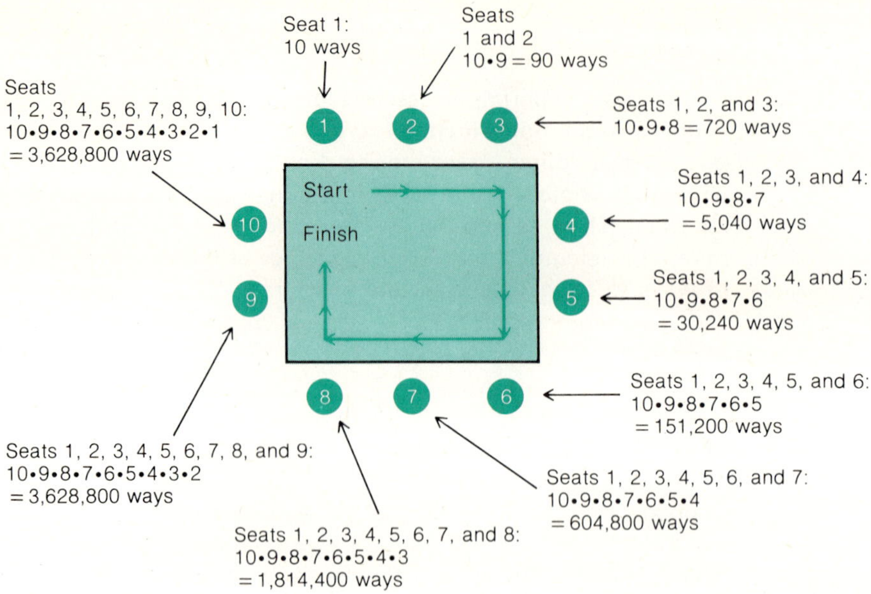

Figure 2.2

multiplication rule, we see that the total number of seating arrangements would have to be

$$10 \cdot 9 \cdot 8 \cdot 7 \cdot 6 \cdot 5 \cdot 4 \cdot 3 \cdot 2 \cdot 1 = 3{,}628{,}800.$$

Incredible as it may seem, there are over 3 million different ways to seat 10 people around a dinner table! If our family of 10 tried a different seating arrangement every night, never repeating an arrangement, it would take over 9,935 years for them to run through all the possibilities! The sample space of this experiment is a list of all 3,628,800 seating arrangements—hardly a sample space that we would want to write out!

The number

$$10 \cdot 9 \cdot 8 \cdot 7 \cdot 6 \cdot 5 \cdot 4 \cdot 3 \cdot 2 \cdot 1,$$

in which we use consecutively decreasing integers and "multiply all the way down to 1," is often referred to as **"10-factorial,"** written

$$10!$$

using an exclamation point—not to suggest excitement, though the size reached by such factorial computations may indeed occasion some excitement. In general,

$n!$ ("n-factorial") means

$$n \cdot (n-1) \cdot (n-2) \cdot \ldots \cdot 3 \cdot 2 \cdot 1$$

where the consecutive integers decreasing from n down to 1 are multiplied together. As further examples:

$6! = 6 \cdot 5 \cdot 4 \cdot 3 \cdot 2 \cdot 1 = 720$

$4! = 4 \cdot 3 \cdot 2 \cdot 1 = 24$

$8! = 8 \cdot 7 \cdot 6 \cdot 5 \cdot 4 \cdot 3 \cdot 2 \cdot 1 = 40{,}320.$

And as the example of the dinner table problem suggests:

The number of different ways of arranging n objects in order is always $n!$ (n-factorial).

EXAMPLES

8. Suppose you are going to read *six* books in some order. Then the number of different orders in which you may read the six books is

$6! = 6 \cdot 5 \cdot 4 \cdot 3 \cdot 2 \cdot 1 = 720.$

9. Suppose *four* telephone calls have to be made in some order. Then the number of different orders in which the calls can be made is

$4! = 4 \cdot 3 \cdot 2 \cdot 1 = 24.$

10. Suppose *eight* people are going to stand in line in front of a theater. Then the number of different orders in which they may line themselves up is

$8! = 8 \cdot 7 \cdot 6 \cdot 5 \cdot 4 \cdot 3 \cdot 2 \cdot 1 = 40{,}320.$

An arrangement of objects *in some order* is called a **permutation.** Thus we may say that

The number of permutations of n objects (taken all at a time) is $n!$ (n-factorial).

For instance, there are $4! = 24$ possible permutations (arrangements in order) of four objects. (Try arranging the letters A, B, C, D in order in various ways, and see if you do not arrive at exactly 24 such arrangements: 24 permutations.)

You will notice that we say there are $n!$ permutations of n objects taken *all at a time,* that is, taken all together, leaving none out. To ask how many ways eight people can stand in line at the theater and to answer by computing $8! = 40{,}320$, we are assuming that *all eight* people do indeed line up (and nobody decides to go bowling instead).

But we may also ask how many permutations there are for a set of objects if *some* but not necessarily *all* of them are to be arranged in order. There is a formula for this.

Given n objects, the number of permutations of the n objects taken r at a time (that is, the number of ways in which r out of the n objects can be taken and arranged in some order) is

$$\frac{n!}{(n-r)!}.$$

EXAMPLES

11. In a party of eight people ($n=8$), the number of ways in which *some three* of them ($r=3$) can go stand *in order* in line at a theater (while whatever remaining five there are do not join them) is

$$\frac{n!}{(n-r)!}=\frac{8!}{(8-3)!}=\frac{8!}{5!}=\frac{8\cdot7\cdot6\cdot5\cdot4\cdot3\cdot2\cdot1}{5\cdot4\cdot3\cdot2\cdot1}$$

$$=\frac{8\cdot7\cdot6}{1}=336.$$

You should understand this means that if, in the party of eight people, someone should say, "Let's pick some three of us to go line up, and then whatever three we pick, go line up in some order," the number of ways in which that can be done is 336. (First, there would be many ways of picking three; then for each such choice, many other ways of lining them up.)

12. The number of ways in which *some two* ($r=2$) out of a total of *six* books ($n=6$) can be selected and read in some order is:

$$\frac{n!}{(n-r)!}=\frac{6!}{(6-2)!}=\frac{6!}{4!}=\frac{6\cdot5\cdot4\cdot3\cdot2\cdot1}{4\cdot3\cdot2\cdot1}$$

$$=\frac{6\cdot5}{1}=30.$$

Notice, in the previous examples, that in computing

$$\frac{n!}{(n-r)!}$$

we generally do not actually have to multiply out all of $n!$ and $(n-r)!$ because the factors of $(n-r)!$ in the denominator will cancel out some of the factors of $n!$ in the numerator.

As a special case in the factorial notation we define 0! (zero-factorial) to be 1:

$$0! = 1.$$

This perhaps strange-looking definition is necessary, because (among other reasons) it is true in general that

$$(n+1) \cdot n! = (n+1)!$$

because if we have, already, $n!$ and multiply $(n+1)$ onto it, we have all the consecutive factors from $n+1$ down to 1, and this product is exactly what we mean by $(n+1)!$. Now if the statement

$$(n+1) \cdot n! = (n+1)!$$

is to remain true when $n=0$, we have:

$$(0+1) \cdot 0! = (0+1)!$$

or $\quad 1 \cdot (0!) = 1!$

or $\quad 0! = 1$

so that 0! *must* be defined to be 1.

The definition $0! = 1$ also works out well in expressing numbers of permutations. For example, in the formula

$$\frac{n!}{(n-r)!}$$

for the number of permutations of n things taken r at a time, if r happens to equal n (that is, if we wish to take the n objects *all* at a time), then the formula reduces to

$$\frac{n!}{(n-r)!} = \frac{n!}{(n-n)!} = \frac{n!}{0!} = \frac{n!}{1} = n!$$

consistent with the fact that we already know n things taken *all* at a time have $n!$ permutations. The latter formula, then, is a special case.

We have been discussing the problem of selecting r objects out of a total pool of n objects and arranging the r objects *in some order*. Now, suppose the order did *not* matter. That is, suppose we only wish to make a *selection* of r objects out of n available ones but do *not* care what order they are in. Such a selection without regard to order is called a **combination,** and the formula needed for the number of combinations possible is as follows.

Given n objects, the number of different *combinations* (selections without regard to order) of these n objects taken r at a time is

$$\frac{n!}{r!(n-r)!}$$

which differs from the permutations formula in that it has a factor of $r!$ in the denominator. Also, we have a special symbol for this new formula, because it is used so often. We write

$$\frac{n!}{r!(n-r)!}=\binom{n}{r}.$$

Thus $\binom{n}{r}$ stands for the number of combinations of n things taken r at a time.

EXAMPLES

13. A club contains eight members ($n=8$). Suppose the experiment consists of selecting some three of the members ($r=3$) to serve on a committee. The sample space, then, consists of *all* such possible three-person committees, and the number of such possible selections is

$$\binom{8}{3}=\frac{8!}{3!(8-3)!}=\frac{8!}{3!\,5!}$$

$$=\frac{8\cdot7\cdot6\cdot5\cdot4\cdot3\cdot2\cdot1}{(3\cdot2\cdot1)(5\cdot4\cdot3\cdot2\cdot1)}=\frac{8\cdot7\cdot6}{3\cdot2\cdot1}$$

$$=8\cdot7=56.$$

There are 56 different three-person committees that can be chosen; there are 56 combinations of eight things taken three at a time.

14. There are seven different movies ($n=7$) playing in town, and we wish to select some two of them ($r=2$) to see, without concern for the order in which we see them. The number of ways this selection of some two movies can be made is

$$\binom{7}{2}=\frac{7!}{2!(7-2)!}=\frac{7!}{2!\,5!}$$

$$=\frac{7\cdot6\cdot5\cdot4\cdot3\cdot2\cdot1}{(2\cdot1)(5\cdot4\cdot3\cdot2\cdot1)}=\frac{7\cdot6}{2\cdot1}=21.$$

There are 21 combinations of seven things taken two at a time. It would be possible for 21 people to choose to see two of the movies, without any two people choosing the same two movies.

15. The number of ways we may choose, from a total of six books ($n=6$), some two books ($r=2$) to read (without regard to reading them in any particular order) is

$$\binom{6}{2}=\frac{6!}{2!(6-2)!}=\frac{6!}{2!\,4!}=\frac{6\cdot 5\cdot 4\cdot 3\cdot 2\cdot 1}{(2\cdot 1)(4\cdot 3\cdot 2\cdot 1)}=\frac{6\cdot 5}{2\cdot 1}=15.$$

Compare this result with Example 12, in which we saw that there are 30 different ways to choose some two books (out of six) and read them *in a particular order.* That is, for six things taken two at a time, there are 30 *permutations* but only 15 *combinations.* In general, permutations are more numerous than combinations, because for combinations, order does not matter and therefore we do not get to count so many situations as different. With permutations, we count a lot of things as different from each other merely on the basis of the order.

A good way to think of the distinction between permutations (order important) and combinations (order unimportant) is to remember that if a *slate of officers* (such as president, vice-president, and secretary) is chosen from a group of people, the selection is a *permutation,* because (say)

Jones for president
Smith for vice-president
Brown for secretary

is genuinely different from (say)

Brown for president
Jones for vice-president
Smith for secretary

—but, if a *committee* of three people is chosen, then (the order in which we name its members being unimportant) the selection is a *combination.* That is, we really do not care if a committee is described as

Jones, Smith, and Brown
or Jones, Brown, and Smith
or Smith, Jones, and Brown
or Smith, Brown, and Jones
or Brown, Jones, and Smith
or Brown, Smith, and Jones

because all of these represent one and the same committee.

NEW TERMS IN THIS SECTION

experiment	**factorial**
sample space	**permutation**
multiplication rule	**combination**

EXERCISES 2-A

1. An experiment consists of ordering a ham sandwich in a certain cafeteria. In this establishment, a ham sandwich can be ordered on white bread, whole wheat, or rye; with or without cheese; with or without tomato; with or without lettuce; with or without pickle; and with mayonnaise, mustard, or plain. How many different ham sandwiches are possible? That is, how many outcomes are present in the experiment's sample space?

2. A vacationer living in the central United States has determined that there are 15 different states he might like to travel to, to spend his vacation (and he will spend it in only one state); he might go by car, bus, train, or air. Also, he can stay in his chosen vacation state for one week, two weeks, or three weeks, and might go alone or with a friend. How many different vacation plans are possible?

3. A cartoonist claims that she can construct cartoon character faces out of a stock of four face-shapes, six types of mouth, four types of nose, seven types of eyes, three types of ears, and five hair styles. If the experiment is to construct a cartoon face, among how many such faces in the sample space can the cartoonist choose?

4. In a certain state, auto license plates consist of two letters of the alphabet followed by a four-digit number, the first digit of which is never zero. How many such plates are possible? (Assume that any two-letter prefix is permissible.)

5. In the beginning of a chess game, White moves first, then Black moves. For White, each of eight pawns can move forward one square or two squares; also, each of the two knights has two possible moves. Then Black, in response, has all these possibilities too. In how many different ways can the first-move sequence (White moves, Black moves) be made? (Assuredly, some of these *possible* moves would be idiotic.)

6. An experiment consists of flipping a coin three times in succession, each time to see whether heads or tails turns up. How many outcomes are there in the sample space? List them, using the notation of Example 2.

7. Do the same with four flips of the coin.

8. In how many different ways (different orders) can
   (a) 5 people stand in a line?
   (b) 7 people stand in a line?
   (c) 9 people stand in a line?
   (d) 11 people stand in a line?
   (e) 12 people stand in a line?
   (f) 15 people stand in a line?

9. A college English instructor wants to assign, in a novels course, *David Copperfield, The Return of the Native, Vanity Fair, The Way of All Flesh, The Scarlet Letter,* and *Wuthering Heights.* In how many different orders can the novels be assigned? In how many ways if the instructor insists on assigning *David Copperfield* first?

10. A television scheduler has 10 film clips and has to select some 6 of them and show them in some order; different orders count as different schedules. In how many ways can it be done?

11. A club consisting of 14 members wishes to elect a slate of officers from among its membership: president, vice-president, secretary, and treasurer. (No person can hold more than one office.) How many different slates are possible?

12. An ice cream parlor has 20 flavors of ice cream. We wish to choose 3 flavors and sample them *in some order.* In how many ways can it be done? If this flavor-sampling is a daily ritual, 7 days a week year-round, then about how long would it take to run through all possible ways of tasting some 3 flavors in order?

13. In Exercise 2-A.12, if it does not matter what order the flavors are tasted in, then how many ways are there of selecting some 3 flavors out of the 20 to taste? If the selection is done daily, about how long would it take to run through all the possible selections?

14. Compute:

(a) $\binom{7}{5}$ (b) $\binom{7}{2}$ (c) $\binom{8}{3}$ (d) $\binom{8}{5}$ (e) $\binom{9}{4}$ (f) $\binom{9}{5}$

(g) $\binom{10}{3}$ (h) $\binom{10}{7}$ (i) $\binom{10}{10}$ (j) $\binom{10}{0}$ (k) $\binom{11}{10}$ (l) $\binom{11}{1}$.

15. A club that consists of 14 members wishes to select, from among its membership, 4 people to constitute a committee. How many different committees are possible? Compare this situation with Exercise 2-A.11.

16. A book club lists 16 books and invites you to select any 5 as an introductory offer. In how many ways can the selection be made?

17. A consumer research group wants to select 4 brands out of a field of 13 brands of cigarettes to test for tar content. In how many ways can the selection be made? What if they not only select four brands but are also interested in the *order* in which the selected brands are tested?

18. A university offers 18 upper-division literature courses. In how many ways can an English major choose 6 of these courses (order unimportant)? In how many ways can a student choose an ordered sequence of 6 courses (assuming all sequences actually possible to enroll in)?

19. A coin is flipped 10 times in succession. How many different heads-tails outcomes are possible? What about 11 times? 12 times? What does each additional flip do to the number of things in the sample space?

20. A student takes an exam containing 11 questions, any 7 of which must be answered. In how many ways can the questions to be answered be chosen? *After* the student chooses some specific seven questions to answer, in how many different *orders* can the student work on the questions, assuming them all independent?

PONDERABLES 2-A

1. Suppose you belong to a club containing 12 members (including yourself). A committee of 5 members is to be chosen.
   (a) Totally, how many committees can be chosen?
   (b) How many committees with you on them are possible?
   (c) How many committees without you on them are possible?

2. At cards (the deck containing 52 cards) how many different 5-card hands are there? How many of these hands contain the ace of spades?

3. An 11-person team wishes to choose a 3-member task force, and then have the task force select some 4 out of a field of 9 possible projects to work on. In how many ways can all this be done?

4. Prove that

$$\binom{n}{r}=\binom{n}{n-r}.$$

5. If n people stand in a line, then $n!$ represents the number of ways (orders) in which they can do so. What would

$$(n!)!$$

stand for? What would this be for $n=3$?

6. Some intuitive idea can be gained of the relation between the formulas

$$\frac{n!}{(n-r)!}$$ (the number of *permutations* of n objects taken r at a time)

and

$$\frac{n!}{r!(n-r)!}$$ (the number of *combinations* of n objects taken r at a time)

by considering the following example. Let $n=5$ and $r=3$. Thus we will look at the numbers of permutations and combinations of five objects taken three at a time.

Suppose the five objects are

$$\{A, B, C, D, E\}.$$

The number of *combinations* of these five objects, taken three at a time, must be

$$\frac{n!}{r!(n-r)!}=\frac{5!}{3!(5-3)!}=\frac{5!}{3!\,2!}=\frac{5\cdot4\cdot3\cdot2\cdot1}{(3\cdot2\cdot1)(2\cdot1)}=\frac{5\cdot4}{2\cdot1}=\frac{20}{2}=10$$

and, in fact, the ten combinations are:

A,B,C   A,B,D   A,B,E   A,C,D   A,C,E

A,D,E   B,C,D   B,C,E   B,D,E   C,D,E.

Now, *each* of these three-object combinations gives rise to *six* (3! = 6) different *permutations* as follows:

A,B,C:	*ABC*	*ACB*	*BAC*	*BCA*	*CAB*	*CBA*
A,B,D:	*ABD*	*ADB*	*BAD*	*BDA*	*DAB*	*DBA*
A,B,E:	*ABE*	*AEB*	*BAE*	*BEA*	*EAB*	*EBA*
A,C,D:	*ACD*	*ADC*	*CAD*	*CDA*	*DAC*	*DCA*
A,C,E:	*ACE*	*AEC*	*CAE*	*CEA*	*EAC*	*ECA*
A,D,E:	*ADE*	*AED*	*DAE*	*DEA*	*EAD*	*EDA*
B,C,D:	*BCD*	*BDC*	*CBD*	*CDB*	*DBC*	*DCB*
B,C,E:	*BCE*	*BEC*	*CBE*	*CEB*	*EBC*	*ECB*
B,D,E:	*BDE*	*BED*	*DBE*	*DEB*	*EBD*	*EDB*
C,D,E:	*CDE*	*CED*	*DCE*	*DEC*	*ECD*	*EDC*

Thus there are 60 permutations, *six* times as many (6 = 3!) as there are combinations. To get from the number of permutations

$$\frac{n!}{(n-r)!}=\frac{5!}{(5-3)!}=\frac{5}{2!}=60$$

to the number of combinations

$$\frac{n!}{r!(n-r)!}=\frac{5!}{3!(5-3)!}=\frac{5!}{3!\,2!}=\frac{1}{3!}\cdot\frac{5!}{2!}=\frac{1}{3!}(60)=\frac{1}{6}(60)=10$$

it is natural to divide by 3! = 6. And in general, dividing the number of *permutations* by $r!$ gives the smaller number of *combinations*, because it is always true that the number of objects (r) in any one combination can be *permuted* $r!$ different ways (but these $r!$ arrangements are still only one *combination*). Try it with $n=6$ and $r=4$, or other values of your own choosing.

B EVENTS AND PROBABILITIES

As we saw in the previous section, given an experiment of some kind, there is always a set of all possible *outcomes* for the experiment, and we call this set the *sample space*.

For example, if the experiment is to flip a coin three times in succession, the sample space (see Exercise 2-A.6 in the previous section) is:

$\{HHH, HHT, HTH, THH, HTT, TTH, THT, TTT\}$.

Suppose we look at a *subset* of this sample space, that is, select *some* but not necessarily all of the elements of the sample space. Such a subset is called an *event*.

An **event** is any subset of a sample space.

For instance, we might take only the elements

$E = \{HHT, HTH, THH\}$.

Considering the elements chosen to make up this subset E, we see that the *event E* is the event that when the coin is flipped three times, heads turns up *exactly twice*. (In E, all those elements with exactly two H's are present, and nothing else is.) Note that we do have to be careful to say *exactly* twice, because the event "When a coin is flipped three times, heads turns up (at least) twice" would be a different event:

$F = \{HHT, HTH, THH, \mathbf{HHH}\}$.

EXAMPLES

1. From the same sample space as just given, the event

$G = \{HHT, THH\}$

would be the event that heads turns up exactly twice *in a row*. Note that (for instance) *HTH* is not included because the H's are not consecutive; and *HHH* is not included because we said *exactly* two times in a row, not three.

It should be noted that finding appropriate *wording* for an event calls for working in the context of a specific given sample space. Thus, for example, in the sample space assumed here, we may also characterize G as the event that "tails turns up exactly once, either on the first flip or on the last." But if the experiment had been to flip a coin *four* times, the two descriptions we have seen would not describe the same event—for instance, the outcome *THHT* would satisfy the first description but not the second.

2. In the experiment of flipping a coin three times, the event

$J = \{HHH\}$

is the event that heads turns up all three times. Such an event, consisting of a subset of the sample space with only one element, is called a **simple event.**

Given an experiment, its sample space, and an *event* E drawn from the sample space, we would like to assign a **probability** to the event E, and denote this probability $P(E)$. That is, we would like to determine a numerical measure of how likely the event E is to occur.

To do this, we first agree that the probability $P(E)$ of an event E must always be a number between 0 and 1 inclusive:

$$0 \leq P(E) \leq 1.$$

An event will have a probability $P(E)=0$ when it is certain *not* to occur. For example, if E is the event that we will get heads *four* times when flipping a coin three times, then clearly we should assign this impossible event the probability $P(E)=0$. On the other extreme, an event E is assigned a probability $P(E)=1$ when the event is *certain* to occur. For example, if E is the event that we will get heads zero, one, two, or three times when flipping a coin three times, then clearly this event is certain and its assigned probability ought to be $P(E)=1$. (In effect, the event E is the whole sample space.)

More often, though, an event E will have a probability that is somewhere *between* the extremes 0 and 1. Frequently such probabilities are stated in terms of percentages: "The probability of rain tomorrow is 75%." In statistics, however, we would prefer to write $P(E)=0.75$, expressing the probability as a pure number between 0 and 1 instead of as a percentage.

To assign probabilities to events, we first think of each single element in the sample space as having a certain appropriate "weight" or intuitive likelihood, depending on the nature of the sample space. If the sample space is such that it is reasonable to assign *equal* "weights" to the elements, then we simply do so by dividing up the total probability of 1 evenly among the elements involved.

For example, in the experiment of flipping a coin three times, with sample space

$$\{HHH, HHT, HTH, THH, HTT, TTH, THT, TTT\}$$

because there are eight elements—and intuitively all are equally likely, assuming the coin is fair—we would assign each element a probability or "weight" of $\frac{1}{8}$. Thus the probability of getting (say) HTH is $\frac{1}{8}$, and we write

$$P(HTH)=\frac{1}{8}.$$

EXAMPLE

3. In the experiment of tossing one die (one of a pair of dice) to see what number turns up, the sample space is

$$\{1, 2, 3, 4, 5, 6\}$$

and assuming the die to be fair so that all these outcomes are equally likely, we would assign to each outcome a "weight" or probability of $\frac{1}{6}$. So, for instance, the probability of getting a 4 is $\frac{1}{6}$, and we may write

$$P(4)=\frac{1}{6}.$$

To assign probabilities to other more inclusive events, we would simply *add* the "weights" involved.

EXAMPLES

4. In the experiment of flipping a fair coin three times, where the sample space has been observed to contain eight elements, each with a "weight" or probability of $\frac{1}{8}$, the probability of the event

$$E=\{HHT, HTH, THH\}$$

would be $\frac{1}{8}+\frac{1}{8}+\frac{1}{8}=\frac{3}{8}$; that is,

$$P(E)=\frac{3}{8}.$$

The probability of getting heads exactly twice out of three tries is $\frac{3}{8}$. Another way to look at it is that *three out of eight* outcomes are describable in that manner, and the **relative frequency** with which the event E can be expected to occur is $\frac{3}{8}$ (three times out of eight) when the experiment is done repeatedly.

It should be pointed out that in studying such probabilities, we are not necessarily obsessed with such gambling-oriented topics as coin flipping. The idea of flipping a fair coin is only an example—a mathematical *model*—of *any* process that has a two-outcome, 50–50 chance nature.

5. In the experiment of tossing one fair die, where the sample space is

$$\{1, 2, 3, 4, 5, 6\}$$

and the probability weight assigned to each element is $\frac{1}{6}$, suppose the event F

is

$F = \{3, 6\}$.

That is, F is the event that the number turning up on the die is either 3 or 6. (The number turning up is a multiple of 3.) Then the probability of F is

$$P(F) = \frac{1}{6} + \frac{1}{6} = \frac{2}{6} = \frac{1}{3}.$$

6. Let the experiment be the drawing of one card at random from a deck of 52 playing cards, so that the sample space is simply a list of all 52 cards and we would assign a probability of $\frac{1}{52}$ to each card. Now suppose the event G is

$$G = \{K♤, K♧, K♡, K♢\},$$

that is, G is the event that a king is drawn. Then the probability of G (the probability that a king is drawn) is

$$P(G) = \frac{4}{52} = \frac{1}{13}.$$

7. An organization consists of 10 people, including Jones, Smith, and Brown. A committee of 3 people is to be drawn from the membership, *completely at random*. What is the probability that the committee chosen will be Jones, Smith, and Brown?

To answer this, we first note that the sample space contains $\binom{10}{3}$ possibilities (the number of different combinations of 10 things taken 3 at a time), and $\binom{10}{3}$ is evaluated as

$$\frac{10!}{3!\,(10-3)!} = \frac{10!}{3!\,7!} = \frac{10 \cdot 9 \cdot 8}{3 \cdot 2 \cdot 1} = 120.$$

And since Jones-Smith-Brown is 1 out of 120 equally likely combinations, the probability required is

$$\frac{1}{120}.$$

Suppose it is impossible for two events E and F (drawn from an experiment's sample space) to occur at the same time (i.e., they cannot *both* occur).

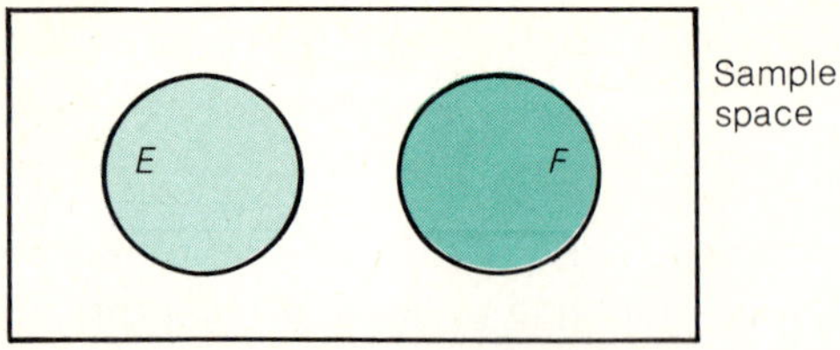

Figure 2.3 *E* and *F* are mutually exclusive events.

Then we say that the two events are **mutually exclusive.** (They literally exclude each other: if *E* happens, then *F* cannot happen, and vice versa). Formally,

Two events *E* and *F* are *mutually exclusive* when $P(E \text{ and } F)=0$.

(In terms of set theory, we may also say that the *intersection* of the sample space subsets *E* and *F*—the set of elements that *E* and *F* share—is the empty set: $E \cap F = \emptyset$.) See Figure 2.3.

EXAMPLES

8. Again, let the experiment be the drawing of one card from a deck of 52, so that the sample space is a list of all 52 cards. Let the events *E* and *F* be

E = {J♧, Q♧, K♧}

and *F* = {J◇, Q◇, K◇}.

(That is, *E* is the event that the card drawn is a club face card, and *F* is the event that the card drawn is a diamond face card.) Then since the two subsets *E* and *F* share *no* elements—*no* card drawn could be *both* a club face card and a diamond face card—we conclude that the events *E* and *F* are mutually exclusive: $P(E \text{ and } F)=0$. *E* and *F* cannot occur at the same time.

9. With still the same experiment, let the events *G* and *H* be:

G = {all black cards},

H = {A♤, A♧, A◇, A♡}.

(*G* is the event that a black card is drawn, and *H* is the event that an ace is drawn.) Then *G* and *H* are *not* mutually exclusive, because they *can* both occur at the same time. It is *not* true that $P(G \text{ and } H)$ is 0, because *G* and *H* do share, in fact, two elements: the ace of spades and the ace of clubs: the two cards that *both* are black and are aces. (See Figure 2.4.) If someone tells us (say) that *G* has occurred, we cannot rule out the possibility that *H* has also occurred. In fact, the probability that they will *both* occur at the same time

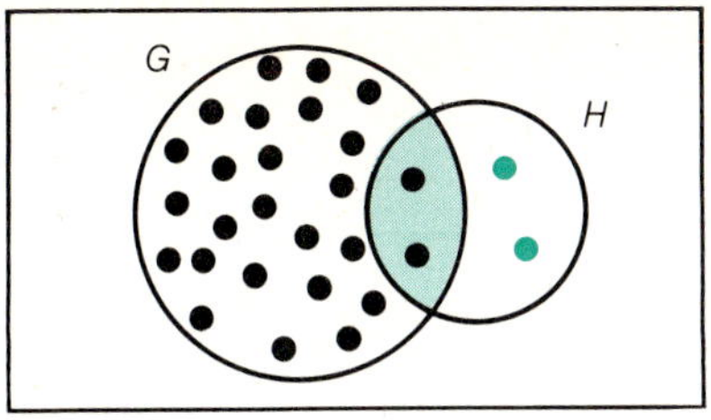

Figure 2.4 *G* and *H* are not mutually exclusive events.

(because there are exactly 2 cards out of 52 answering to both descriptions) is $P(G \text{ and } H) = \frac{2}{52} = \frac{1}{26}$.

When two events are mutually exclusive, we can compute the probability that one event *or* the other will occur, by simply *adding* the individual probabilities.

When events E and F are mutually exclusive, then $P(E \text{ or } F) = P(E) + P(F)$.

EXAMPLES

10. For the situation in Example 8, in the draw-one-card experiment with events

$E = \{J♣, Q♣, K♣\}$

$F = \{J♢, Q♢, K♢\}$,

because E and F are *mutually exclusive,* we can compute the probability that either E *or* F will occur by adding the probabilities of E and F:

$$P(E \text{ or } F) = P(E) + P(F) = \frac{3}{52} + \frac{3}{52} = \frac{6}{52} = \frac{3}{26}.$$

Note that the event E *or* F is the event

$\{J♣, Q♣, K♣, J♢, Q♢, K♢\}$

(the event that a club face card *or* a diamond face card is drawn), where the probability is verifiably $\frac{6}{52} = \frac{3}{26}$. (The set shown is really the *union* of the given sets: $E \cup F$, the result of "pooling" the elements of E together with the elements of F.)

11. For again the same experiment (drawing a playing card) let the events E

and F be

$E = \{A\spadesuit, A\clubsuit, A\heartsuit, A\diamondsuit\}$,

$F = \{$all 48 cards that are *other* than aces$\}$.

Clearly E and F are mutually exclusive, because no card drawn could both be and *not* be an ace. Thus we may say that the probability that E *or* F will occur is

$$P(E \text{ or } F) = P(E) + P(F) = \frac{4}{52} + \frac{48}{52} = \frac{52}{52} = 1.$$

Not a surprising result, when you think about it. (The card drawn *must* absolutely either be an ace or not be an ace.) Such events, which are mutually exclusive and have probabilities adding up to 1, are called **complementary events;** each event is the *complement* of the other. In a complementary pair such as this, each event is essentially a statement that the other event does *not* occur. (Note, though, that in Example 10 we have two events which, though they are mutually exclusive, are not complementary, because their probabilities do not "cover the whole sample space"; that is, they do not add up to 1.)

But suppose two events E and F are *not* mutually exclusive. Then we *cannot* compute $P(E \text{ or } F)$ by just adding $P(E) + P(F)$.

We must first clarify what "E *or* F" means, exactly, in case the events E and F are not mutually exclusive, that is, in case they *can* both occur simultaneously. The mathematical usage of the word "or" is the "inclusive *or*"—the statement "E *or* F" is taken to mean "E *or* F *or both.*" (That is, "E *or* F" means that at least one of the events E, F occurs—and we do not rule out that they *both* occur.) With this understanding, the way to compute the probability $P(E \text{ or } F)$ is by a new formula:

Given two events E and F that are *not necessarily mutually exclusive,* the probability that E occurs *or* F occurs (or both) is

$$P(E \text{ or } F) = P(E) + P(F) - P(E \text{ and } F).$$

Thus we do not just add the individual probabilities—we also subtract the probability that the two events would *both* occur at the same time. The reasoning behind this can be seen by thinking about Figure 2.5. If we *did* just add $P(E) + P(F)$, we would be counting some things *twice*: those items belonging *both* to E and to F; the double shading of the subsets' intersection shows this. Because we do not want to count anything twice, we look at the portion of the probability that *has* been counted twice, in the intersection, corresponding to the statment "E *and* F", and, to compensate, subtract that portion back off again *once.*

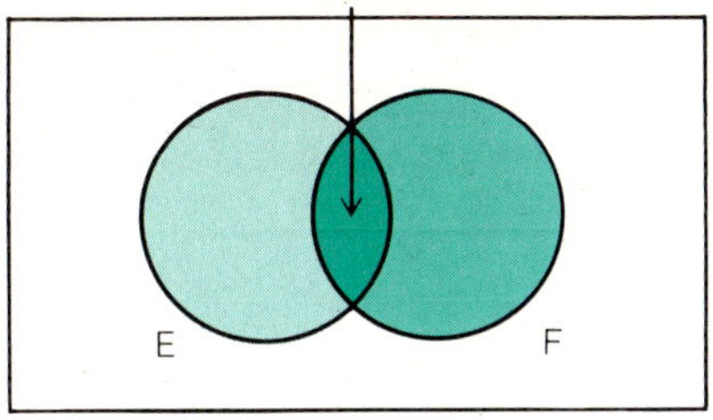

Figure 2.5 Everything that is shaded: "*E* or *F* (or both)."
$P(E \text{ or } F) = P(E) + P(F) - P(E \text{ and } F)$.

EXAMPLES

12. For the experiment of drawing one playing card, let the events E and F (which are *not* mutually exclusive) be

$E = \{Q\spadesuit, Q\clubsuit, Q\heartsuit, Q\diamondsuit\},$

$F = \{\text{all 26 red cards}\},$

and let us compute $P(E \text{ or } F)$, the probability that the card drawn is a queen *or* a red card (or both). We *cannot* just add $P(E)+P(F) = \frac{4}{52}+\frac{26}{52}=\frac{30}{52}$, because this procedure would, in effect, count some cards *twice* (the ones that answer *both* descriptions: queen of hearts and queen of diamonds). Because the probability of E and F happening together, simultaneously, is

$$P(E \text{ and } F) = \frac{2}{52},$$

the correct probability is

$$P(E \text{ or } F) = P(E) + P(F) - P(E \text{ and } F)$$

$$= \frac{4}{52}+\frac{26}{52}-\frac{2}{52}=\frac{30}{52}-\frac{2}{52}$$

$$= \frac{28}{52}=\frac{7}{13}.$$

And we may corroborate this result by noticing how many cards there are, in fact, that are *either* queens *or* red (or both): 28, because we have 26 red cards (including the two red queens) and also the two black queens ($Q\spadesuit$ and $Q\clubsuit$).

13. Suppose you and a friend belong to a club in which there are 12 members. A committee of 4 is to be chosen at random. What is the probability that *you* will be on the committee? What is the probability that your friend will be on the committee? What is the probability that *both* of you will be? The probability that either you *or* your friend (or both) will be?

Let us first see how many committees (combinations) of 4 out of 12 members are possible:

$$\binom{12}{4} = \frac{12!}{4!\,(12-4)!} = \frac{12!}{4!\,8!}$$

$$= \frac{12 \cdot 11 \cdot 10 \cdot 9}{4 \cdot 3 \cdot 2 \cdot 1} = 495.$$

Now: in how many ways can a committee be chosen with *you* on it? If you are to be on the committee, then in some manner 3 more people (out of a field of 11) must be chosen, and the number of ways of doing that would be

$$\binom{11}{3} = \frac{11!}{3!\,(11-3)!} = \frac{11!}{3!\,8!} = \frac{11 \cdot 10 \cdot 9}{3 \cdot 2 \cdot 1} = 165.$$

That is, there are 165 (out of a total of 495) committees possible that would include you. Thus the *probability* that you will be on the chosen committee is the ratio

$$\frac{\text{number of committees with you belonging}}{\text{total number of committees}} = \frac{165}{495} = \frac{1}{3}.$$

(Hardly what one could call a "common sense" result or intuitive result—it had to be computed.) It should be pointed out that this probability (using the idea of the ratio just examined) can also be computed directly (making use of available cancellations) by forming the ratio:

$$\frac{\binom{11}{3}}{\binom{12}{4}} = \frac{\dfrac{11!}{3!\,8!}}{\dfrac{12!}{4!\,8!}} = \frac{11!}{3!\,8!} \cdot \frac{4!\,8!}{12!}$$

$$= \frac{11!\,4!}{12!\,3!} = \frac{4}{12} = \frac{1}{3}.$$

Clearly, the probability that your *friend* will be on the committee is computed in the same way as for you, and is also $\frac{1}{3}$. The probability that *both* you and your friend will be chosen is computed as follows. If both of you are chosen, then the remaining 2 slots must be filled in some fashion from among the remaining 10 people; and the number of ways of doing so is

$$\binom{10}{2} = \frac{10!}{2!\,(10-2)!} = \frac{10!}{2!\,8!} = \frac{10 \cdot 9}{2 \cdot 1} = 45.$$

And the probability that both you and your friend are chosen is given by the ratio

$$\frac{\text{number of committees including both of you}}{\text{total number of committees possible}}=\frac{45}{495}=\frac{1}{11}.$$

Finally, to answer the question, "What is the probability that you *or* your friend (or both) will be on the committee?" we follow the formula

$$P(E \text{ or } F)=P(E)+P(F)-P(E \text{ and } F)$$

(with E the event that you are selected, and F the event that your friend is selected) obtaining

$$P(E \text{ or } F)=\frac{1}{3}+\frac{1}{3}-\frac{1}{11}=\frac{2}{3}-\frac{1}{11}$$

$$=\frac{22}{33}-\frac{3}{33}=\frac{19}{33}.$$

NEW TERMS IN THIS SECTION

event
simple event
probability
relative frequency
mutually exclusive events
complementary events
***odds**
***conditional probability**
***independent events**

EXERCISES 2-B

1. Write out the sample space for the experiment of flipping a coin *four* times, and write out each of the following events as subsets of the sample space:
   (a) Heads turns up exactly two times.
   (b) Heads turns up exactly two times, and in a row.
   (c) Heads turns up exactly three times.
   (d) Heads turns up exactly three times in a row.
   (e) Heads turns up at least three times.
   (f) Heads turns up at least three times in a row.
   (g) Heads and tails alternate.
   (h) Tails turns up first.
   (i) Heads turns up, but first tails turns up at least twice.
   (j) It never happens that tails occurs as much as twice in a row.

2. What is the probability of each event in Exercise 2-B.1? Let the event in part (a) be called A, the event in (b) be called B, and so on, so that you are evaluating $P(A)$, $P(B)$, and so on. Reduce all fractions to lowest terms.

3. What can be said about the event J in the previous exercise and the event K in which tails occurs at least three times?

4. A bag contains a number of colored cards: 17 red cards, 12 green cards, 15 yellow cards, 21 blue cards, and 19 orange cards. One card is drawn at random. Find the probability that the card drawn is

(a) green
(b) orange
(c) blue
(d) red
(e) yellow
(f) blue or orange
(g) blue and orange
(h) yellow or green
(i) yellow, orange, or red
(j) not green

5. What can be said about the events in Exercise 2-B4 (a) and (j) (aside from the fact that they are mutually exclusive)? Look at their probabilities.

6. If 23% of the students at a certain college are English majors and 11% are chemistry majors, and if no students are double-majoring in both English and chemistry, then what is the probability (write your answers as decimals) that a student chosen at random will be
(a) a chemistry major?
(b) an English major?
(c) a chemistry major *or* an English major?
(d) other than a chemistry major?
(e) other than an English major?
(f) neither a chemistry major nor an English major?

7. In Exercise 2-B.6 look again at event D in part (d) and event E in part (e). Are they mutually exclusive? How would you compute $P(D \text{ or } E)$? What is the result? What does it mean?

8. A fair pair of dice is tossed, to see what *two* numbers turn up (not just to see their sum); think of the dice as falling first one, then the other, so that (for example) getting 2 and then 3 is counted as different from getting 3 and then 2.
(a) What is the sample space, written out?
(b) What is the probability that the sum of the numbers will be

12? 11? 10? 9? 8? 7? 6? 5? 4? 3? 2? 1?

(c) What is the probability that the sum of the numbers will be even?
(d) What is the probability that the sum of the numbers will be odd?
(e) Even *or* odd? Even *and* odd?
(f) What is the probability that the sum of the numbers will be greater than 4? Greater than or equal to 4?

9. An English instructor teaching a poetry course and looking for possible textbooks has available paperback poetry collections by Robert Frost, Emily Dickinson, W. H. Auden, Percy Bysshe Shelley, John Keats, Algernon Swinburne, John Milton, Carl Sandburg, John Greenleaf

Whittier, John Masefield, Walt Whitman, Elizabeth Barrett Browning, and Edgar Allan Poe. The instructor wishes to choose some five of these collections to use in class.

(a) In how many ways can the choice be made?

(b) In how many ways can the choice be made with the Poe collection included?

(c) What would be the probability, then, that the Poe collection would be included if the selection were made *at random*? (Admittedly, not the best way, generally, to choose textbooks!)

(d) What (again, assuming random selection) would be the probability that the five collections chosen would be Frost, Dickinson, Poe, Whittier, and Sandburg?

(e) What would be the probability that the five chosen would *not* be those five exactly?

(f) Assuming random selection, what would be the probability that Whittier, Swinburne, and Whitman would all be chosen?

(g) What would be the probability that Keats, Browning, Milton, and Auden would all be chosen?

(h) What would be the probability that neither Shelley nor Masefield would be chosen?

10. Alice, Jane, Ann, Betty, and Carol go to the movies and stand in line at the box office.

(a) In how many different orders can they line up?

(b) What is the probability, if the order is random, that they will line up in the order Alice, Jane, Ann, Betty, Carol?

(c) What is the probability that they will *not* line up in that order?

(d) In how many ways can they line up in some order with Betty first?

(e) What is the probability (again, assuming randomness) that Betty will be first?

(f) What is the probability that Betty would *not* be first?

(g) In how many ways can they line up with Jane first and Alice last?

(h) What is the probability that Jane will end up first and Alice last?

(i) What is the probability that Alice will end up first and Jane last?

(j) What is the probability that *either* they will end up Jane first with Alice last, or Alice first with Jane last?

(k) What is the probability that neither Alice nor Jane will be either first or last (that is, Alice and Jane will both be at "internal" positions in the line)?

PONDERABLES 2-B

1. Is the formula

$$P(E \text{ or } F) = P(E) + P(F) - P(E \text{ and } F)$$

valid when E and F *are* mutually exclusive? Explain.

2. Given two events E and F, what is represented, conceptually, by the probability

$$P(E) + P(F) - 2 \cdot P(E \text{ and } F)?$$

Refer to Figure 2.4.

3. When the probability of an event E is $P(E) = \frac{a}{b}$, then we may express $P(E)$ in terms of **odds**

by saying that the odds of E's occurrence are a to $(b-a)$. For example, if $P(E)=\frac{2}{5}$, then the *odds* that E will occur would be 2 to $(5-2)$, that is, 2 to 3. Express each of the following probabilities in terms of odds:

(a) $P(E)=\frac{3}{7}$ (b) $P(E)=\frac{5}{8}$ (c) $P(E)=\frac{2}{9}$ (d) $P(E)=\frac{11}{12}$

(e) $P(E)=\frac{6}{7}$ (f) $P(E)=\frac{1}{4}$ (g) $P(E)=\frac{7}{10}$ (h) $P(E)=\frac{17}{20}$

(i) $P(E)=\frac{19}{100}$ (j) $P(E)=\frac{9}{109}$

4. Conversely, if we say, "The odds for E are s to t," then we are really saying:

$$P(E)=\frac{s}{s+t}.$$

For example, if the odds are 7 to 13 that an event E will occur, then

$$P(E)=\frac{7}{7+13}=\frac{7}{20}.$$

Find $P(E)$ if the odds that E will occur are

(a) 5 to 8 (b) 7 to 9 (c) 4 to 15 (d) 3 to 7
(e) 15 to 16 (f) 16 to 15 (g) 1 to 9 (h) 11 to 3
(i) 8 to 9 (j) 14 to 5

5. Beside talking about the simple probability $P(E)$ of an event E, we may talk about **conditional probability.** A conditional probability $P(E \mid F)$ ("the probability of E, *given* F") is the probability that one would assign to E *given* that F has occurred, or is certain to occur. For example, in the experiment of drawing one playing card from a deck of 52, let the events E and F be as follows:

E: the ace of spades is drawn.
F: a black card is drawn.

Without any reference to the event F, if we have drawn a card but have not looked at it, we would set $P(E)=\frac{1}{52}$. *But* if someone says, "*I* saw the card that you drew, and I can tell you that the card is black," then we have been *given* that the event F has occurred, and the *conditional* probability we would now assign to E would be

$$P(E \mid F)=\frac{1}{26}$$

because (given F) we know that we only need consider the 26 black cards, and the ace of spades is one of them. For the events

A: a red card is drawn
B: a face card (J, Q, K) is drawn
C: an ace is drawn
D: a card in the suit of diamonds is drawn

(together with the events E and F already described), find:

(a) $P(E)$	(b) $P(E \mid C)$	(c) $P(E \mid A)$	(d) $P(E \mid D)$
(e) $P(B)$	(f) $P(B \mid F)$	(g) $P(F)$	(h) $P(F \mid B)$
(i) $P(D)$	(j) $P(D \mid A)$	(k) $P(C)$	(l) $P(C \mid E)$
(m) $P(E \mid B)$	(n) $P(A \mid F)$	(o) $P(B \mid A)$	(p) $P(B \mid D)$
(q) $P(D \mid C)$	(r) $P(D \mid B)$	(s) $P(A \mid D)$	(t) $P(D \mid F)$

6. Given two events J and K, the probability that J and K will *both* occur can be expressed as

$$P(J \text{ and } K) = P(J \mid K) \cdot P(K).$$

For example, again in the card-drawing experiment, for the events

J: the card drawn is a numerical card (2 through 10) of rank greater than 4
K: the card drawn is a numerical card of *even* rank,

we would have

$$P(J \text{ and } K) = P(J \mid K) \cdot P(K)$$
$$= \frac{12}{20} \cdot \frac{20}{52} = \frac{12}{52} = \frac{3}{13},$$

and we may check this result directly by noting that there are 12 cards in the deck that are *both* between 5 and 10 inclusive *and* even-ranked, so that $P(J \text{ and } K)$ should indeed by direct count be $\frac{12}{52}$. Use this new formula and your results from the previous problem to compute each of the following, and check your result by direct count:
(a) $P(B \text{ and } F)$ (b) $P(D \text{ and } C)$ (c) $P(D \text{ and } F)$

7. It *may* happen, for two events J and K, that $P(J \mid K) = P(J)$. That is, it may happen that the probability of J, *given* K, is the same as the simple probability of J with no mention of whether K has occurred or not, so that K does not affect J's probability at all. In this case, J is said to be **independent** of K. By using the formula in the previous problem in the form

$$P(J \mid K) = \frac{P(J \text{ and } K)}{P(K)} \quad \text{or} \quad P(K \mid J) = \frac{P(K \text{ and } J)}{P(J)}$$

(presuming $P(K)$ and $P(J)$ to be nonzero) prove that if $P(J \mid K) = P(J)$, then $P(K \mid J) = P(K)$. (That is, prove that if J is independent of K, then K is independent of J. Thus the events do not affect each other's probabilities.) We simply say, then, that J and K are independent events.

8. Can mutually exclusive events be independent? Explain.

9. Show that events A and B of Ponderable 2-B.5 are independent. Intuitively, why does this happen?

10. Suppose 45 people apply for a job and all have exactly the same apparent qualifications, so that the employer decides to choose the person for the job at random. Some of the applicants are men and some are women, and some are married and some single, as follows:

	Men	Women
married	12	15
single	8	10

Let A be the event that the person chosen is a man, and let B be the event that the person chosen is married. Show that the events A and B in this situation are *independent* by showing that $P(A \mid B) = P(A)$. (Remember, $P(A \mid B)$ is the probability that we would assign to the probability that the chosen person is a man if we somehow *know* that the person chosen is married.)

11. If two events J and K are independent, so that $P(J \mid K) = P(J)$, the formula

$$P(J \text{ and } K) = P(J \mid K) \cdot P(K)$$

becomes

$$P(J \text{ and } K) = P(J) \cdot P(K).$$

That is, when two (or in fact more) events do not affect each other's probabilities, we can find the probability that they *both* (or all) happen by multiplying the individual probabilities. Remembering this, consider the following. An experiment consists of first tossing a fair die (getting 1, 2, 3, 4, 5, or 6) and then drawing a playing card at random from a deck of 52. Clearly we would expect the die result and the card result to be independent. Find, then, the probability that:

(a) The die will be 3 and the card will be red.
(b) The die will be 5 and the card will be hearts.
(c) The die will show an even number and the card will be a face card.
(d) The die will show a number larger than 2 and the card will be an ace.
(e) The die will show something other than 4 and the card will not be clubs.

C BINOMIAL DISTRIBUTION

Let us suppose that an experiment had the following form.

1. The experiment consists of a number of "trials." This may entail a sequence in which something is done repeatedly, time after time, or a group of separate things being done at the same time.

2. Each trial will lead to exactly one of two possible outcomes: "success" or "failure," where the meanings of these words depend upon the particular experiment; each trial is simply a yes-or-no situation.
3. The trials are independent of each other; that is, one trial does not affect the outcome of another trial, so that for all trials there is the same constant probability that any one trial will result in success. We use the symbol π (the Greek letter *pi*, having nothing whatever to do with the geometric π encountered in studying circles) to denote the probability that any single trial will be a success.

If an experiment has this form, then it is called a **binomial experiment,** because (among other reasons algebraically too far-flung to go into here) *bi-* means "two," and each trial has exactly two possible outcomes (success or failure).

EXAMPLES

1. Consider the experiment of flipping a fair coin three times, each time to see whether we get heads (which we will arbitrarily designate as "success") or tails (which we designate as "failure"). The experiment is a binomial experiment. Each flip of the coin is a trial, so that in this case there are three trials, each with two possible outcomes. The trials are independent, because it is reasonable to assume that no flip of the coin is affected by any other flip. (Do not fall into the trap of thinking that if tails has just turned up twice then it is time to get heads on the next flip—a form of the Gambler's Fallacy.) And the probability of success (heads) each time is $\pi = \frac{1}{2}$.

2. A bag contains 10 cards: 3 green cards and 7 blue cards. The experiment is to draw a card at random five times in succession (*replacing* the drawn card each time before drawing again), each draw being arbitrarily considered successful if the card drawn is green; blue, then, is failure. The experiment is binomial, because each of the five independent trials (independent because of *replacing* the card each time) has exactly two possible outcomes, and the probability of success each time is a constant $\pi = \frac{3}{10} = 0.3$.

3. In the previous example, suppose we do *not* replace the card before drawing again. Then the experiment is *not* binomial in nature, because the trials are *not independent:* one trial will affect another, and the probability of success changes from one trial to the next. For example, after the first trial has taken place, we have on the second trial a $\frac{2}{9}$ or $\frac{3}{9}$ chance of getting a green card, depending upon whether the first trial produced a green card or not. (See Ponderable 2-C.2 at the end of this section: the hypergeometric distribution.)

When we do have a binomial experiment, we are often interested in knowing the probability that, out of a total of n trials, some number x of successes will be obtained. For example, in the experiment of flipping a coin eight times, identifying heads as success, we may want to find out the probability that (say) five out of the eight flips will be successes.

We compute such probabilities by using the **binomial probability formula,** as follows.

Given a binomial experiment, with n trials and a probability π of success on any one trial, the probability that exactly x successes out of n trials will be obtained is

$$P(x \text{ successes}) = \binom{n}{x}\pi^{x}(1-\pi)^{n-x}.$$

Remember, $\binom{n}{x}$ is the number of combinations of n things taken x at a time:

$$\binom{n}{x} = \frac{n!}{x!\,(n-x)!}.$$

EXAMPLES

4. Suppose a binomial experiment consists of flipping a fair coin eight times ($n=8$) and we want to know the probability of getting heads (success) exactly five times ($x=5$). Because the probability of success is $\pi=\frac{1}{2}$ each time, the desired probability is

$$\begin{aligned} P(5 \text{ successes}) &= \binom{8}{5}\left(\frac{1}{2}\right)^{5}\left(1-\frac{1}{2}\right)^{8-5} \\ &= \frac{8!}{5!\,(8-5)!}\left(\frac{1}{2}\right)^{5}\left(\frac{1}{2}\right)^{3} = \frac{8!}{5!\,3!}\left(\frac{1}{2}\right)^{8} \\ &= \frac{8\cdot7\cdot6}{3\cdot2\cdot1}\left(\frac{1}{2^{8}}\right) = 8\cdot7\left(\frac{1}{256}\right) = \frac{56}{256} = \frac{7}{32} \approx 0.219. \end{aligned}$$

5. Suppose the experiment consists of administering an experimental drug independently to each of eight patients ($n=8$), where the drug is known to have a probability of succeeding (producing physical improvement) of $\pi=0.5$ in any one case. If we want to know the probability that the drug will succeed in exactly five of the eight cases, then we note that this experimental situation is exactly equivalent to the "eight flips of the coin" problem in Example 4. Thus the desired probability (computed here exactly as it was there) is $\frac{7}{32}\approx 0.219$.

6. Suppose in Example 5 the drug had only a $\frac{1}{10}$ chance ($\pi = 0.1$) of succeeding in any one case. Then in being administered to eight patients, its probability of succeeding in exactly five of the eight cases would be

$$\begin{aligned}
P(5 \text{ successes}) &= \binom{8}{5}(0.1)^5(1-0.1)^{8-5} \\
&= \binom{8}{5}(0.1)^5(0.9)^3 = \frac{8!}{5!\,(8-5)!}(0.1)^5(0.9)^3 \\
&= \frac{8!}{5!\,3!}(0.00001)(0.729) \\
&= \frac{8 \cdot 7 \cdot 6}{3 \cdot 2 \cdot 1}(0.00000729) = 8 \cdot 7(0.00000729) \\
&= 56(0.00000729) = 0.00040824 \approx 0.0004
\end{aligned}$$

making exactly five out of eight successes an extremely unlikely event: less than 1 chance in 1,000.

7. Let an experiment consist of the following. It is known that one-fourth of the executives in a certain industry develop an identifiable neurosis within a given amount of time on the job. A psychologist plans to examine six of these executives, six who work and live independently of each other. (Why is that important?) What is the probability that exactly two of the six executives will prove to be neurotic? (Note that we call *having* a neurosis a "success" in this analysis; the word "success" in probability theory does not necessarily carry any connotation of desirability.) With $n = 6$, $x = 2$, and $\pi = \frac{1}{4}$, we have

$$\begin{aligned}
P(2 \text{ successes}) &= \binom{6}{2}\left(\frac{1}{4}\right)^2\left(1-\frac{1}{4}\right)^{6-2} = \frac{6!}{2!\,4!}\left(\frac{1}{4}\right)^2\left(\frac{3}{4}\right)^4 \\
&= \frac{6 \cdot 5}{2 \cdot 1}\left(\frac{1}{16}\right)\left(\frac{81}{256}\right) = 15\left(\frac{81}{4{,}096}\right) = \frac{1{,}215}{4{,}096} \approx 0.297.
\end{aligned}$$

8. In the following experiment, one-third of the potted plants of a certain variety are known to bloom under given conditions of soil, temperature, and humidity in a greenhouse. Seven of these plants are set out, and we wish to know the probability that the exact number that will bloom will be (a) 0; (b) 1; (c) 2; (d) 3; (e) 4; (f) 5; (g) 6; (h) all 7. We may compute these probabilities, in sequence, using the binomial formula with $n = 7$, $\pi = \frac{1}{3}$, and x successively 0, 1, 2, 3, 4, 5, 6, and 7:

(a)

$$P(0 \text{ successes}) = \binom{7}{0}\left(\frac{1}{3}\right)^0\left(1-\frac{1}{3}\right)^{7-0}$$

which requires knowing that *a nonzero number* $\left(\text{like } \frac{1}{3}\right)$ *raised to the power 0 equals 1.* (Remember also: $0! = 1$.) Thus we proceed with:

$$P(0 \text{ successes}) = \frac{7!}{0!\,(7-0)!}\left(\frac{1}{3}\right)^0\left(\frac{2}{3}\right)^7 = \frac{7!}{0!\,7!}(1)\left(\frac{2}{3}\right)^7$$

$$= \frac{7!}{(1)7!}(1)\left(\frac{2}{3}\right)^7 = (1)(1)\left(\frac{2}{3}\right)^7 = \frac{2^7}{3^7}$$

$$= \frac{128}{2{,}187} \approx 0.059.$$

(b)

$$P(1 \text{ success}) = \binom{7}{1}\left(\frac{1}{3}\right)^1\left(1-\frac{1}{3}\right)^{7-1}$$

$$= \frac{7!}{1!\,(7-1)!}\left(\frac{1}{3}\right)^1\left(\frac{2}{3}\right)^6 = \frac{7!}{6!}\left(\frac{1}{3}\right)\left(\frac{2}{3}\right)^6$$

$$= 7\left(\frac{1}{3}\right)\left(\frac{2^6}{3^6}\right) = \frac{7}{3}\cdot\frac{64}{729} = \frac{448}{2{,}187} \approx 0.205.$$

(c)

$$P(2 \text{ successes}) = \binom{7}{2}\left(\frac{1}{3}\right)^2\left(1-\frac{1}{3}\right)^{7-2}$$

$$= \frac{7!}{2!\,(7-2)!}\left(\frac{1}{3}\right)^2\left(\frac{2}{3}\right)^5 = \frac{7!}{2!\,5!}\left(\frac{1}{9}\right)\left(\frac{32}{243}\right)$$

$$= \frac{7\cdot 6}{2\cdot 1}\left(\frac{1}{9}\right)\left(\frac{32}{243}\right) = \frac{21}{9}\left(\frac{32}{243}\right) = \frac{672}{2{,}187}$$

$$\approx 0.307.$$

(d)

$$P(3 \text{ successes}) = \binom{7}{3}\left(\frac{1}{3}\right)^3\left(1-\frac{1}{3}\right)^{7-3} = \frac{7!}{3!\,(7-3)!}\left(\frac{1}{3}\right)^3\left(\frac{2}{3}\right)^4$$

$$= \frac{7!}{3!\,4!}\left(\frac{1}{27}\right)\left(\frac{16}{81}\right) = \frac{7\cdot 6\cdot 5}{3\cdot 2\cdot 1}\left(\frac{1}{27}\right)\left(\frac{16}{81}\right)$$

$$= 35\left(\frac{16}{2{,}187}\right) = \frac{560}{2{,}187} \approx 0.256.$$

(e)

$$P(4 \text{ successes}) = \binom{7}{4}\left(\frac{1}{3}\right)^4\left(1-\frac{1}{3}\right)^{7-4}$$

$$= \frac{7!}{4!\,(7-4)!}\left(\frac{1}{3}\right)^4\left(\frac{2}{3}\right)^3 = \frac{7!}{4!\,3!}\left(\frac{1}{81}\right)\left(\frac{8}{27}\right)$$

$$= \frac{7\cdot 6\cdot 5}{3\cdot 2\cdot 1}\left(\frac{1}{81}\right)\left(\frac{8}{27}\right) = 35\left(\frac{8}{2{,}187}\right)$$

$$= \frac{280}{2{,}187} \approx 0.128.$$

(f)

$$P(5 \text{ successes}) = \binom{7}{5}\left(\frac{1}{3}\right)^5\left(1-\frac{1}{3}\right)^{7-5}$$

$$= \frac{7!}{5!\,(7-5)!}\left(\frac{1}{3}\right)^5\left(\frac{2}{3}\right)^2 = \frac{7!}{5!\,2!}\left(\frac{1}{243}\right)\left(\frac{4}{9}\right)$$

$$= \frac{7\cdot 6}{2\cdot 1}\left(\frac{4}{2{,}187}\right) = 21\left(\frac{4}{2{,}187}\right) = \frac{84}{2{,}187}$$

$$\approx 0.038.$$

(g)

$$P(6 \text{ successes}) = \binom{7}{6}\left(\frac{1}{3}\right)^6\left(1-\frac{1}{3}\right)^{7-6} = \frac{7!}{6!\,(7-6)!}\left(\frac{1}{3}\right)^6\left(\frac{2}{3}\right)^1$$

$$= \frac{7!}{6!\,1!}\left(\frac{1}{729}\right)\left(\frac{2}{3}\right) = 7\left(\frac{1}{729}\right)\left(\frac{2}{3}\right)$$

$$= \frac{14}{2{,}187} \approx 0.006.$$

(h)

$$P(7 \text{ successes}) = \binom{7}{7}\left(\frac{1}{3}\right)^7\left(1-\frac{1}{3}\right)^{7-7} = \frac{7!}{7!\,(7-7)!}\left(\frac{1}{3}\right)^7\left(\frac{2}{3}\right)^0$$

$$= \frac{7!}{7!\,0!}\left(\frac{1}{2{,}187}\right)(1) = \frac{7!}{7!\,(1)}\left(\frac{1}{2{,}187}\right)$$

$$= \frac{1}{2{,}187} \approx 0.000. \quad \text{(Actually, about 0.00046.)}$$

We may add these various probabilities to verify that, as they should, they add up to 1:

$$P(0 \text{ successes}) + P(1 \text{ success}) + P(2 \text{ successes}) + P(3 \text{ successes})$$
$$+ P(4 \text{ successes}) + P(5 \text{ successes}) + P(6 \text{ successes}) + P(7 \text{ successes})$$
$$= \frac{128}{2{,}187} + \frac{448}{2{,}187} + \frac{672}{2{,}187} + \frac{560}{2{,}187} + \frac{280}{2{,}187} + \frac{84}{2{,}187} + \frac{14}{2{,}187} + \frac{1}{2{,}187}$$
$$= \frac{2{,}187}{2{,}187} = 1.$$

(Adding in decimal form gives only 0.999 because of cumulative round-off error.)

9. Using the results of Example 8 we may compute various other probabilities by *adding:*

(a) The probability that the number of blooms will be *four or more* (that is, four, five, six, or seven) is $P(4)+P(5)+P(6)+P(7)$, which is:

$$\frac{280}{2{,}187} + \frac{84}{2{,}187} + \frac{14}{2{,}187} + \frac{1}{2.187} = \frac{379}{2{,}187} \approx 0.173.$$

We can obtain the correct result by adding in this manner for the reason that any two of the events concerned here (for example, getting four successes or getting five successes) are *mutually exclusive* of each other.

(b) The probability that the number of blooms will be *five or less* is $P(5)+P(4)+P(3)+P(2)+P(1)+P(0)$, which is:

$$\frac{84}{2{,}187} + \frac{280}{2{,}187} + \frac{560}{2{,}187} + \frac{672}{2{,}187} + \frac{448}{2{,}187} + \frac{128}{2{,}187} = \frac{2{,}172}{2{,}187} \approx 0.993,$$

a very high probability. One could bet confidently on five blooms or less.

The binomial probabilities obtained in Example 8 may be summarized in a table called a **probability distribution:**

x: number of successes	0	1	2	3	4	5	6	7
$P(x)$: probability	$\frac{128}{2{,}187}$ ≈ 0.059	$\frac{448}{2{,}187}$ ≈ 0.205	$\frac{672}{2{,}187}$ ≈ 0.307	$\frac{560}{2{,}187}$ ≈ 0.256	$\frac{280}{2{,}187}$ ≈ 0.128	$\frac{84}{2{,}187}$ ≈ 0.038	$\frac{14}{2{,}187}$ ≈ 0.006	$\frac{1}{2{,}187}$ ≈ 0.000

The variable x in such a display (in this case, the number of blooms) is often called a **random variable.** (Strictly speaking, mathematically, a random variable is a rule of correspondence that assigns a probability value to each element of an experiment's sample space; thus in this case assigning 0 the probability 0.059, assigning 1 the probability 0.205, and so on.)

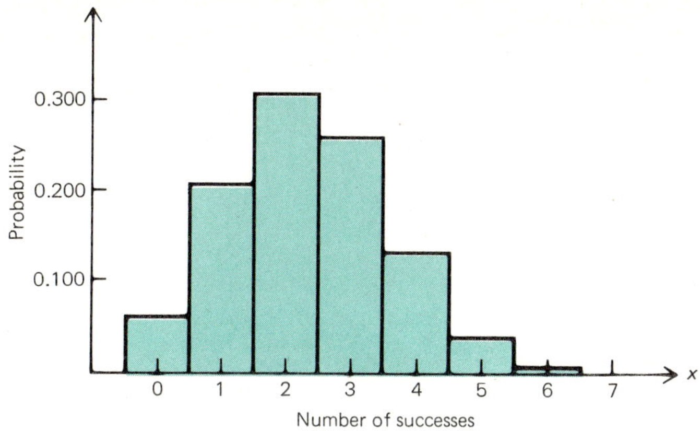

Figure 2.6

We may display the probability distribution visually by means of a *histogram* as shown in Figure 2.6.

Note that the histogram in Figure 2.6 is not symmetric; the probabilities tend to "pile up" more to the left, less to the right. (With such a lack of symmetry, the distribution is said to be **skewed.**) This is because the probability of success on each trial is only $\pi = \frac{1}{3}$: Specifically, π is different from $\frac{1}{2}$. If π had been $\frac{2}{3}$, the situation would be reversed, with the high probabilities on the right and the low ones on the left.

When $\pi = \frac{1}{2}$ exactly, in a binomial experiment, then the resulting distribution of probabilities has a histogram that is exactly symmetric, left and right.

EXAMPLE

10. Consider the experiment of flipping a fair coin four times ($n = 4$), success being heads, so that for each trial $\pi = \frac{1}{2}$. The number of successes can be $x = 0, 1, 2,$ 3, or 4, with the following probabilities:

(a) $P(0) = \binom{4}{0}\left(\frac{1}{2}\right)^0\left(1-\frac{1}{2}\right)^{4-0} = \frac{4!}{0!\,(4-0)!}\left(\frac{1}{2}\right)^0\left(\frac{1}{2}\right)^4$

$$= \frac{4!}{0!\,4!}(1)\left(\frac{1}{16}\right) = \frac{4!}{(1)4!}\left(\frac{1}{16}\right) = (1)\left(\frac{1}{16}\right)$$

$$= \frac{1}{16}.$$

(b) $P(1)=\binom{4}{1}\left(\frac{1}{2}\right)^1\left(1-\frac{1}{2}\right)^{4-1}=\frac{4!}{1!\,(4-1)!}\left(\frac{1}{2}\right)^1\left(\frac{1}{2}\right)^3$

$$=\frac{4!}{1!\,3!}\left(\frac{1}{2}\right)\left(\frac{1}{8}\right)=\frac{4}{1}\left(\frac{1}{16}\right)=\frac{4}{16}=\frac{1}{4}.$$

(c) $P(2)=\binom{4}{2}\left(\frac{1}{2}\right)^2\left(1-\frac{1}{2}\right)^{4-2}=\frac{4!}{2!\,(4-2)!}\left(\frac{1}{2}\right)^2\left(\frac{1}{2}\right)^2$

$$=\frac{4!}{2!\,2!}\left(\frac{1}{4}\right)\left(\frac{1}{4}\right)=\frac{4\cdot 3}{2\cdot 1}\left(\frac{1}{16}\right)=\frac{6}{16}=\frac{3}{8}.$$

(d) $P(3)=\binom{4}{3}\left(\frac{1}{2}\right)^3\left(1-\frac{1}{2}\right)^{4-3}=\frac{4!}{3!\,(4-3)!}\left(\frac{1}{2}\right)^3\left(\frac{1}{2}\right)^1$

$$=\frac{4!}{3!\,1!}\left(\frac{1}{8}\right)\left(\frac{1}{2}\right)=\frac{4}{1}\left(\frac{1}{16}\right)=\frac{4}{16}=\frac{1}{4}.$$

(e) $P(4)=\binom{4}{4}\left(\frac{1}{2}\right)^4\left(1-\frac{1}{2}\right)^{4-4}=\frac{4!}{4!\,(4-4)!}\left(\frac{1}{2}\right)^4\left(\frac{1}{2}\right)^0$

$$=\frac{4!}{4!\,0!}\left(\frac{1}{16}\right)(1)=\frac{4!}{4!\,(1)}\left(\frac{1}{16}\right)=(1)\left(\frac{1}{16}\right)=\frac{1}{16}.$$

In summary, the probability distribution for this binomial experiment is:

x	0	1	2	3	4
$P(x)$	$\frac{1}{16}$	$\frac{4}{16}$	$\frac{6}{16}$	$\frac{4}{16}$	$\frac{1}{16}$

with a *symmetric* histogram as shown in Figure 2.7 (symmetric because $\pi=\frac{1}{2}$; the probability, for example, of getting three heads is the same as the

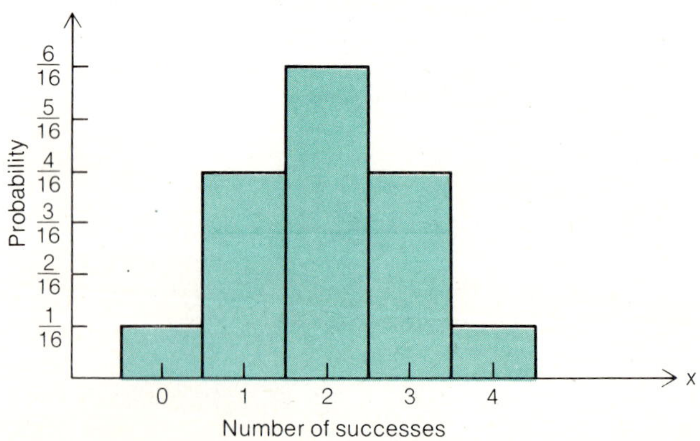

Figure 2.7

probability of getting one tails, which because $\pi = \frac{1}{2}$ is the same as the probability of getting heads once).

Obviously, the arithmetic involved in computing some bionomial probabilities can be considerable. Tables have been compiled in which some such probabilities can be looked up, but even such tables are of no avail if n is large or if π is not some common convenient value. But there are excellent ways to *approximate* such probabilities with little actual computation, as we see in the next chapter.

NEW TERMS IN THIS SECTION

binomial experiment
binomial probability formula
probability distribution
random variable
skewed distribution
***hypergeometric distribution**

EXERCISES 2-C

1. A jar contains 30 green beans and 20 white beans, and the experiment is to choose a bean at random, six times in succession, with the chosen bean being replaced each time before the next selection. What is n? What is π? Find the probability (as a fraction and as a decimal correct to three places) that the number of green beans ("successes") that will be chosen is
(a) exactly 0 (b) exactly 1 (c) exactly 2 (d) exactly 3 (e) exactly 4
(f) exactly 5 (g) exactly 6 (h) 4 or less (i) less than 4 (j) 3 or more
(k) more than 3 (l) at least 5 (m) at most 2 (n) from 2 to 4 inclusive
(o) at least 1

2. Verify by direct addition that the probabilities in Exercise 2-C.1(a) through (g) add up to 1, and draw a histogram representing this probability distribution. Would you have expected symmetry?

3. A tranquilizing drug is known to produce results in 80% of the cases. An experimental psychologist decides to use the drug in the laboratory, administering it to four subjects, if there is at least an 85% chance that the drug will be effective in at least three of the four cases. Should the drug be declared effective enough for use in the laboratory? Explain.

4. On a certain college campus, half the students are women. From time to time a student is chosen at random to serve on the student court. Once chosen, a student is not removed from the "pool" and conceivably could be chosen again, at any time. During the school year, the choice is made seven times. Find the probability (expressing your answers as fractions and as decimals correct to three places) that out of seven times, the choice will be a woman student
(a) 0 times (b) 1 time (c) 2 times (d) 3 times (e) 4 times (f) 5 times
(g) 6 times (h) 7 times

5. In Exercise 2-C.4 (which is, you should notice, equivalent to a coin-flipping problem), could some of the results have been anticipated from earlier answers? Why? Do your probabilities add up to 1? Draw a histogram representing the probability distribution.

6. In Exercise 2-C.4, what is the probability that the choice for the student court will be a woman student
(a) at least 4 times? (b) more than 4 times? (c) at least 3 times?
(d) 3 times or less? (e) from 2 to 5 times inclusive? (f) at least once?

7. Do the same as in Exercise 2-C.4 in the event that the student choice is made eight times instead of seven. That is, what is the probability that a woman student will be chosen exactly
(a) 0 times? (b) 1 time? (c) 2 times? (d) 3 times? (e) 4 times? (f) 5 times?
(g) 6 times? (h) 7 times? (i) 8 times?

8. Verify that the probabilities obtained in Exercise 2-C.7 add up to 1, and draw a histogram representing the probability distribution.

9. If one-third of the people listed in a certain telephone directory are registered voters, and if a name is chosen from the directory at random on each of five different occasions (with previously chosen names still eligible to be chosen again), what is the probability (as a fraction and a three-place decimal each time) that a registered voter will be chosen
(a) 0 times? (b) 1 time? (c) 2 times? (d) 3 times? (e) 4 times? (f) all 5 times?

10. Using the probabilities obtained in Exercise 2-C.9 (verify that they do add up to 1), draw a histogram. How does it differ from the distributions of Exercises 2-C.5 and 2-C.8? Why?

11. In Exercise 2-C.9, from the probabilities you computed, find the probability that the number of times the choice is a registered voter, will be
(a) at least 1 (b) more than 1 (c) 3 or less (d) less than 3 (e) at least 4
(f) at most 4 (g) less than 4 (h) between 1 and 3 inclusive

12. In a certain city it has been noted that one-fourth of the cars on the streets can be expected to have luggage racks on top. *Not* knowing this, a group of office employees decide to observe the first eight cars that pass by on the street running past their window, starting at noon, and let people in the group bet money in a pool beforehand as to the number of cars that will have luggage racks. Jones puts a dollar on *three* cars, and Anderson puts a dollar on *one* car. Which of the two stands the better chance of winning? Explain.

PONDERABLES 2-C

1. Suppose there are five work areas, including the assembly line area, in an industrial plant. A safety inspector who visits the plant once a month decides to choose, at random, three work areas each month to inspect. Areas already chosen can be chosen again later. Over a four-month period, find the probability (correct to two decimal places) that the assembly line area will be inspected
(a) at least once (b) all four times (c) exactly twice (d) more than twice

2. Recall that for an experiment consisting of several trials (each culminating in success or failure) to be *binomial*, the trials must be *independent*. For example, in choosing a card from a deck several times in a row, the chosen card always has to be *replaced* before the next drawing to keep the trials independent of each other. In such an experiment, if the chosen card were *kept out* of the deck each time, then the trials would not be independent, and the experiment, strictly speaking, would not be binomial. For such nonbinomial experiments, the correct way to compute the probability of getting a certain number (x) of successes is to use the **hypergeometric distribution formula** (which gives probabilities that are said to have "hypergeometric distribution") as follows.

In an experiment consisting of n trials, each ending in either "success" or "failure," where there are A ways of "succeeding" and B ways of "failing" in the sample space, the probability of getting exactly x successes out of n trials is

$$\frac{\binom{A}{x}\binom{B}{n-x}}{\binom{A+B}{n}}.$$

For example, in choosing a card at random from a bag containing six blue cards and four red cards (success being considered a blue card) and making the choice four times ($n=4$) *without* replacement, the probability of getting exactly *three* ($x=3$) successes out of four trials is

$$\frac{\binom{6}{3}\binom{4}{1}}{\binom{10}{4}}=\frac{\frac{6!}{3!\,3!}\cdot\frac{4!}{1!\,3!}}{\frac{10!}{4!\,6!}}$$

$$=\frac{6!}{3!\,3!}\cdot\frac{4!}{1!\,3!}\cdot\frac{4!\,6!}{10!}$$

$$=\frac{8}{21}\approx 0.381$$

but if we had (incorrectly) used the binomial formula with $\pi=\frac{6}{10}=\frac{3}{5}$, the result would have been

$$\binom{4}{3}\left(\frac{3}{5}\right)^3\left(\frac{2}{5}\right)^1=\frac{216}{625}\approx 0.346.$$

Then, using the hypergeometric formula (where necessary; some answers can be arrived at by common sense): if a box contains seven intact ("successful") light bulbs and three faulty ones, and if we choose a bulb at random from the box five times in succession, *not* replacing the bulb chosen (after all, why would we replace the bulb after we have determined whether or not it is one of the good ones?) find the probability (as a fraction and as a three-place decimal) that out of five trials, the number of intact bulbs chosen will be

(a) exactly 0 (b) exactly 1 (c) exactly 2 (d) exactly 3 (e) exactly 4
(f) exactly 5 (g) at least 1 (h) at most 1 (i) more than 3 (j) less than 3

3. In problem 2, if (inappropriately) we had used the binomial formula to answer parts (c) and (d), what would have been the results? Compare them with the correct hypergeometric results. (*Remark:* If n is large, the binomial formula can be used as an approximation to the hypergeometric. However, the present comparisons, with n small, should show that for small n the two are not necessarily very similar.)

4. Verify that the results of parts (a) through (f) of problem 2 do add up to 1 as they should, and draw a histogram representing this hypergeometric probability distribution.

SUGGESTED READINGS

Elements of Finite Probability, by William J. Adams (Kaedmon Publishing Company, 1971). This book offers a thoroughgoing discussion of probability, with elementary but fascinating examples of applications to such diverse areas as genetics, marketing, and particle physics.

Stonehenge Decoded, by Gerald S. Hawkins (Dell Publishing Co., Inc., 1965). Chapter 9 provides an interesting discussion of the probabilities relating to the astronomical alignments of the stones at Stonehenge.

Introduction to Applied Statistics, by Marvin Lentner (Prindle, Weber & Schmidt, Incorporated, 1975). Chapters 6 and 7 provide many interesting examples in elementary probability theory.

An Introduction to Mathematical Models in the Social and Life Sciences, by Michael Olinick (Addison-Wesley Publishing Company, 1978). Chapters 9 and 10 provide somewhat more advanced topics in probability theory, including Markov processes.

3 Normal Distribution

For Art may err, but Nature cannot miss.—Dryden

A NORMAL CURVES

Suppose the managers of a supermarket decide to count, during each of 335 one-hour periods, the number of shoppers entering the supermarket, and suppose the results are as follows:

Shoppers	*One-hour periods*
86	7
87	9
88	8
89	11
90	12
91	14
92	13
93	16
94	19
95	23
96	26
97	29
98	24
99	21
100	18
101	15
102	17
103	14
104	10
105	10
106	8
107	6
108	5

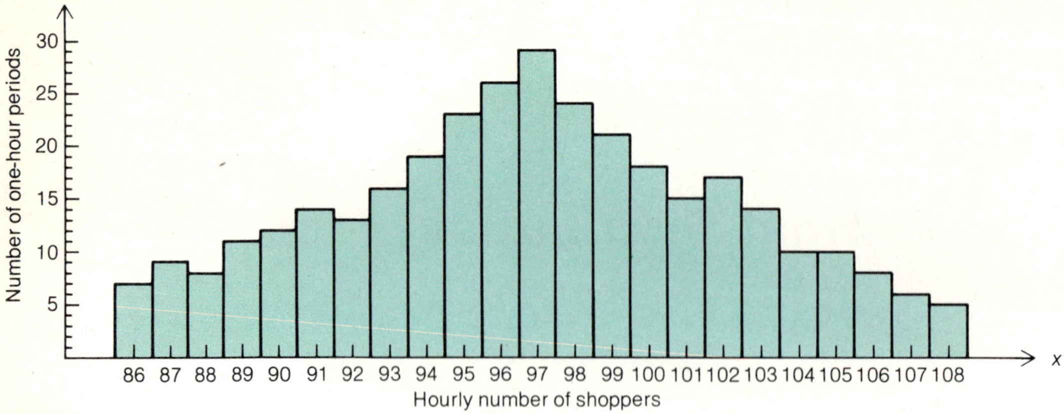

Figure 3.1

This information can more clearly be displayed as a histogram with the hourly numbers of shoppers (x) measured along the horizontal axis and the numbers of one-hour periods (for which each number of shoppers was observed) measured along the vertical axis, as shown in Figure 3.1.

This variable x, the hourly number of supermarket shoppers, is an example of what statisticians call a **discrete variable:** it can assume only certain *isolated* values (in this case, whole-number values) but not the values in between. Clearly, unless something pretty bizarre is going on, the hourly number of shoppers in a supermarket is a whole number. We can have 102 shoppers or 103 shoppers, but not 102.4 shoppers. A discrete variable x can thus be thought of as being like Figure 3.2, able to assume the values represented by the isolated dots but not actually running through other values between them.

For such a discrete variable x in a histogram like Figure 3.1, because we could not find instances of hourly periods where x took on values like 102.4, we could *not* have other rectangles *between* the ones already established (see Figure 3.3).

However, some variables are *not* discrete; these variables do allow the building of such intermediate rectangles. That is to say, some variables take on not only such isolated values as the whole numbers, but also values in between. Such variables are said to be **continuous,** and Figure 3.4 shows the nature of a continuous variable: it takes on a solid spectrum of values, *not* just hopping (for example) from one whole number to the next.

For example, suppose x stands for the heights of people, measured in centimeters. A person's height may be 175 centimeters, may be 176 centimeters, *and* may be something in between, such as 175.4 or 175.09 or even 175.387 centimeters. Consider, in fact, that a person who *grows* from 175 to 176 centimeters does go through all such intermediate values. (We are now

Figure 3.2

A discrete variable

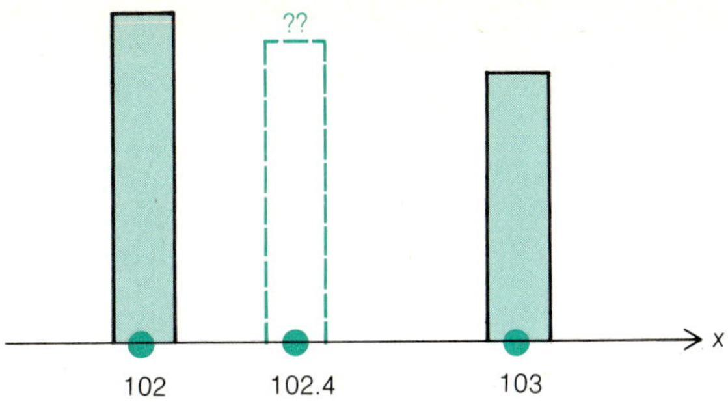

Figure 3.3

talking about a *measure* situation rather than a *count* situation.) The height variable x would take on a continuous spectrum of values, and therefore would be classified as continuous rather than discrete. If we were observing how many people had various heights, we could very well not only find people 175 centimeters tall or 176 centimeters tall, but also people with heights in between.

Figure 3.4

A continuous variable

Suppose we have measured every adult in the world at a given time and have recorded the heights (admittedly, a horrendous task, a task performable only in the imagination). Let us say first that we have measured the heights to the nearest 5 centimeters and have recorded, in round numbers, how many people occur at each such height value. We display our results in a histogram, as shown in Figure 3.5.

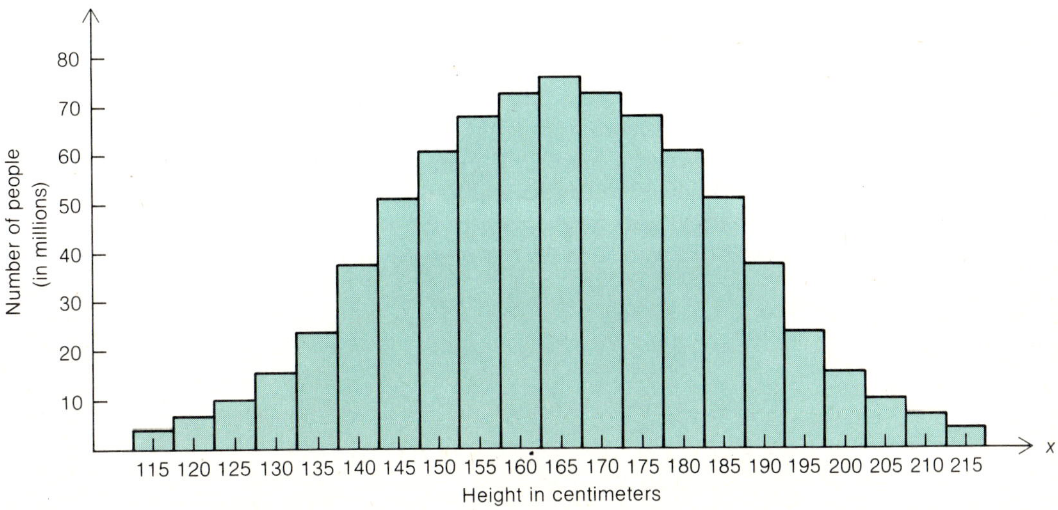

Figure 3.5

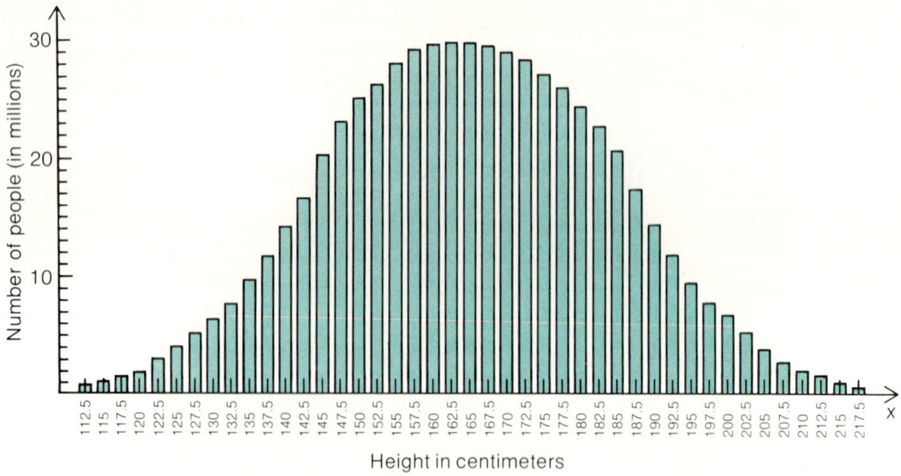

Figure 3.6

The visual display in Figure 3.5 can be refined by more closely measuring the heights and increasing the number of rectangles. Remember, the height variable x is *continuous* and takes on a solid line of values, not just the values 115, 120, 125, and so on. For example, if we measure everyone's height to the nearest 2.5 centimeters instead of the nearest 5, the effect would be to change Figure 3.5 by inserting a new rectangle *between* each of the two x-values previously considered, as in Figure 3.6. That is, we might have taken some of the people who had been *counted* as 115 centimeters tall or 120 centimeters tall, and counted them as 117.5 centimeters if that were closer to their true heights. Such more careful measurement would create and fill new height categories.

Then such a refinement procedure could continue. Perhaps we might *more* closely measure the heights by counting the people not just as occurring at 112.5, 115.0, 117.5, and so on, but also at values in between (such as 113.75, 116.25, and the like), inserting even more rectangles *between* the places where rectangles exist in Figure 3.6.

As we continue to measure the heights x more and more closely, partitioning the horizontal axis into finer and finer, narrower and narrower intervals, increasing the number of rectangles and making them thinner and thinner, the "jagged skyline" effect seen at the tops of the rectangles *tends to smooth out into a solid curve,* a curve such as the one shown in Figure 3.7.

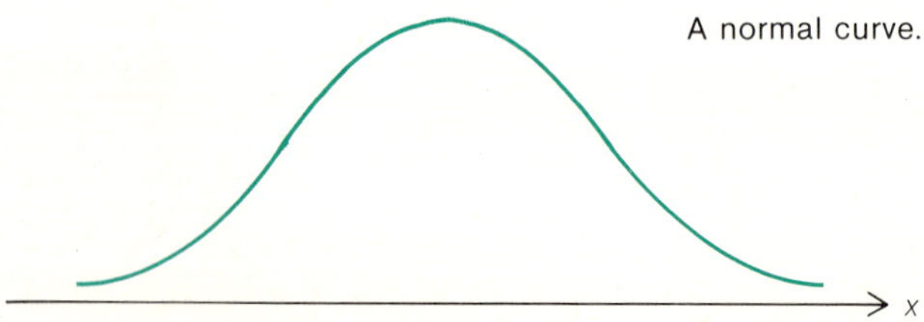

A normal curve.

Figure 3.7

Such a symmetric curve turns out to be the type of curve which mathematicians call a **normal curve.** Although it is often popularly called a *bell-shaped curve* or *bell curve*, the correct statistical term is *normal curve.* (A normal curve is actually the graph of a certain type of mathematical function, the theoretical explanation of which is beyond the scope of this book.) The heights of all these people, because they tend to suggest a normal curve when more and more closely analyzed, are said to be **normally distributed.** The variable x has a *normal distribution* (also sometimes called a *Gaussian* distribution, in honor of Karl Friedrich Gauss, who did pioneer work with the theory of the normal curve).

Normal curves are of extreme importance in statistics. Many things, including many things in nature, tend to be normally distributed and tend to suggest this "bell curve" type of graphical behavior in their distributions. If we were to pluck all the leaves off 1,000 maple trees and weigh them in milligrams, we should find that there is some "central" or "typical" weight around which the weights would tend to gather, and leaves would tend to be *over* or *under* that weight in a more or less symmetric fashion—that is, leaves would tend to weigh a given amount *less* than the "typical" weight about as often as they were seen to be that much *over* the typical weight. The distribution of the weights should look like the symmetric curve of Figure 3.7, and it would not be surprising to learn that the weights of maple leaves are normally distributed.

Let us look at some of the important attributes of a normal curve. The most obvious attribute is the curve's *symmetry*: One could draw a vertical line down the center of the curve, and the left and right portions would be identical mirror images of each other. (See Figure 3.8.) Also, the x-value (on the horizontal axis) at which we would draw such a vertical centerline is the *mean* of the distribution, denoted by the Greek letter μ (*mu*). (In general, the mean of a whole population of x's is an example of a population *parameter*, which should be denoted by a Greek letter, in contrast with a sample *statistic* such as a sample mean $\bar{x}$, denoted by a Roman letter. Because we are talking here about the mean of a whole population of normally distributed x-values, the mean of such values is a population parameter and will be written μ.)

Another feature of a normal distribution is the *standard deviation* of the variable x. This is the standard deviation of a whole population of x-values and

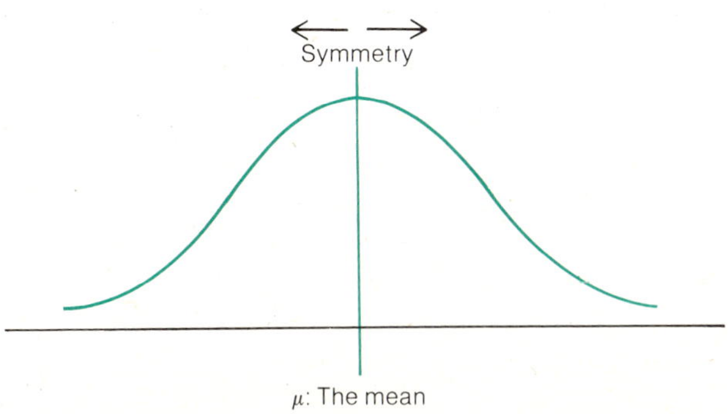

Figure 3.8

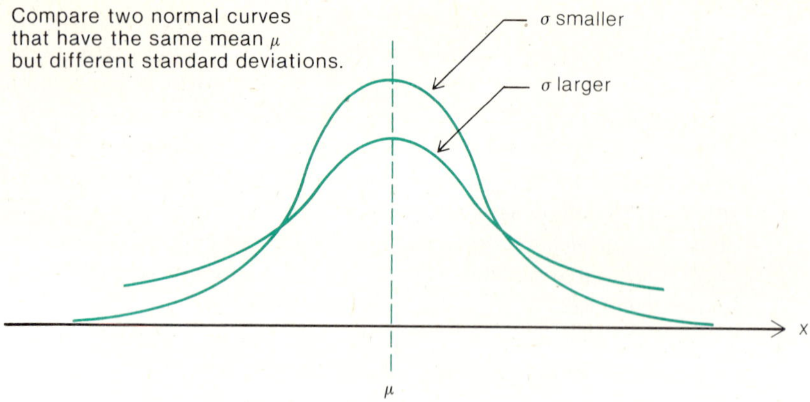

Figure 3.9

is therefore a parameter; it is denoted σ (Greek *sigma*, in contrast with a *sample* standard deviation, which would be denoted s). Recall that standard deviation measures *dispersion*, or the extent to which the x-values are scattered or dissimilar from each other. The more dispersed the values, the larger σ is. If σ is small, then there is little dispersion, and the x-values "pile up" greatly around the mean, as shown in Figure 3.9. But if σ is large, then there is much dispersion, and the values are less tightly massed at the mean, occurring more out under the tails of the curve away from the mean. (Again, see Figure 3.9.) For any fixed mean μ and fixed standard deviation σ, there is exactly one normal curve; but changing the standard deviation affects the shape of the curve, whereas changing the mean shifts the curve left or right.

We come now to the computationally most important feature of the normal curve, the *area* under the curve. First, because of the theoretical nature of the normal curve, *the total area under the curve is 1 square unit.* (See Figure 3.10.) Even though the tails of the curve taper off toward the horizontal axis indefinitely, the area under the curve can be shown mathematically to be exactly 1.

An immediate consequence of this fact, considering the perfect symmetry of the curve, is that the area *on each side* of the mean is exactly $\frac{1}{2} = 0.5$. (See Figure 3.11.) This is an important fact to remember, because we often work with half of the curve at a time.

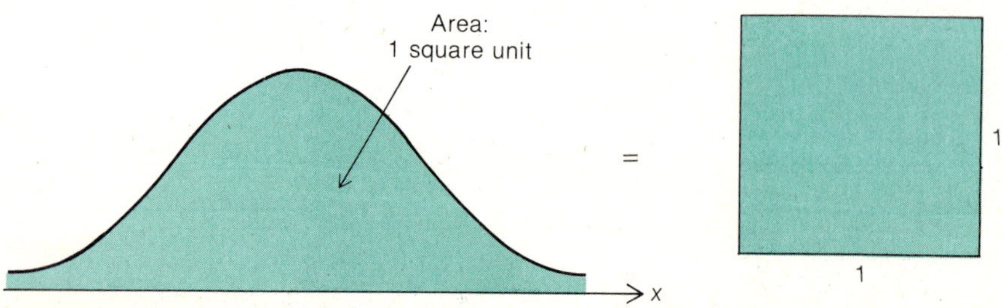

Figure 3.10

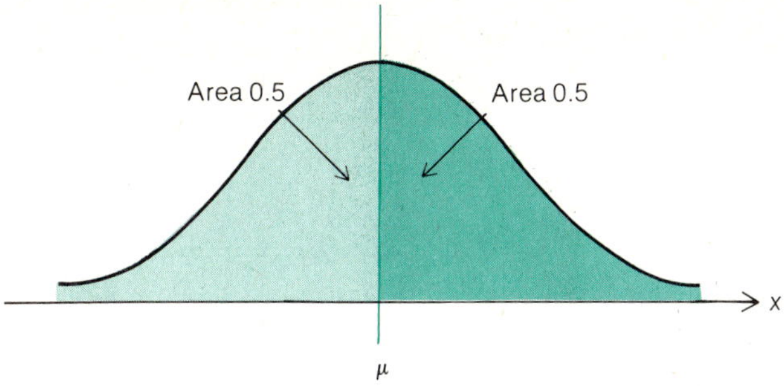

Figure 3.11

And the chief interpretation which we attach to areas under the normal curve is this. Suppose we look at two x-values a and b as shown in Figure 3.12, drawing vertical lines through them and considering the area bounded by those lines, the x-axis, and the normal curve itself. That area (a number between 0 and 1) will be *the probability that an* **x** *chosen at random would lie between* **a** *and* **b**.

For example, if the area in question turns out to be 0.2475, then that means that if an x-value were chosen at random, the probability that it would lie between the values a and b is 0.2475. In other words, we would expect 24.75% of the normally distributed x-values to lie between a and b.

If these areas under the normal curve measure certain important probabilities, then it must be essential to know, numerically, what such areas are. But then the question is: How do we evaluate them?

If we had to measure these areas directly, from scratch, we would have to haul out some pretty heavy mathematical artillery, because the process would require advanced methods in the calculus. Fortunately, we do not have to resort to such heavy-handed methods, because there are *tables* in which the areas can be looked up. But we need to inquire into how the tables are used, and this requires some explanation.

Remember, for a given mean μ and standard deviation σ, there is a unique normal curve; but if we change either μ or σ, then we have a different normal curve. Thus one might think that for every different μ and σ, different tables would be needed to determine the areas under the normal curve. But luckily

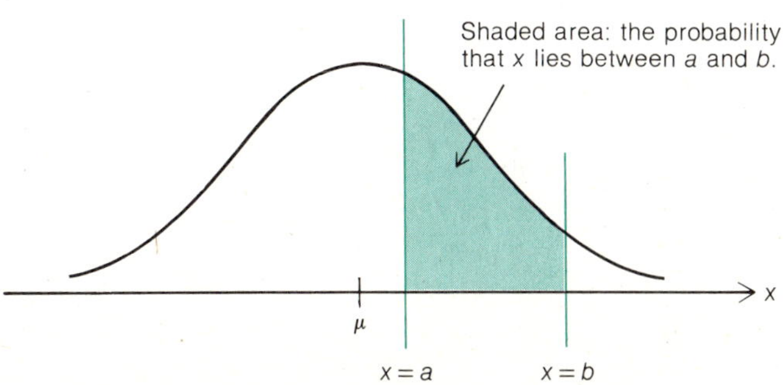

Figure 3.12

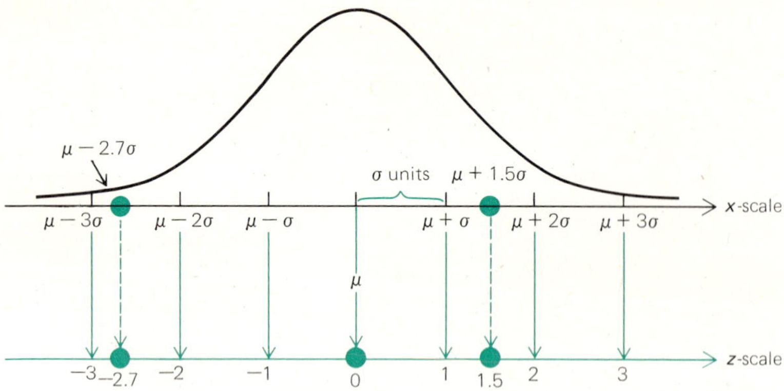

Figure 3.13

this is not the case, because we may always *standardize* a given normal curve by choosing *standard units* to express the x-values.

To do this, we set up a new horizontal scale of numbers, which we may write below the given x-axis as shown in Figure 3.13, and we call this new horizontal axis the **z-scale.** The z-scale works as follows. We first define the *mean* on the z-scale (corresponding to μ on the x-scale) to be 0. Then we take *the standard deviation itself* (σ, whatever its value may happen to be) as defining *1 unit* on the z-scale. That is, if on the x-scale we go σ units to the right of the mean, then on the z-scale we simply record that location as 1. If we go 2σ units to the right (from the mean) on the x-scale, then on the z-scale we arrive at the point 2, and so on. The same idea holds for noninteger multiples of σ. If we travel 1.5 σ's to the right, the z-value reached is 1.5. If we go to the *left* of the mean, and travel σ units, we arrive at −1 on the z-scale; likewise, 2σ units to the left of the mean is −2 on the z-scale, 2.7 standard deviations to the left of the mean is −2.7 on the z-scale, and so on. Again, see Figure 3.13.

In general, to change a value on the given x-scale into **standard units,** that is, z-values on the z-scale, the formula is

$$z = \frac{x - \mu}{\sigma}.$$

Basically z represents *how many standard deviations the raw x-score is away from the mean of the distribution.*

EXAMPLES

1. Given a normal curve with mean $\mu = 75$ and standard deviation $\sigma = 4$, to convert the x-value 80 into a z-score: we use the formula with $x = 80$, $\mu = 75$, and $\sigma = 4$:

$$z = \frac{x - \mu}{\sigma} = \frac{80 - 75}{4} = \frac{5}{4} = 1.25.$$

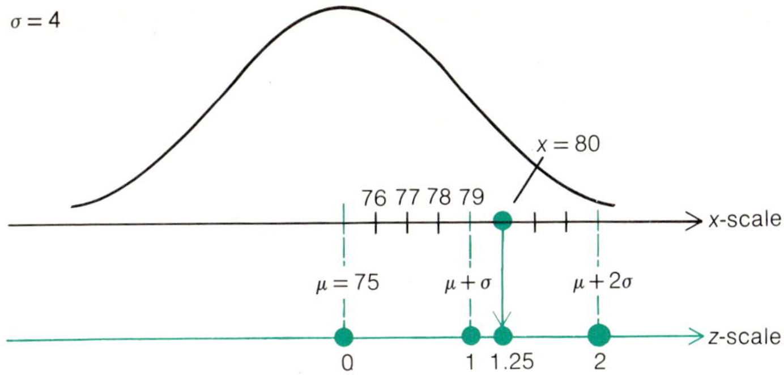

Figure 3.14

This result means that on the original x-scale, the value $x = 80$ lies $1\frac{1}{4}$ standard deviations to the right of the mean. And you can verify directly that this is so, since in this case σ is 4. (The score $x = 80$ lies 5 units out; how many 4's out does it lie? Clearly, $1\frac{1}{4}$.) See Figure 3.14.

2. Given a normal curve with mean $\mu = 48$ and standard deviation $\sigma = 7$, to convert the x-value 43 into a z-score:

$$z = \frac{x - \mu}{\sigma} = \frac{43 - 48}{7} = \frac{-5}{7} = -\frac{5}{7} \approx -0.71.$$

This result (a negative z-score this time) means that the raw score $x = 43$ lies $\frac{5}{7}$ of a standard deviation to the *left* of the mean. (That makes sense, since $\frac{5}{7}$ of σ in this case is $\frac{5}{7}$ of 7, which is 5, and certainly 43 lies 5 units to the left of $\mu = 48$.)

To use the table for normal curve areas, we must first change any x-values involved into z-scores in the manner just illustrated. The fundamental fact about the table "Normal Curve Areas Between Mean and z" in the back of the book is that for a given x-value changed into a z-score, looking up the z-score in the table gives *the area from the mean out to the given score.* Work through the following examples very carefully—they are chosen to illustrate the various situations that you will encounter when using the table for normal curve areas.

EXAMPLES

3. Given a normal curve with mean $\mu = 75$ and standard deviation $\sigma = 4$, to use the normal table to find the area under the curve from the mean out to the score $x = 80$.

We first change $x=80$ into a z-score, and as we saw in Example 1, this produces

$$z=\frac{x-\mu}{\sigma}=\frac{80-75}{4}=\frac{5}{4}=1.25.$$

(If necessary we would have rounded this off to the *nearest hundredth.*) Then in the table we look up $z=1.25$ like this:

```
z  . . . . . . . .  0.05
.                    |
.                    |
.                    ↓
1.2 ───────────→  .3944
```

and the resulting number 0.3944 is the area under the curve from the mean out to $x=80$ (see Figure 3.15). Thus for such a normally distributed variable x, the probability that an x would at random lie between the mean $\mu=75$ and the value 80 is determined to be 0.3944. That is, we would expect about 39.44% of the x's to lie between 75 and 80.

4. For a normal curve with mean $\mu=62.8$ and standard deviation $\sigma=5.0$, find the area under the curve *beyond* $x=66.4$. We first change $x=66.4$ into a z-score:

$$z=\frac{x-\mu}{\sigma}=\frac{66.4-62.8}{5}=\frac{3.6}{5}=0.72.$$

Then we proceed as suggested in Figure 3.16. We look up the area *out to* $z=0.72$ (the blank area shown in Figure 3.16) and then *subtract* this area from the total area under the right half of the curve, which is 0.5. In the table, the area *out to* $z=0.72$ is:

```
z  . . . . . . . .  0.02
.                    |
.                    |
.                    ↓
0.7 ───────────→  .2642
```

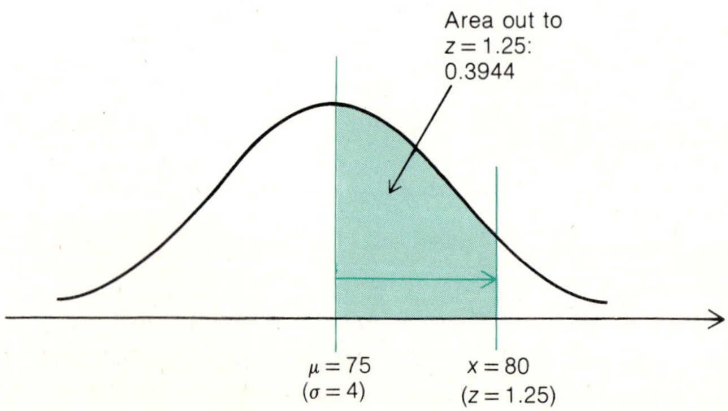

Figure 3.15

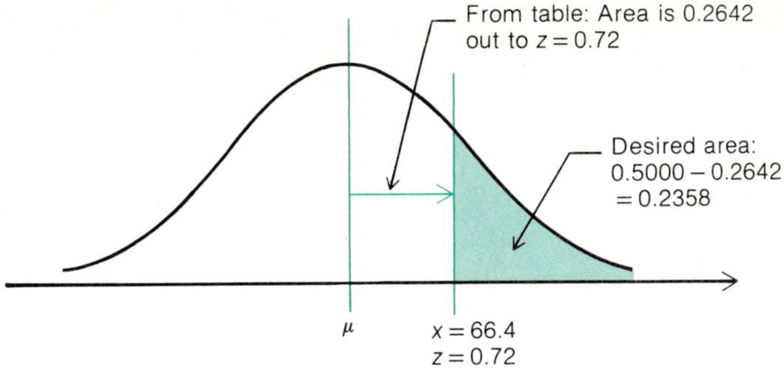

Figure 3.16

and the subtraction from 0.5 gives

$0.5000 - 0.2642 = 0.2358,$

the desired shaded area out under the tail of the curve as shown. The probability that a randomly chosen x would lie beyond 66.4 is 0.2358. That is, 23.58% of the x's could be expected to exceed 66.4.

5. Given a normal curve again with mean $\mu = 62.8$ and standard deviation $\sigma = 5.0$, find the area under the curve from the mean out to $x = 56.7$.
As a z-score, $x = 56.7$ is

$$z = \frac{x-\mu}{\sigma} = \frac{56.7-62.8}{5}$$

$$= \frac{-6.1}{5} = -1.22,$$

naturally a *negative* z-score because $x = 56.7$ is *less* than the mean. Now, when we look up $z = -1.22$ in the table, we look it up just as if it were a positive $z = 1.22$:

```
z   . . . .  0.02
.              |
.              |
.              ↓
1.2 ———————→ .3888
```

because the *symmetry* of the normal curve ensures that working on the left side of the curve is equivalent to working on the right side. Thus the area from the mean out to $x = 56.7$ is 0.3888. (See Figure 3.17.)

6. Given a normal curve with mean $\mu = 88.0$ and standard deviation $\sigma = 9.0$, find the area under the curve *between* $x = 90$ and $x = 95$ (see Figure 3.18.)

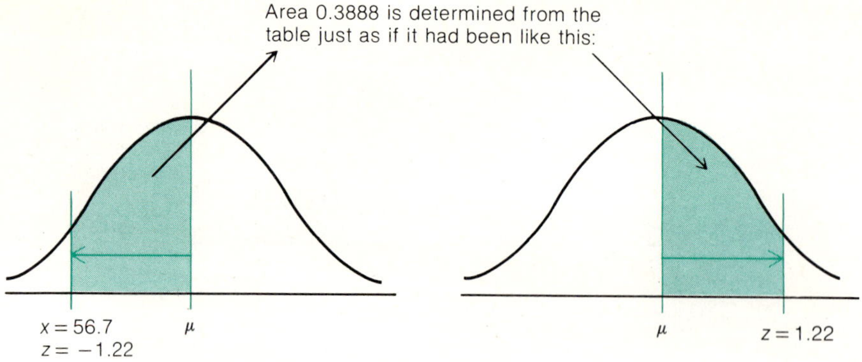

Figure 3.17

We first change *both* the x-scores into z-scores:

For $x=95$, $z=\dfrac{95-88}{9}=\dfrac{7}{9}\approx 0.78.$

For $x=90$, $z=\dfrac{90-88}{9}=\dfrac{2}{9}\approx 0.22.$

Then we look up the area out to the further z:

Area out to $z=0.78$ is 0.2823

and the area out to the nearer z:

Area out to $z=0.22$ is 0.0871.

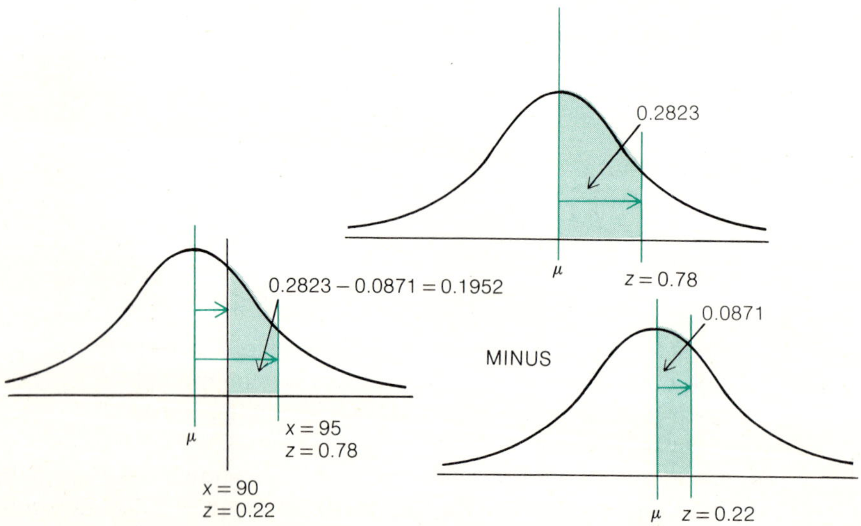

Figure 3.18

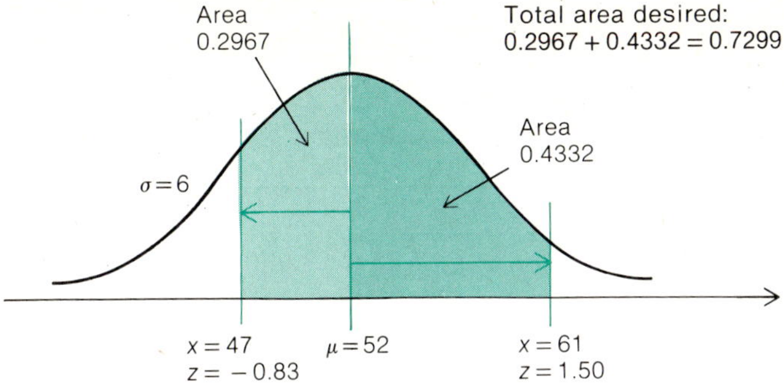

Figure 3.19

And we *subtract* these areas:

$0.2823 - 0.0871 = 0.1952.$

Thus the probability that an x randomly chosen would lie between 90 and 95 is 0.1952. About 19.52% of the x-values should lie between 90 and 95.

7. Given a normal curve with mean $\mu = 52.0$ and standard deviation $\sigma = 6$, find the area under the curve between $x = 47$ and $x = 61$.

We first change the x-values involved into z-scores:

For $x = 47$, $z = \dfrac{47-52}{6} = \dfrac{-5}{6} \approx -0.83.$

For $x = 61$, $z = \dfrac{61-52}{6} = \dfrac{9}{6} = \dfrac{3}{2} = 1.50.$

Now look at Figure 3.19. The desired area this time is the *sum* of two areas, instead of the difference as in the previous example. We find in the table:

Area from μ out to $z = -0.83$: 0.2967.

Area from μ out to $z = 1.50$: 0.4332.

And we *add* these two areas:

$0.2967 + 0.4332 = 0.7299$

so that about 72.99% (very nearly 73%) of the x-values can be expected to lie between $x = 47$ and $x = 61$.

Using the normal table, we may determine a few basic facts about the normal curve. Specifically, we can tell what percentages of the normally distributed x-values lie within certain proximities of the mean.

Out to $z = 1.00$, the area is 0.3413 from the table; out to $z = -1.00$, then, the area would again be 0.3413.

Thus *between* $z = -1.00$ and $z = 1.00$, the area would be $2(0.3413) = 0.6826$. That is, for a *normally distributed* variable x,

68.26% of the values lie *within one standard deviation of the mean.*

Similarly, you may verify that for x normally distributed,

95.44% of the values lie within *two* standard deviations of the mean.

and

99.74% of the values lie within *three* standard deviations of the mean.

Thus virtually *all* the data could be expected to lie between $\mu - 3\sigma$ and $\mu + 3\sigma$. (See Figure 3.20.)

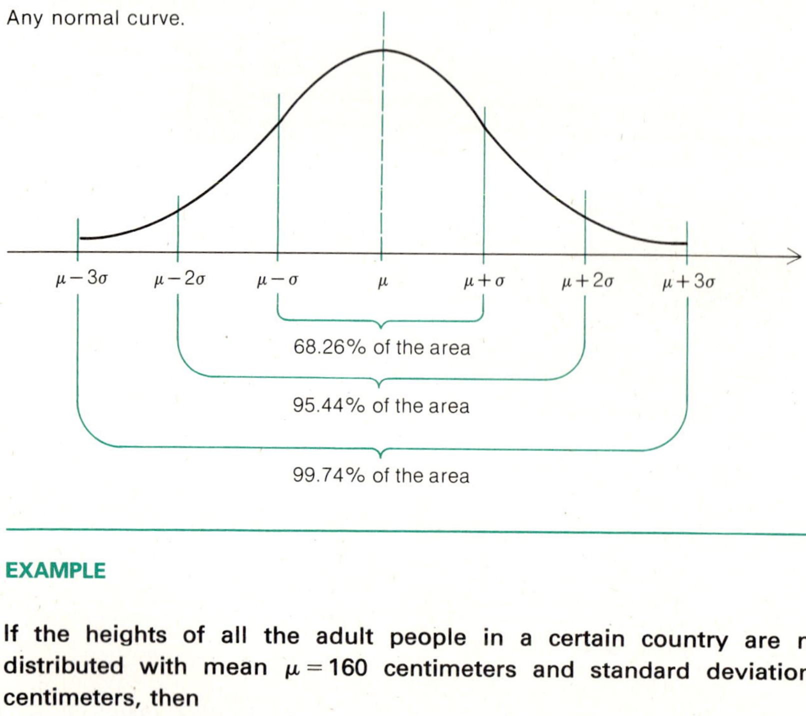

Figure 3.20

EXAMPLE

8. If the heights of all the adult people in a certain country are normally distributed with mean $\mu = 160$ centimeters and standard deviation $\sigma = 10$ centimeters, then

68.26% of the people can be expected to have heights between $\mu - \sigma$ and $\mu + \sigma$, that is, between 150 and 170 centimeters.

95.44% of the people can be expected to have heights between $\mu - 2\sigma$ and $\mu + 2\sigma$, that is, between 140 and 180 centimeters.

99.74% of the people can be expected to have heights between $\mu - 3\sigma$ and $\mu + 3\sigma$, that is, between 130 and 190 centimeters.

Remember from the beginning of this section that we entertain the idea of a possible normal curve when the variable x is *continuous*, taking on a solid spectrum of values—as contrasted with a *discrete* variable (for example, a variable capable of being only a whole number, and hopping suddenly from one whole number value to the next without taking on the values in between). Again, look at Figures 3.2 and 3.4 to see the difference.

But even though we should, strictly speaking, associate a possible normal-curve distribution only with a continuous variable, there are times when we may wish to regard a *discrete* variable as being *approximately* normally distributed. The necessary computations then require making an adjustment called the **continuity correction.** Read the following examples carefully to see how this works.

EXAMPLES

9. The daily number of traffic tickets handed out by the police department of a certain large city is regarded as being approximately normally distributed with mean $\mu = 240$ and standard deviation $\sigma = 11$. Find the probability that on a given day the number of tickets will be between 250 and 260 inclusive.

We first note that the variable x (the daily number of tickets) is *discrete*, because it must always be a whole number—the police cannot, for instance, hand out 256.37 tickets today. Thus we will make use of the *continuity correction*. This consists of "spreading" each integer involved on the axis and thinking of it as an *interval*: 250 is thought of as being not just a point on the line but rather an interval running from 249.5 to 250.5, and similarly 260 is thought of as an interval running from 259.5 to 260.5. (Each number is "spread" a half unit on each side of itself, as it were.) This procedure makes it possible to associate some *area* under the curve with each number, as suggested in Figure 3.21: The area associated with 250 is the area under the curve from 249.5 to 250.5. Thus to find the desired area we must measure from 249.5 all the way over to 260.5 (using the "outside" endpoints to make sure that the intervals representing $x = 250$ and $x = 260$ are included, because the problem did say from 250 to 260 *inclusive*). We change 249.5 and 260.5 to z-scores in the usual way and look up the areas out to them:

$$\text{For } x = 260.5, \quad z = \frac{260.5 - 240}{11} = \frac{20.5}{11} \approx 1.86, \text{ with area } 0.4686.$$

$$\text{For } x = 249.5, \quad z = \frac{249.5 - 240}{11} = \frac{9.5}{11} \approx 0.86, \text{ with area } 0.3051.$$

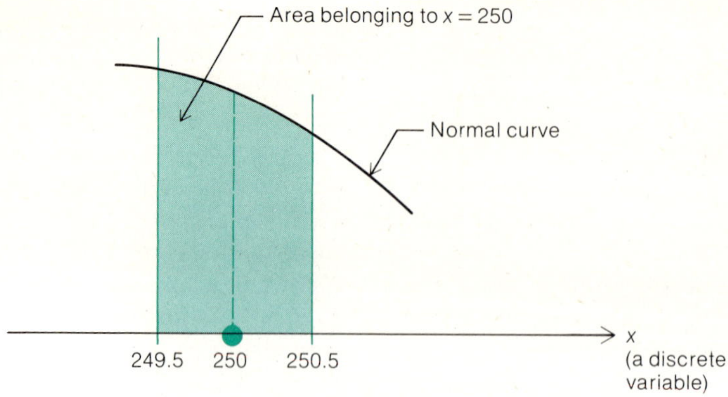

Figure 3.21

Then (see Figure 3.22) the desired area is

$0.4686 - 0.3051 = 0.1635.$

Thus we would expect the police department in the city in question to give out daily from 250 to 260 tickets, inclusive, about 16.35% of the time.

Sometimes we use much the same technique even when the variable is continuous, if we want to describe the probability that x will have one specific value. For instance, if we want the probability that the diameter of a tree trunk will be 37 centimeters, we may think of 37 as meaning 36.5 to 37.5.

10. For the situation in Example 9, find the probability that on a given day the police department will hand out *more than* 248 tickets.

Again using the continuity correction we think of 248 as being "spread" from 247.5 to 248.5, and because we want the area strictly *beyond* 248 (the problem says *more than* 248) we measure the area *from* 248.5 *outward* (see

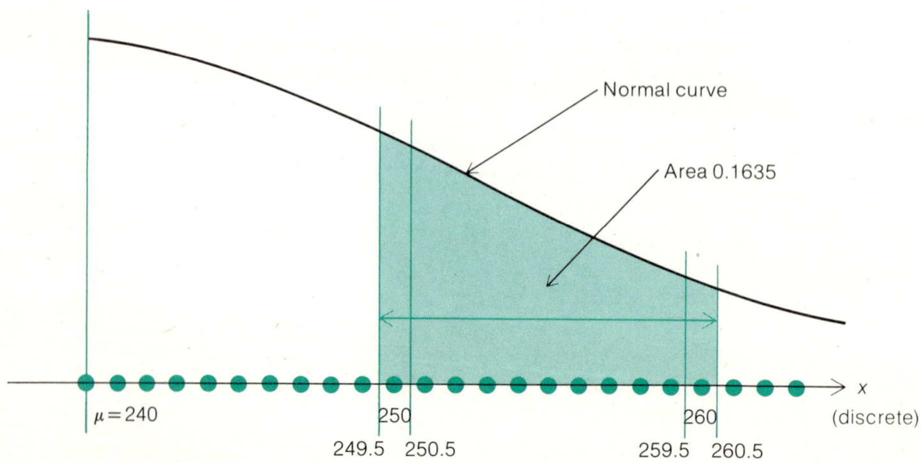

Figure 3.22

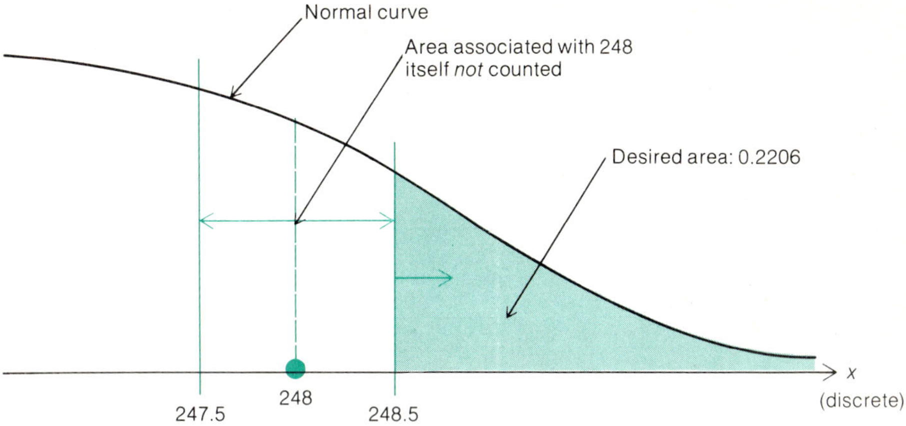

Figure 3.23

Figure 3.23): we change $x = 248.5$ to a z-score:

$$z = \frac{248.5 - 240}{11} = \frac{8.5}{11} \approx 0.77$$

then look up the area *out to* $z = 0.77$ (getting 0.2794) and subtract it from 0.5000 to get the area shown shaded in Figure 3.23:

$$0.5000 - 0.2794 = 0.2206.$$

Roughly speaking, we would expect the police department on a daily basis to exceed 248 tickets about 22% of the time.

NEW TERMS IN THIS SECTION

discrete variable	***z*-scale**
continuous variable	**standard (*z*) units**
normal curve	**continuity correction**
normally distributed variable	***Tchebycheff's Theorem**

EXERCISES 3-A

(Assume the variable x to be continuous unless otherwise indicated.)

1. In each of the following situations, given a value of μ and σ, change the indicated x-value into a z-score (to the nearest hundredth):

(a) $\mu = 70, \sigma = 4; x = 77$ (b) $\mu = 70, \sigma = 4; x = 63$
(c) $\mu = 70, \sigma = 4; x = 80$ (d) $\mu = 70, \sigma = 4; x = 60$
(e) $\mu = 70, \sigma = 4; x = 70$ (f) $\mu = 81.0, \sigma = 5; x = 88.3$
(g) $\mu = 81.0, \sigma = 5; x = 73.7$ (h) $\mu = 45, \sigma = 7; x = 54$

(i) $\mu = 45$, $\sigma = 7$; $x = 36$
(j) $\mu = 67$, $\sigma = 6$; $x = 71$
(k) $\mu = 67$, $\sigma = 6$; $x = 63$
(l) $\mu = 28.5$, $\sigma = 2$; $x = 31.8$
(m) $\mu = 28.5$, $\sigma = 2$; $x = 25.2$
(n) $\mu = 57$, $\sigma = 12$; $x = 72$
(o) $\mu = 57$, $\sigma = 12$; $x = 42$
(p) $\mu = 57$, $\sigma = 12$; $x = 58$
(q) $\mu = 57$, $\sigma = 12$; $x = 56$
(r) $\mu = 39$, $\sigma = 2.5$; $x = 45$
(s) $\mu = 39$, $\sigma = 2.5$; $x = 33$
(t) $\mu = 39$, $\sigma = 2.5$; $x = 39$

2. Given a normal curve with mean $\mu = 92.7$ and standard deviation $\sigma = 5$, find the area under the curve

(a) from the mean out to $x = 100.0$
(b) from the mean out to $x = 96.0$
(c) from the mean out to $x = 93.8$
(d) from the mean out to $x = 81.0$
(e) from the mean out to $x = 85.4$
(f) from the mean out to $x = 77.5$
(g) beyond (to the right of) $x = 95.2$
(h) beyond (to the right of) $x = 99.1$
(i) beyond (to the right of) $x = 105.9$
(j) beyond (to the left of) $x = 83.3$
(k) beyond (to the left of) $x = 79.2$
(l) beyond (to the left of) $x = 90.7$

3. For the situation in Exercise 3-A.2, find the following areas under the normal curve, drawing a diagram to help in each case:

(a) between $x = 95.8$ and $x = 103.4$
(b) between $x = 92.8$ and $x = 106.9$
(c) between $x = 97.6$ and $x = 97.7$
(d) between $x = 90.1$ and $x = 94.6$
(e) between $x = 90.0$ and $x = 95.4$
(f) between $x = 82.8$ and $x = 93.1$
(g) between $x = 83.5$ and $x = 101.9$
(h) between $x = 82.2$ and $x = 89.1$
(i) between $x = 92.0$ and $x = 92.1$
(j) between $x = 78.8$ and $x = 91.8$
(k) between $x = 77.7$ and $x = 107.7$
(l) between $x = 92.6$ and $x = 92.8$

4. Suppose a normal curve is given, with mean $\mu = 165.3$ and standard deviation $\sigma = 17.8$. Remembering that the area under each *half* of the curve is 0.5, and drawing a diagram each time, find the area under the curve

(a) to the left of $x = 185.0$
(b) to the left of $x = 171.5$
(c) to the left of $x = 177.2$
(d) to the left of $x = 191.4$
(e) to the left of $x = 182.7$
(f) to the right of $x = 146.9$
(g) to the right of $x = 153.6$
(h) to the right of $x = 132.1$
(i) to the right of $x = 162.8$
(j) to the right of $x = 111.9$

5. The heights of all the trees in a vast forest are normally distributed with mean $\mu = 16.4$ meters and standard deviation $\sigma = 3.2$ meters. Find the probability that the height of a tree chosen at random would be

(a) between 16.4 and 21.6 meters
(b) between 14.3 and 16.4 meters
(c) between 15.0 and 17.0 meters
(d) between 13.4 and 19.4 meters
(e) between 10.0 and 22.8 meters
(f) between 20.7 and 22.7 meters
(g) between 12.5 and 14.9 meters
(h) between 16.5 and 26.2 meters
(i) more than 17.5 meters
(j) less than 10.8 meters
(k) more than 12.1 meters
(l) less than 25.4 meters

6. Reinterpret your answers to Exercise 3-A.5 in terms of percentages. That is, what percentage of the forest could be expected to be in each height range described?

7. The reaction times of subjects to a psychological test stimulus are regarded as normally distributed with mean $\mu = 2.8$ seconds and standard deviation $\sigma = 0.4$ seconds. Find the percentage of test subjects whose reaction times could be expected to be

(a) between 3.0 and 3.5 seconds
(b) between 2.5 and 2.7 seconds
(c) between 2.3 and 3.3 seconds
(d) between 2.0 and 3.6 seconds
(e) between 1.6 and 4.0 seconds
(f) more than 3.7 seconds
(g) more than 1.9 seconds
(h) less than 2.2 seconds
(i) less than 3.4 seconds
(j) less than 3.9 seconds

8. A market analyst for a clothing manufacturer claims that men's sleeve lengths for long-sleeved shirts are normally distributed with mean $\mu = 54$ centimeters and standard deviation $\sigma = 7$ centimeters. The manufacturer has produced shirts with sleeves as short as 39 centimeters and shirts with sleeves as long as 69 centimeters. Assuming the market analyst's normality assumption to be correct, what percentage of men shoppers may still find that their arms are too short or too long to find a good fit among the available shirts?

9. The weights of all the automobiles on the road in a certain region are said to be normally distributed with mean 2,045 kilograms and standard deviation 225 kilograms. Also, in this region, the ancient and rickety Gauss River Bridge bears a warning sign: "One car at a time; maximum safe load 2,400 kilograms." Assuming that automobile drivers in the region all use the bridge regardless of the warning, what percentage of the time, when a car is on the bridge, could the bridge be considered too heavily loaded for safety?

10. Following the idea of Example 3-A.8: if the weights of all the adult people in a certain state are normally distributed with mean 70 kilograms and standard deviation 9 kilograms, between what weights would one expect to find 68.26% of the people? 95.44% of the people? 99.74% of the people?

11. The daily number of telephone calls received at a switchboard is regarded as approximately normally distributed with mean 580 and standard deviation 10. Remembering to use the *continuity correction* (why?) find the probability that on a given day the number of calls will be

(a) between 585 and 590 inclusive
(b) between 575 and 585 inclusive
(c) between 581 and 595 inclusive
(d) between 560 and 600 inclusive
(e) exactly 580 calls
(f) more than 584 calls
(g) 584 calls or more
(h) less than 590 calls
(i) 590 calls or less
(j) exactly 582 calls
(k) exactly 578 calls
(l) less than 572 calls
(m) 572 calls or less
(n) more than 572 calls
(o) 572 calls or more

12. The hourly number of letters mailed at a certain post office is regarded as approximately normally distributed with mean 208 and standard deviation 19. Using the continuity correction (why?), find the probability that in any given hour the number of letters mailed will be

(a) between 200 and 216 inclusive
(b) between 191 and 225 inclusive
(c) between 180 and 236 inclusive
(d) between 158 and 258 inclusive
(e) exactly 208
(f) exactly 207

(g) exactly 209
(h) between 210 and 230 inclusive
(i) between 209 and 264 inclusive
(j) between 168 and 205 inclusive
(k) more than 260
(l) 260 or more
(m) less than 263
(n) more than 153
(o) less than 160

PONDERABLES 3-A

1. It is possible to use the relation

$$z = \frac{x - \mu}{\sigma}$$

backward, in a manner of speaking. Usually, one starts with a raw score x and changes it directly into a z-score. However, it is possible to start with a predetermined z-score and use the same formula, in another form, to find out what the corresponding x-score would have been. We rewrite the z formula, solving it for x:

$$z = \frac{x - \mu}{\sigma}$$

$$z\sigma = x - \mu$$

$$z\sigma + \mu = x$$

$$x = z\sigma + \mu.$$

Use this form of the formula (and the normal table) to do the following problems:

(a) If the proper trouser lengths of adults in a certain community are normally distributed with mean 102 centimeters and standard deviation 11 centimeters, above what trouser length (to the nearest centimeter) would one find the most long-legged 4% of the people?

(b) Below what trouser length would one find the most short-legged 4% of the people?

(c) Between what trouser lengths (values symmetric around the mean) would one find about 94% of the people?

2. The completion times of subjects of psychological testing reacting to a certain test item are normally distributed with mean 94 seconds and standard deviation 14 seconds.

(a) Below what completion time (to the nearest second) would one find the fastest 1.5% of the subjects?

(b) Above what completion time would one find the slowest 1.5% of the subjects?

(c) Between what two completion times (symmetric values about the mean) would one find about 80% of the subjects?

(d) About 95% of the subjects?

3. If the lengths of the shoes worn by adult women in a certain city are normally distributed with mean 29.0 centimeters, and if 98% of the lengths lie between 21.8 and 36.2 centimeters, find the standard deviation σ correct to the nearest tenth.

4. **Tchebycheff's Theorem.** As we have seen, when a variable x is normally distributed, we may predict how much of the time the values of x will be within certain proximities of the mean. For example, we have seen that 95.44% of the values can be expected to lie within two standard deviations of the mean (that is, between $\mu-2\sigma$ and $\mu+2\sigma$).

But what if the variable x is *not* known to be normally distributed?

It turns out that, even in that event, we can still say something about the fraction or percentage of the time the values should lie within certain proximities of the mean. We make use of

Tchebycheff's Theorem: No matter how the values of a variable x are distributed, the *least* fraction of the values that will lie within k standard deviations of the mean (where $k \geq 1$) is equal to

$$1-\frac{1}{k^2}.$$

For example, when $k=2$,

$$1-\frac{1}{k^2}=1-\frac{1}{2^2}=1-\frac{1}{4}$$
$$=\frac{3}{4}$$

and this means that *at least* $\frac{3}{4}$ of the values (75% of the values) must lie within *two* standard deviations of the mean, whatever the standard deviation happens to be. Typically, Tchebycheff's Theorem gives a very cautious, conservative estimate. It may say, for instance, that *at least* $\frac{3}{4}$ of the values must lie within two standard deviations of the mean, while it may turn out in actuality that much more than $\frac{3}{4}$ of the values do so. In fact, when $k=1$, what the theorem says is true but worthless: at least *no* values lie within *one* σ of the mean. But Tchebycheff's Theorem is all we have when the nature of the distribution is unknown.

At least what fraction of the x-values in any distribution, then, must lie within

(a) $1\frac{1}{4}$ standard deviations of the mean?

(b) $1\frac{1}{2}$ standard deviations of the mean?

(c) $2\frac{1}{4}$ standard deviations of the mean?

(d) $2\frac{1}{2}$ standard deviations of the mean?

(e) 3 standard deviations of the mean?

(f) $3\frac{1}{2}$ standard deviations of the mean?

(g) 4 standard deviations of the mean?

(h) $4\frac{1}{2}$ standard deviations of the mean?

(i) 5 standard deviations of the mean?

(j) $5\frac{1}{2}$ standard deviations of the mean?

(k) $5\frac{3}{4}$ standard deviations of the mean?

(l) 6 standard deviations of the mean?

5. Express each of your answers in the previous problem in four-place decimal form and compare it with what the decimal would be if the distribution of x were known to be *normal.*

6. Using Tchebycheff's Theorem, how many standard deviations would you have to go out on each side of the mean to be sure to include at least 99% of the x-values?

7. If the prices of a certain commodity, as sold in various stores, have mean \$100.00 and standard deviation \$1.25, but are *not* known to be normally distributed, what is the least fraction of the time (expressed as a fraction and as a two-place decimal) one would expect the price to lie between

(a) \$97.50 and \$102.50? (b) \$97.25 and \$102.75?
(c) \$96.25 and \$103.75? (d) \$94.00 and \$106.00?
(e) \$92.50 and \$107.50?

B APPROXIMATING BINOMIAL PROBABILITIES

Recall from Chapter 2 that in a *binomial* experiment, in which a number (n) of independent trials are done, each with possible outcomes "success" and "failure," with a probability of success π on each individual trial, the probability of getting exactly x successes out of n trials is found by the binomial probability formula

$$P(x \text{ successes}) = \binom{n}{x}\pi^x(1-\pi)^{n-x}$$

where $\binom{n}{x}$ is the number of combinations of n things taken x at a time:

$$\binom{n}{x} = \frac{n!}{x!(n-x)!}.$$

At times, this formula is quite easy to use, as when n is small and π is a value not difficult to work with. For instance, if we want to roll a die three times ($n=3$) and compute the probability that a 5 will turn up exactly twice ($x=2$), we observe that $\pi = \frac{1}{6}$ (presuming the die is fair) and the desired probability is

$$P(2 \text{ successes}) = \binom{3}{2}\left(\frac{1}{6}\right)^2\left(1-\frac{1}{6}\right)^{3-2} = \frac{3!}{2!(3-2)!}\left(\frac{1}{36}\right)\left(\frac{5}{6}\right)$$

$$= 3\left(\frac{1}{36}\right)\left(\frac{5}{6}\right) = \frac{1}{12}\cdot\frac{5}{6} = \frac{5}{72} \approx 0.0694.$$

But much of the time the values for n, x, and π are such that the direct use of the binomial formula is difficult. For instance, suppose we draw a random card from a deck of 52 playing cards, 48 times in succession (always replacing the drawn card before drawing again), and suppose we want to know the probability of getting diamonds exactly 14 times. Then $n = 48$, $x = 14$, and $\pi = \frac{1}{4}$, so that the probability computation would call for evaluating

$$\binom{48}{14}\left(\frac{1}{4}\right)^{14}\left(1-\frac{1}{4}\right)^{48-14} = \frac{48!}{14!(48-14)!}\left(\frac{1}{4}\right)^{14}\left(\frac{3}{4}\right)^{34}$$

$$= \frac{48!}{14!34!}\cdot\frac{3^{34}}{4^{14}\cdot 4^{34}} = \frac{48!}{14!34!}\cdot\frac{3^{34}}{4^{48}},$$

a computation that is almost unthinkable actually to carry out. We react: There must be a better way!

And there is.

Remember, the probabilities obtained in a binomial experiment for all the various x-values involved may be displayed in a *histogram*. For example, in the experiment of flipping a fair coin 12 times ($n = 12$) and considering heads a successful outcome $\left(\text{so that } \pi = \frac{1}{2}\right)$, the various numbers of successes possible (0 through 12) can be assigned probabilities, with some difficulty, as follows:

$$P(0 \text{ successes}) = \binom{12}{0}\left(\frac{1}{2}\right)^{0}\left(1-\frac{1}{2}\right)^{12-0} = \frac{1}{4{,}096} \approx 0.0002.$$

$$P(1 \text{ success}) = \binom{12}{1}\left(\frac{1}{2}\right)^{1}\left(1-\frac{1}{2}\right)^{12-1} = \frac{12}{4{,}096} \approx 0.0029.$$

$$P(2 \text{ successes}) = \binom{12}{2}\left(\frac{1}{2}\right)^{2}\left(1-\frac{1}{2}\right)^{12-2} = \frac{66}{4{,}096} \approx 0.0161.$$

$$P(3 \text{ successes}) = \binom{12}{3}\left(\frac{1}{2}\right)^{3}\left(1-\frac{1}{2}\right)^{12-3} = \frac{220}{4{,}096} \approx 0.0537.$$

$$P(4 \text{ successes}) = \binom{12}{4}\left(\frac{1}{2}\right)^{4}\left(1-\frac{1}{2}\right)^{12-4} = \frac{495}{4{,}096} \approx 0.1208.$$

$$P(5 \text{ successes}) = \binom{12}{5}\left(\frac{1}{2}\right)^{5}\left(1-\frac{1}{2}\right)^{12-5} = \frac{792}{4{,}096} \approx 0.1934.$$

$$P(6 \text{ successes}) = \binom{12}{6}\left(\frac{1}{2}\right)^{6}\left(1-\frac{1}{2}\right)^{12-6} = \frac{924}{4{,}096} \approx 0.2256.$$

$$P(7 \text{ successes}) = \binom{12}{7}\left(\frac{1}{2}\right)^{7}\left(1-\frac{1}{2}\right)^{12-7} = \frac{792}{4{,}096} \approx 0.1934.$$

$$P(8 \text{ successes}) = \binom{12}{8}\left(\frac{1}{2}\right)^{8}\left(1-\frac{1}{2}\right)^{12-8} = \frac{495}{4{,}096} \approx 0.1208.$$

$$P(9\text{ successes}) = \binom{12}{9}\left(\frac{1}{2}\right)^{9}\left(1-\frac{1}{2}\right)^{12-9} = \frac{220}{4{,}096} \approx 0.0537.$$

$$P(10\text{ successes}) = \binom{12}{10}\left(\frac{1}{2}\right)^{10}\left(1-\frac{1}{2}\right)^{12-10} = \frac{66}{4{,}096} \approx 0.0161.$$

$$P(11\text{ successes}) = \binom{12}{11}\left(\frac{1}{2}\right)^{11}\left(1-\frac{1}{2}\right)^{12-11} = \frac{12}{4{,}096} \approx 0.0029.$$

$$P(12\text{ successes}) = \binom{12}{12}\left(\frac{1}{2}\right)^{12}\left(1-\frac{1}{2}\right)^{12-12} = \frac{1}{4{,}096} \approx 0.0002.$$

And a histogram showing this distribution of probabilities is shown in Figure 3.24.

It is clear in the computation of these probabilities and in the drawing of the histogram that because $\pi = \frac{1}{2}$, the probabilities (and their graph) exhibit *symmetry:* the probabilities of 0 successes and 12 successes are equal, the probabilities of 1 success and 11 successes are equal, and so on. Because of this symmetry and the resulting appearance of the histogram, it requires little or no stretching of the imagination to imagine a *normal curve* running "through" the histogram's rectangle tops—a normal curve *approximating* the histogram, as it were (see Figure 3.25).

That is, we may use the normal curve as a sort of stand-in that approximates the actual distribution of the probabilities. We are using the normal curve, then, as a **mathematical model** to describe, at least approximately, the reality of the situation.

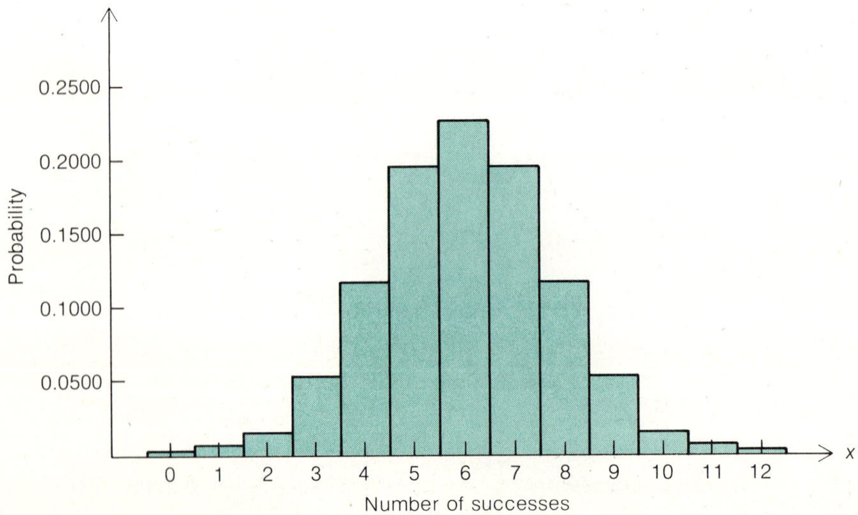

Figure 3.24

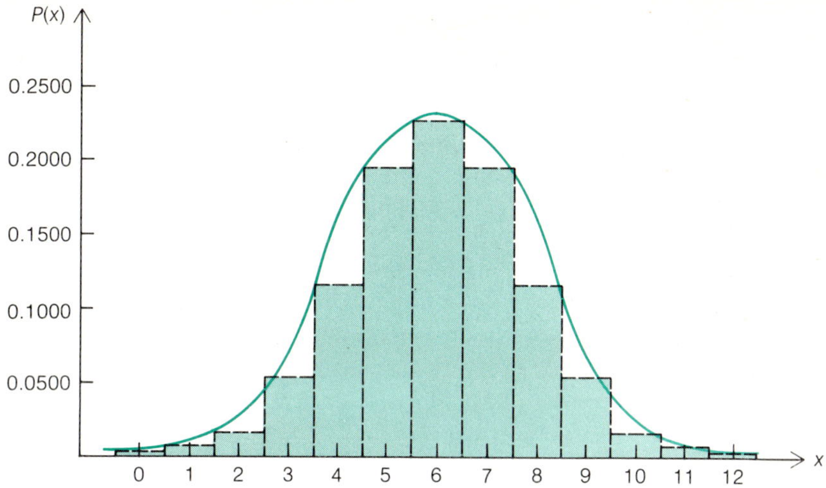

Figure 3.25

Statistical theory shows that the nature of this approximation is considered good, provided two conditions are met:

$$n\pi > 5$$

and $n(1-\pi) > 5.$

That is, *both* products $n\pi$ and $n(1-\pi)$ must exceed 5 before the normal curve model is guaranteed to be a good model for the actual probability distribution. Essentially, these conditions require that

1. π should be reasonably close to $\frac{1}{2}$, or
2. The number n of trials should be reasonably large

before we attempt the approximation. For example, if $\pi = \frac{1}{5}$ and $n = 20$, then $n\pi = 4$, which does *not* exceed 5. Similarly, if $\pi = \frac{4}{5}$ and $n = 20$, then $n\pi$ exceeds 5 *but* $n(1-\pi) = 20\left(1 - \frac{4}{5}\right) = 4$, which does not exceed 5. The problem in both cases is that π is too far away from $\frac{1}{2}$ (and the sample size is too small) to guarantee the sort of *symmetric* histogram that $\pi = \frac{1}{2}$ ensures, as was seen in Figure 3.25. On the other hand, if $\pi = \frac{3}{5}$ $\left(\text{a bit closer to } \frac{1}{2} \text{ this time}\right)$ but $n = 8$, then $n\pi = 8\left(\frac{3}{5}\right) = 4.8$, which does not exceed 5. The problem this time is that n is too small for a good normal approximation to be guaranteed. (In "Ponderables" at the end of this section we explore some not-so-good approximations in which the basic conditions are not met.)

Then, if a normal curve is to be used as a model representing approximate binomial probabilities, we need the usual two items required for working with a normal curve: a mean μ and a standard deviation σ. These parameters and the operating rules are supplied as follows:

To use a **normal curve approximation** of binomial probabilities, given a binomial experiment consisting of n trials with probability of success π on each trial, the appropriate mean and standard deviation for the normal curve computations are given by the formulas

$$\mu = n\pi$$

$$\sigma = \sqrt{n\pi(1-\pi)}$$

provided both $n\pi$ and $n(1-\pi)$ exceed 5. The *continuity correction* is always used, since the variable x (the whole-number value standing for number of successes) is discrete.

EXAMPLES

1. Returning to the situation in which a fair coin is flipped 12 times (success being heads), we will approximate the probability of getting heads exactly 7 times. We will use a normal curve.

   First, $n = 12$ and $\pi = \frac{1}{2}$, so that checking to see that the necessary conditions are met, we see that they are:

$$n\pi = 12\left(\frac{1}{2}\right) = 6 > 5,$$

$$n(1-\pi) = 12\left(1-\frac{1}{2}\right) = 12\left(\frac{1}{2}\right) = 6 > 5.$$

   Now, we compute the necessary μ and σ:

$$\mu = n\pi = 12\left(\frac{1}{2}\right) = 6.0$$

$$\sigma = \sqrt{n\pi(1-\pi)} = \sqrt{12\left(\frac{1}{2}\right)\left(1-\frac{1}{2}\right)}$$

$$= \sqrt{12\left(\frac{1}{2}\right)\left(\frac{1}{2}\right)} = \sqrt{3} \approx 1.7.$$

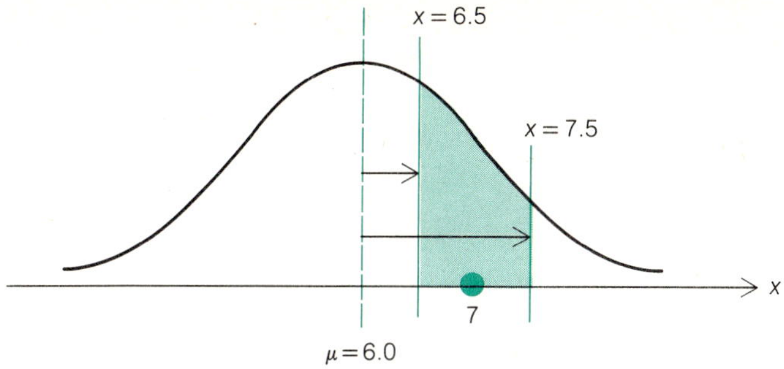

Figure 3.26

To find the probability that exactly $x=7$ successes will be obtained, we employ the continuity correction and think of 7 as being spread out as an interval from 6.5 to 7.5 (see Figure 3.26), and we find the area *between* $x=6.5$ and $x=7.5$ by the usual procedure: change the x's to z-scores, look up the areas in the normal table, and subtract them:

For $x=7.5$, $z=\dfrac{7.5-6.0}{1.7}=\dfrac{1.5}{1.7}\approx 0.88$:

Area out to $z=0.88$ is 0.3105.

For $x=6.5$, $z=\dfrac{6.5-6.0}{1.7}=\dfrac{0.5}{1.7}\approx 0.29$:

Area out to $z=0.29$ is 0.1141.

And subtracting, we have

$0.3105-0.1141=0.1964.$

Comparing this result with the same probability previously computed directly from the binomial formula (0.1934, which is also a slight approximation due to division round-off) we see that the normal curve approximation has provided a result only about 0.003 units away from the true value. Not a bad approximation, for n as small as 12. Actually, it could have been improved by using a better approximation for $\sqrt{3}$ than 1.7. You may verify that if $\sqrt{3}\approx 1.73$ is used, the resulting normal approximation is 0.1937, only about 0.0003 units away from the true value.

2. Suppose we wish to know, in the same situation (flipping a fair coin 12 times) the probability that the number of heads will be *7 or more*. That is, we want to know the likelihood of getting 7, 8, 9, 10, 11, or 12 successes. By the binomial formula, computing from scratch, we would be required to compute

$$\binom{12}{7}\left(\frac{1}{2}\right)^{7}\left(1-\frac{1}{2}\right)^{12-7}+\binom{12}{8}\left(\frac{1}{2}\right)^{8}\left(1-\frac{1}{2}\right)^{12-8}+\binom{12}{9}\left(\frac{1}{2}\right)^{9}\left(1-\frac{1}{2}\right)^{12-9}$$
$$+\binom{12}{10}\left(\frac{1}{2}\right)^{10}\left(1-\frac{1}{2}\right)^{12-10}+\binom{12}{11}\left(\frac{1}{2}\right)^{11}\left(1-\frac{1}{2}\right)^{12-11}+\binom{12}{12}\left(\frac{1}{2}\right)^{12}\left(1-\frac{1}{2}\right)^{12-12},$$

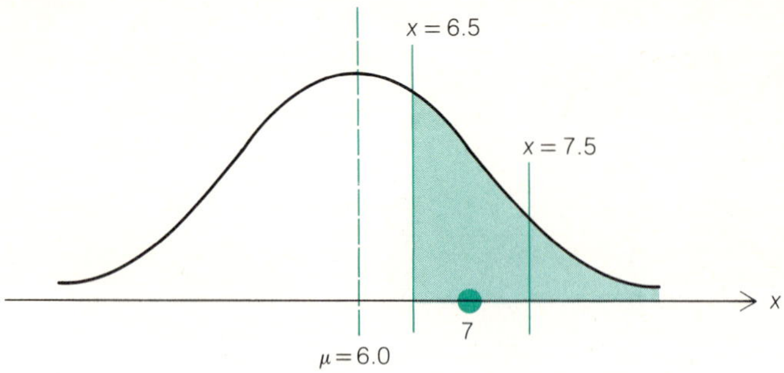

Figure 3.27

surely a computation whose delights are surpassed only by hanging by one's thumbs. Fortunately, we may avoid all this by using the normal approximation (as before, $\mu = 6.0$ and $\sigma = \sqrt{3} \approx 1.7$), again thinking of the value $x = 7$ as being spread into an interval from 6.5 to 7.5, and finding the area from $x = 6.5$ *out*, as shown in Figure 3.27. The idea is that we start at the left endpoint 6.5 to include the area belonging to 7 itself, that is, the area from 6.5 to 7.5. The total desired area is everything to the right of 6.5, from the wording of the problem, "*7 or more.*"

Thus we change $x = 6.5$ into a z-score:

$$z = \frac{6.5 - 6.0}{1.7} = \frac{0.5}{1.7} \approx 0.29,$$

and we find the area from $z = 0.29$ outward: the area *out to* $z = 0.29$ is 0.1141, and the area beyond (shaded in Figure 3.27) is

$$0.5000 - 0.1141 = 0.3859.$$

We may note that from the computations previously done, just before Example 1, using the binomial formula directly, the direct (and laborious) result, *not* appealing to the normal approximation, would have been

$$P(7) + P(8) + P(9) + P(10) + P(11) + P(12)$$

$$= 0.1934 + 0.1208 + 0.0537 + 0.0161 + 0.0029 + 0.0002$$

$$= 0.3871 \text{ (with some accumulated round-off error)}$$

so that our normal approximation 0.3859 misses the mark only by about 0.0012 units. Again, not bad!

3. A book collector, two-fifths of whose books are fiction, chooses a book at random every evening to read (by matching random digits against the books' card catalogue numbers). If she makes the selection 54 times, what is the probability that she will choose a fictional work more than 18 times? (It is assumed that once chosen, a book is put back and may be chosen again at any time, so that the experiment is binomial, with $n = 54$ trials, and the probability of success each time—getting a fiction book—is $\frac{2}{5} = 0.4.$)

We will approximate the desired probability with a normal curve, provided the necessary conditions are met, and they indeed are:

$$n\pi = 54\left(\frac{2}{5}\right) = 54(0.4) = 21.6 > 5$$

and $n(1-\pi) = 54(0.6) = 32.4 > 5.$

Thus it is safe to proceed, and we compute the necessary parameters μ and σ:

$$\mu = n\pi = 54(0.4) = 21.6,$$

$$= \sqrt{n\pi(1-\pi)} = \sqrt{54\left(\frac{2}{5}\right)\left(\frac{3}{5}\right)}$$

$$= \sqrt{\frac{324}{25}} = \frac{\sqrt{324}}{\sqrt{25}} = \frac{18}{5} = 3.6.$$

Then we organize the problem as shown in Figure 3.28. We think of $x = 18$ as being spread from 17.5 to 18.5. And because the problem says "*more than* 18 times," we start at the right endpoint, *not* counting the area between 17.5 and 18.5, and measure all the area to the right of $x = 18.5$. (If the problem had said, "18 *or more*," we would have started at $x = 17.5$.) We change 18.5 into a z-score:

$$z = \frac{18.5 - 21.6}{3.6} = \frac{-3.1}{3.6} \approx -0.86.$$

The area from the mean out to $z = -0.86$ is 0.3051, and (as shown in Figure 3.28) the total desired area is

$$0.3051 + 0.5000 = 0.8051.$$

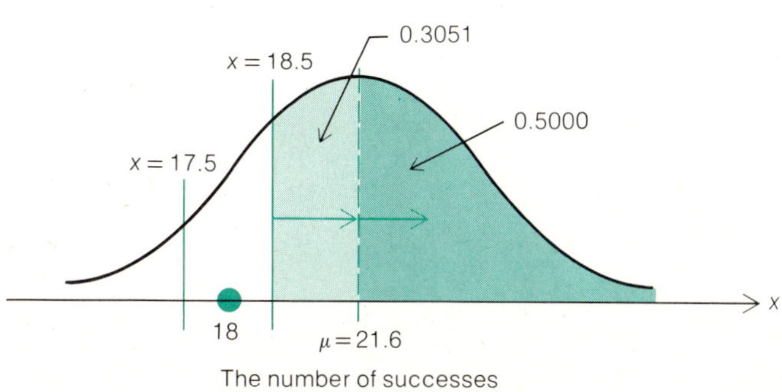

Figure 3.28

Computing this probability directly from the binomial formula would require finding

$$P(19)+P(20)+P(21)+\ldots+P(53)+P(54),$$

and because even the *first* of these 36 terms *by itself* would require evaluating

$$\binom{54}{19}\left(\frac{2}{5}\right)^{19}\left(1-\frac{2}{5}\right)^{54-19}=\frac{54!}{19!\,35!}\left(\frac{2}{5}\right)^{19}\left(\frac{3}{5}\right)^{35}$$

you may imagine what the whole computation would be like. Clearly, the normal curve approximation technique can save an enormous amount of time and trouble.

NEW TERMS IN THIS SECTION

mathematical model
normal approximation

EXERCISES 3-B

1. A file cabinet full of account folders contains 800 folders, 500 of which contain overdue accounts. A folder is chosen at random, 60 times in succession (folders chosen are replaced and become eligible to be chosen again). Using the normal approximation (is it legitimate here?—explain), specify μ and σ (express σ as a fraction or a two-place decimal), and find the probability that out of 60 trials, the number of overdue accounts chosen will be

(a) from 38 to 42 inclusive
(b) from 33 to 42 inclusive
(c) from 38 to 47 inclusive
(d) from 28 to 47 inclusive
(e) more than 40
(f) 40 or more
(g) more than 35
(h) 35 or more
(i) less than 37
(j) 37 or less
(k) less than 44
(l) 44 or less
(m) exactly 38
(n) exactly 37
(o) exactly 41

2. A card is drawn at random from a deck of 52 playing cards, 48 times in succession, the drawn card being replaced each time. Using the normal approximation (if appropriate—explain) find the probability that the number of diamonds chosen out of 48 trials (diamonds being one-fourth of the deck) will be

(a) exactly 12
(b) exactly 14
(c) exactly 10
(d) from 7 to 15 inclusive
(e) from 13 to 20 inclusive
(f) from 5 to 19 inclusive
(g) more than 16
(h) 16 or more
(i) more than 8
(j) 8 or more
(k) less than 18
(l) 18 or less
(m) 11 or less
(n) less than 11
(o) from 9 to 17 inclusive

3. In a certain locality, it is claimed that three days before the state auto inspection deadline, 30% of the cars on the road still do not have new inspection stickers. Assuming this claim to be true, if the first 2,100 cars passing a certain checkpoint on that day are observed, specify μ and σ for a normal curve approximation, and find the probability that the number of cars still lacking a new inspection sticker will be

(a) more than 670
(b) 670 or more
(c) less than 580
(d) 580 or less
(e) from 610 to 650 inclusive
(f) from 590 to 670 inclusive
(g) from 570 to 690 inclusive
(h) less than 665
(i) 665 or less
(j) exactly 630
(k) 605 or more
(l) more than 605
(m) from 615 to 655 inclusive
(n) from 600 to 640 inclusive
(o) from 635 to 645 inclusive

4. Think about what it would take to do Exercise 3-B.3 using the binomial formula directly.

5. A storage bin contains 4,000 fuses, and 1,440 of them are defective. Suppose we choose 100 fuses to take out and test to see whether they work. (For practical reasons we would, of course, refrain from *replacing* a chosen fuse once it was tested, and the experiment is, strictly speaking, not binomial. But the numbers involved are so large that very little error is made if we act as if the experiment *were* binomial, as if the chosen fuses were replaced.) Using a normal approximation, specify μ and σ, and find the probability that the number of defective fuses chosen will be

(a) from 31 to 41 inclusive
(b) from 26 to 46 inclusive
(c) from 22 to 50 inclusive
(d) more than 43
(e) 43 or more
(f) less than 34
(g) 34 or less
(h) exactly 36
(i) less than 45
(j) 45 or less
(k) 26 or more
(l) more than 26
(m) exactly 38
(n) from 30 to 42 inclusive
(o) more than 21

PONDERABLES 3-B

1. Using normal approximations, find the probability that heads will turn up exactly half the time when a fair coin is flipped

(a) 16 times
(b) 36 times
(c) 64 times
(d) 100 times
(e) 144 times
(f) 196 times
(g) 256 times
(h) 400 times
(i) 2,500 times
(j) 10,000 times

2. Remember, the normal approximation to the binomial distribution is guaranteed to be a satisfactory approximation only when $n\pi > 5$ and $n(1-\pi) > 5$. Suppose a bag contains 20 green cards and 80 red cards. A card will be chosen at random, six times in succession, with replacement; success will mean getting a green card, so that $\pi = \frac{20}{100} = \frac{1}{5}$, and we may observe that the basic conditions are *not* met. Compute and compare the following, observing the

comparatively rough nature of the normal approximations under these circumstances:

(a) The probability of getting exactly two green cards (using the binomial formula).
(b) The same thing, using a normal curve.
(c) The probability of getting exactly one green card (using the binomial formula).
(d) The same thing, using a normal curve.
(e) The probability of getting *no* green cards (using the binomial formula).
(f) The same thing, using a normal curve.

SUGGESTED READINGS

Men of Mathematics, by E. T. Bell (Simon & Schuster, 1961). Chapter 14 of this very readable and valuable historical-biographical volume treats the life and work of Karl Friedrich Gauss, after whom the normal distribution is known as "Gaussian distribution."

The Nature and Growth of Modern Mathematics, by Edna Kramer (Fawcett Publications, Inc., 1970, 2 volumes). Chapter 14 provides an illuminating discussion of normal distribution for the student interested in knowing more about its theory and origins.

The World of Mathematics, presented with commentaries and notes by James R. Newman (Simon & Schuster, 1956). Part VIII, Chapter 5 ("On the Average and Scatter," by M. J. Moroney) provides a good discussion of dispersion and distributions in general, including the normal (with reference to distribution of IQ scores).

Statistics: An Applied Approach, by Neil Ullman (Xerox College Publishing, 1972). Chapter 10 gives an excellent discussion of the nature of the normal curve, with a close look at continuous and discrete variables.

4 Estimation of Population Parameters

By a small sample we may judge of the whole piece.—Cervantes

A INTERVAL ESTIMATES OF A MEAN

We come now to the beginning of what may really be called *inferential* statistics: the study of techniques by which information obtained from a *sample* allows us to make certain inferences, draw certain *probable conclusions*, about the population from which the sample was drawn. The ability to do so is one of the chief advantages to be gained from the study of statistics.

First let us review a distinction made earlier.

A computation that measures some aspect of a whole population is called a population *parameter*. For example, given a population of numbers, the *mean* of all the numbers, usually denoted μ, is a parameter of the population, a parameter measuring central tendency. Also, the *standard deviation* of all the numbers, usually denoted σ, is a parameter of the population, a parameter measuring dispersion. If the population contains N numbers, these parameters could theoretically be computed by the formulas

$$\mu = \frac{\Sigma x}{N} \quad \text{and} \quad \sigma = \sqrt{\frac{\Sigma(x-\mu)^2}{N}}.$$

We say "theoretically" computed by those formulas because, in practice, a population is commonly much too large to allow such direct computations to be feasible. More often, these parameters are *estimated*, as we shall see.

In contrast with such computations as μ and σ, a computation which measures some aspect of a *sample* is called a *statistic*. For example, the *mean* of a sample of numbers is a sample statistic, a statistic measuring central tendency and usually denoted $\bar{x}$. Also, the *standard deviation* of a sample is a statistic, a statistic measuring dispersion and usually denoted s. Given a *sample* of n numbers, these statistics are computed by the formulas

$$\bar{x} = \frac{\Sigma x}{n} \quad \text{and} \quad s = \sqrt{\frac{\Sigma(x-\bar{x})^2}{n-1}}.$$

(In general, population parameters are denoted by Greek letters, whereas sample statistics are denoted by Roman letters.)

The fundamental idea here is that *a sample statistic is used for estimating a population parameter.* For instance, if we do not know the true population standard deviation σ, we may take a good-sized random sample, compute the sample standard deviation s, and use the statistic s as an estimator of the parameter σ.

In particular, population *means* (μ) are often estimated by sample means ($\bar{x}$), and we will now examine some of the implications.

Suppose we consider, for example, a population consisting of the times (in seconds) which all people taking a certain intelligence test will require to complete a given test item. Now, because conceivably any number of people could take the test, we are dealing with a very large population. Thus if we wanted to compute the true population mean time μ, we would first have to wait until everyone of concern had taken the test, and then we would have an enormous information-gathering and computation job to do: adding all the observed times (perhaps many thousands of them) and dividing by N. The serious problem here, though, is not just the tedium of computing $\Sigma x/N$; the problem is that we may not have all of the test times because some people who might take the test may not yet have done so. We may wish to know what the true mean μ might be like *if* everyone concerned had taken the test, and we may have to find this out without having the leisure of waiting around until everyone conceivable has been tested.

The procedure, then, would be to select a number of people at random, have them take the test under proper conditions, and (regarding their completion times as a random sample) compute the sample mean $\bar{x}$. This result would then be an initial estimate of μ, at least. Consider, for instance, the following.

EXAMPLE

1. Suppose 50 people are examined at random and their completion times (in seconds), for the test item in question, are as follows:

47	44	49	45	46	64	68	72	59	66
62	52	55	58	50	48	55	60	58	42
42	65	53	63	47	49	43	55	62	65
59	61	68	69	56	57	42	42	46	48
67	65	41	52	64	63	59	47	51	53.

Adding these numbers, we get $\Sigma x = 2{,}700$, so that the *sample mean* is

$$\bar{x} = \frac{\Sigma x}{n} = \frac{2{,}700}{50} = 54.0.$$

This result 54.0 is our initial simple estimate of the true population mean μ, the mean completion time for *all* examinees.

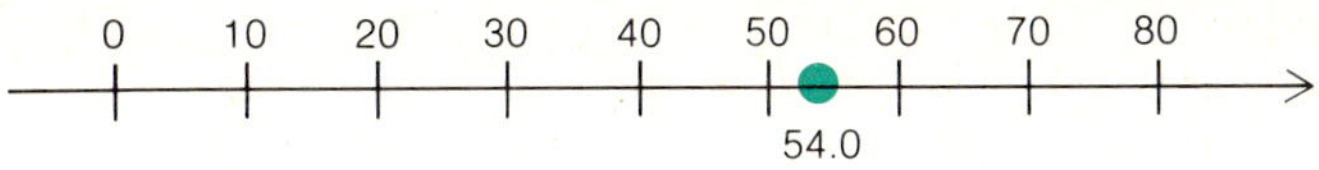

Figure 4.1 A point estimate of μ.

Such an estimate as shown in Example 1, where we estimate μ by coming up with a single number $\bar{x} = 54.0$, is called a **point estimate** of the population parameter—a point estimate because it consists of a single value and would be plotted on a number line as a single point (see Figure 4.1).

The trouble with a point estimate, though, is that it does not carry with it any indication of how *good* an estimate it is. If you ask me to find an estimate of μ, and I take a sample and come back with the point estimate $\bar{x} = 54.0$, you may well ask me, "But how good is this estimate? Could the true value of μ be as far away as, say, 50 or 58?" I get a little nervous and reply, "Well, I suppose it *could* be; I've only looked at a sample of test results and computed a sample mean 54.0." You ask, further: "Well, then, how *likely* is it that the true value of μ is that far away from your point estimate 54.0?" I begin, at this point, to wish in earnest that you were not pressing me with all these embarrassing questions!

The problem is that a point estimate alone does not allow one to answer such questions with any authority. Maybe there is very little chance that the real μ is as far away as 50 or 58. Maybe, on the other hand, there is an excellent chance that the real μ may be that far away from the point estimate. We simply do not know.

Yet.

But there is a way to find out. The point estimate alone is not the last word in parameter estimation, by any means.

There is a much more informative kind of estimate called an **interval estimate.** We can construct an interval, running some number below 54.0 to some number above 54.0, and assign a high probability that the true value of μ should lie within the interval somewhere, that is, within a certain proximity of 54.0 one way or the other. (See Figure 4.2.)

To do this, we make use of a remarkable theorem in the theory of statistics, called the **Central Limit Theorem.**

Suppose, in trying to estimate μ, the true mean completion time for the intelligence test item, we had taken random samples of size 50 *repeatedly*: 50 test results chosen on one day, 50 on another day, 50 on still another day, and so on, over an indefinitely long period of time. Now, it is admittedly unlikely that we would have the leisure actually to indulge in such sampling over and

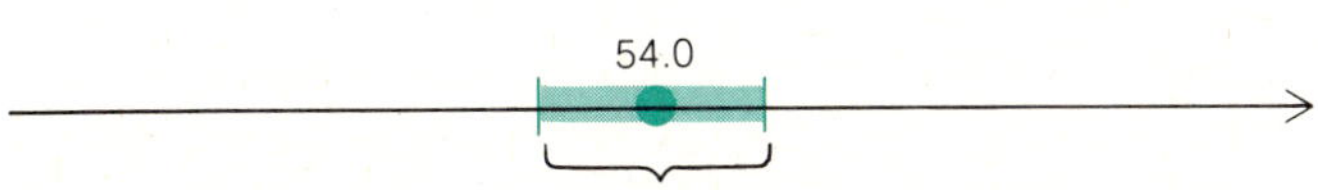

Figure 4.2 An interval estimate of μ.

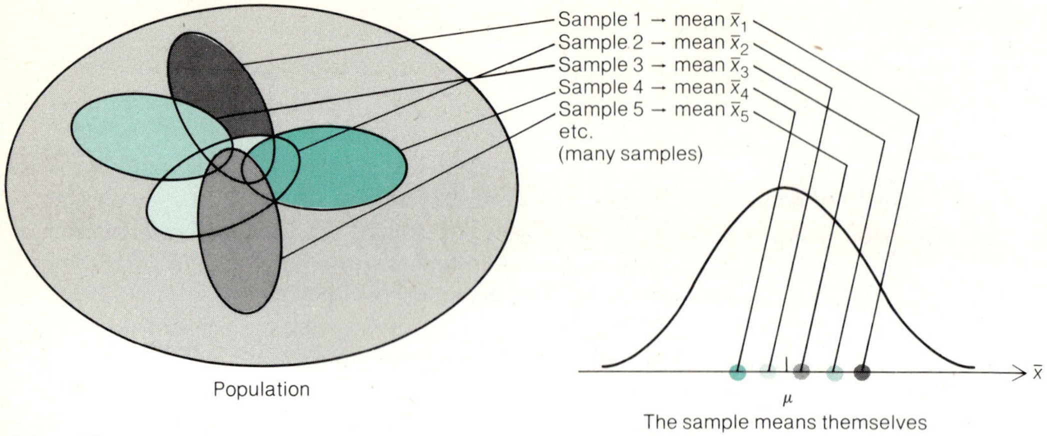

Figure 4.3

over again. But if we *did* do so, statistical theory has a lot to say about what could be expected to happen.

Specifically, the different sample means obtained (and they would be somewhat different, no doubt: perhaps 54.3 one time, 53.8 another time, and so on, because of chance fluctuation in the sampling)—the different sample means obtained at this sample size (50) *would tend to be normally distributed.* This happens, in fact, as long as the sample size is greater than 30. (See Figure 4.3.)

The theorem, formally, is as follows:

Central Limit Theorem. If large random samples of a fixed size (in practice, samples of size $n > 30$) are taken from a numerical population (with parameters μ, σ), and if a sample mean $\bar{x}$ is computed for each sample, the distribution of sample means $\bar{x}$ is closely approximated by a normal curve whose mean (denoted $\mu_{\bar{x}}$) is the mean of the population itself:

$$\mu_{\bar{x}} = \mu$$

and whose standard deviation (denoted $\sigma_{\bar{x}}$) is the population standard deviation σ divided by the square root of the sample size:

$$\sigma_{\bar{x}} = \frac{\sigma}{\sqrt{n}}.$$

The larger the sample size n, the more nearly the sample means $\bar{x}$ are normally distributed. And the remarkable thing about this theorem is that it is true *even if the population itself is not normally distributed.* (However, if the population does happen to be normally distributed, then we can relax the requirement that the sample size n be greater than 30.)

This theorem has some very interesting implications. Notably, if one sampled repeatedly and got sample means $\bar{x}$ that (by the theorem) were approximately normally distributed, the *standard deviation* $\sigma_{\bar{x}}$ of that normal curve would be much *smaller* than the σ of the original population, because in $\sigma_{\bar{x}} = \frac{\sigma}{\sqrt{n}}$ we are dividing the original σ by a quantity $\sqrt{n}$. Remember, too: standard deviation measures dispersion, and $\sigma_{\bar{x}}$ measures the dispersion of all the sample means $\bar{x}$ that one might obtain by sampling at a given sample size. The immediate implication, because $\sigma_{\bar{x}}$ is always smaller than σ, is that *the sample means are always much more similar to each other* (much less dispersed) *than the original numbers in the population are.* In a sense, then, the sample means are "better behaved" and more predictable than the numbers from which they are drawn.

The normal curve in question here, which describes the distribution of the sample means $\bar{x}$, is called the **sampling distribution of the mean,** and its standard deviation $\sigma_{\bar{x}} = \frac{\sigma}{\sqrt{n}}$ is called the **standard error of the mean.**

Let us go back to our data in Example 1, the times (in seconds) required for 50 people sampled at random to complete a certain test item.

In that sample, the mean obtained was $\bar{x} = 54.0$. Now, by the Central Limit Theorem, all such obtainable means $\bar{x}$ (at sample size 50, which exceeds 30 as required) would be normally distributed with mean $\mu_{\bar{x}} = \mu$, the population mean. Intuitively, this means that the $\bar{x}$'s would be expected to be "close to" a central value μ in a predictable way, with their dispersion measured by the standard deviation

$$\sigma_{\bar{x}} = \frac{\sigma}{\sqrt{n}} = \frac{\sigma}{\sqrt{50}}.$$

Now, the problem is that we do not *know* the original population standard deviation σ. In practice, we handle this difficulty in about the only way we can: we compute the *sample* standard deviation s from the sample directly and use s in place of σ.

By returning to the given 50 numbers in the sample, you may verify that their sample standard deviation is

$$s = \sqrt{\frac{\Sigma(x-\bar{x})^2}{n-1}} = \sqrt{\frac{\Sigma(x-54.0)^2}{50-1}} = \sqrt{\frac{3{,}686}{49}}$$
$$\approx \sqrt{75.2} \approx 8.7.$$

Therefore the desired $\sigma_{\bar{x}}$ may be computed as:

$$\sigma_{\bar{x}} = \frac{\sigma}{\sqrt{n}} \approx \frac{s}{\sqrt{n}} = \frac{8.7}{\sqrt{50}},$$

which we may simplify (to avoid dividing by $\sqrt{50}$) as follows:

$$\frac{8.7}{\sqrt{50}} = \frac{8.7}{\sqrt{50}} \cdot \frac{\sqrt{50}}{\sqrt{50}} = \frac{8.7\sqrt{50}}{50}$$

so that

$$\sigma_{\bar{x}} \approx \frac{8.7\sqrt{50}}{50} \approx \frac{8.7(7.071)}{50} = \frac{61.5177}{50} \approx 1.2.$$

We may place a "hat" (^) over the symbol $\sigma_{\bar{x}}$ to indicate that we have *approximated* the standard deviation by substituting s for σ:

$$\hat{\sigma}_{\bar{x}} = \frac{s}{\sqrt{n}}.$$

Note that $\sigma_{\bar{x}}$, with $n = 50$, is less than $\frac{1}{7}$ as much as the original population σ: the sample means are less than $\frac{1}{7}$ as much dispersed from each other as the original numbers x were!

Now that we have an approximate value for $\sigma_{\bar{x}}$, we proceed with examples showing the actual construction of *interval estimates* of μ. The following examples illustrate the basic technique and should be studied carefully.

EXAMPLES

2. Referring to the given sample of 50 numbers as before, we have a *point estimate* for μ, namely $\bar{x} = 54.0$, and we know that such $\bar{x}$'s would be approximately normally distributed with standard deviation $\sigma_{\bar{x}} \approx 1.2$.

   Now, we may construct an *interval* estimate of μ. The particular interval estimate will be the kind known as a "95% **confidence interval.**" That is, we will construct an interval which we can feel is 95% likely to contain the true value of μ. To do so, we first note that *under any normal curve,* 95% *of the area lies between* $z = -1.96$ *and* $z = 1.96$. That is, 95% of the area lies within 1.96 standard deviations of the mean. (This is because the area beyond $z = 1.96$ or -1.96, on either side of the curve, is 0.0250, as you may verify from the normal table; see Figure 4.4.) Remember this z-score 1.96, because it is very frequently used. Our procedure will be, then, to use the *point* estimate

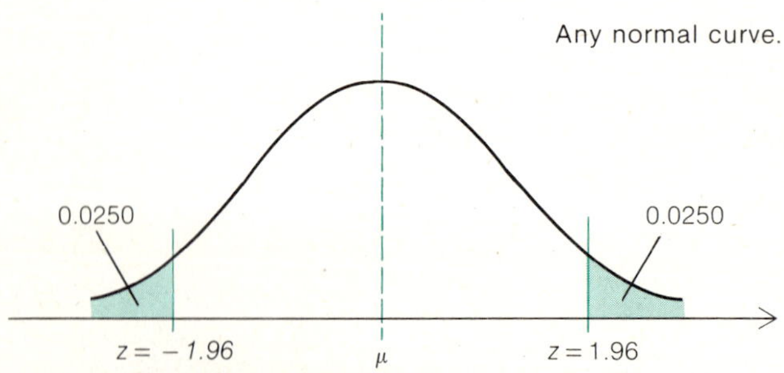

Figure 4.4 Total colored area under the tails: 0.0500. Uncolored area: 0.9500.

$\bar{x} = 54.0$ as the midpoint of the interval to be constructed, and obtain the right and left endpoints of the interval by *adding and subtracting* 1.96 *standard deviations*, where the appropriate standard deviation is $\sigma_{\bar{x}}$. That is, our interval will run from

$$\bar{x} - 1.96\sigma_{\bar{x}} \quad \text{to} \quad \bar{x} + 1.96\sigma_{\bar{x}}.$$

Because, $\sigma_{\bar{x}} \approx 1.2$ here, we get $1.96\sigma_{\bar{x}} \approx 1.96(1.2) = 2.352$, so that the endpoints of the interval would be

$$\bar{x} - 1.96\sigma_{\bar{x}} = 54.0 - 2.352 = 51.648 \quad \text{and}$$

$$\bar{x} + 1.96\sigma_{\bar{x}} = 54.0 + 2.352 = 56.352.$$

To round these endpoints off to tenths, we use (instead of the usual round-off considerations) the principle that *round-off should always be done in such a direction as to widen the interval*, so that we do not endanger the claim that we have 95% confidence in the interval's inclusion of μ. (A narrower interval might be *less* than 95% likely to include μ.) Thus left endpoints are always rounded leftward on the number line, while right endpoints are always rounded rightward:

$$51.648 \to 51.6, \quad 56.352 \to 56.4.$$

Therefore our 95% confidence interval estimate of μ is:

$$51.6 < \mu < 56.4.$$

We can feel 95% confident that the true population mean μ will lie between 51.6 seconds and 56.4 seconds. (Thus, to return to an earlier question: how likely is it that the true mean time μ would lie as far away from $\bar{x}$ as 50 or 58 seconds?—we may answer, less than a 5% chance, because those numbers lie outside our 95% confidence interval.)

3. To construct a **99%** confidence interval estimate of μ, we proceed in much the same way as in Example 2, except that we use the *z*-value 2.58 instead of 1.96, because (as you may verify from the normal table) about 99% of the area under any normal curve lies between $z = -2.58$ and $z = 2.58$. That is to say, 99% of the area lies within 2.58 standard deviations of the mean (see Figure 4.5). Remember, also, this *z*-score 2.58, because it is used quite often.

Thus for a 99% confidence interval (an interval that will, we are "99% sure," contain μ), we follow the same procedure as before, except that 2.58 (not 1.96) is the appropriate *z*-value to use. That is, we elect the point estimate $\bar{x} = 54.0$ as the interval's midpoint, then add and subtract 2.58 standard deviations

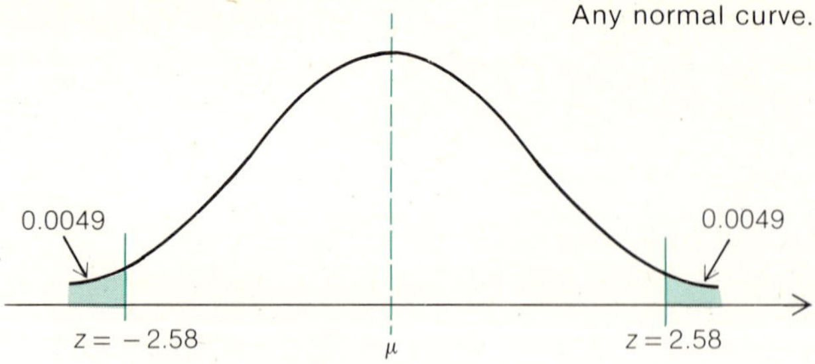

Figure 4.5

Total colored area: 0.0098 ≈ 0.0100.
Uncolored: 0.9902 ≈ 0.9900.

($\sigma_{\bar{x}} \approx 1.2$) to obtain the interval's endpoints:

$$\bar{x} - 2.58\sigma_{\bar{x}} = 54.0 - 2.58(1.2) = 54.0 - 3.096 = 50.904$$

$$\bar{x} + 2.58\sigma_{\bar{x}} = 54.0 + 2.58(1.2) = 54.0 + 3.096 = 57.096$$

and we can round off to tenths in such a way as to *widen* the interval (so that we do not endanger the validity of the claim that the interval is 99% likely to contain μ). Thus the finished 99% confidence interval estimate of μ is:

$$50.9 < \mu < 57.1.$$

These examples have illustrated what is true in general:

A 95% confidence interval for μ is given by

$$\bar{x} - 1.96\sigma_{\bar{x}} < \mu < \bar{x} + 1.96\sigma_{\bar{x}}.$$

A 99% confidence interval for μ is given by

$$\bar{x} - 2.58\sigma_{\bar{x}} < \mu < \bar{x} + 2.58\sigma_{\bar{x}}.$$

Note that the 99% confidence interval obtained in Example 3 is noticeably *wider* than the 95% confidence interval obtained in Example 2 (see Figure 4.6). This comparison is to be expected, because if we want to be "99% sure" instead of only "95% sure," we must *pay* for the privilege; that is, the interval must be *wider* to make it that likely that μ is contained in it. It is, to be sure, *better* to have a narrow interval, because a narrow interval gives a clearer, more specific indication of the real location of μ. But we can't have everything: Either we have to make do with a wider (less specific) interval but be *more* confident in it, or have a narrower (more specific) interval but be *less* confident in it. In any event, however, an interval can be narrowed by *increasing the*

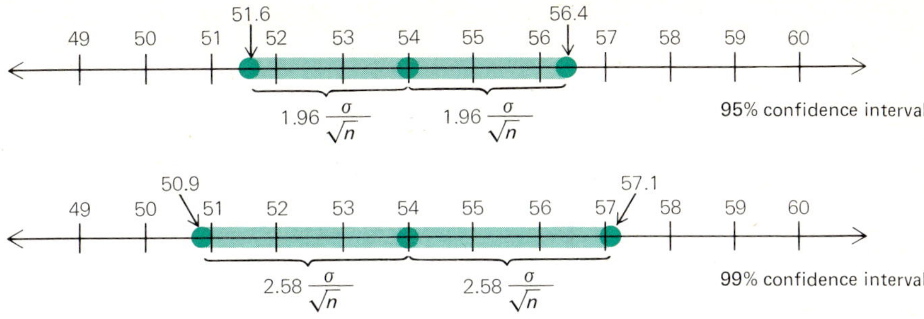

Figure 4.6

sample size, because the larger n is, the smaller $\frac{\sigma}{\sqrt{n}}$ is, and the less we add and subtract to produce the interval endpoints. (Appropriately, when we take the trouble to obtain a larger sample, we are rewarded with a narrower and thus more informative and useful confidence interval.)

EXAMPLES

4. To estimate the true mean (μ) weight of all the automobiles that use a certain large parking garage, a statistician takes a random sample of size 100, obtaining a sample mean $\bar{x} = 2030.0$ kilograms and sample standard deviation 216.0 kilograms. What is the 95% confidence interval estimate for μ?

We first observe that at the sample size $n = 100 > 30$, the obtainable sample means $\bar{x}$ are approximately normally distributed with standard deviation

$$\sigma_{\bar{x}} = \frac{\sigma}{\sqrt{n}} \approx \frac{s}{\sqrt{n}} = \frac{216.0}{\sqrt{100}} = \frac{216.0}{10} = 21.6.$$

And to have a 95% confidence interval we must add and subtract (from the midpoint $\bar{x} = 2030.0$ kilograms) the quantity $1.96\sigma_{\bar{x}}$, that is, approximately

$$1.96(21.6) = 42.336,$$

so that

$$\bar{x} - 1.96\sigma_{\bar{x}} = 2030.0 - 42.336 = 1987.664 \quad \text{and}$$

$$\bar{x} + 1.96\sigma_{\bar{x}} = 2030.0 + 42.336 = 2072.336.$$

And we round off these results in such a direction (outward) as to widen, not narrow, the interval, producing then the confidence interval

$1987.6 < \mu < 2072.4$ (see Figure 4.7).

5. In Example 4, if the results $\bar{x} = 2030.0$ kilograms and $s = 216.0$ kilograms had been obtained from a sample of size $n = 400$ (instead of only 100), then a

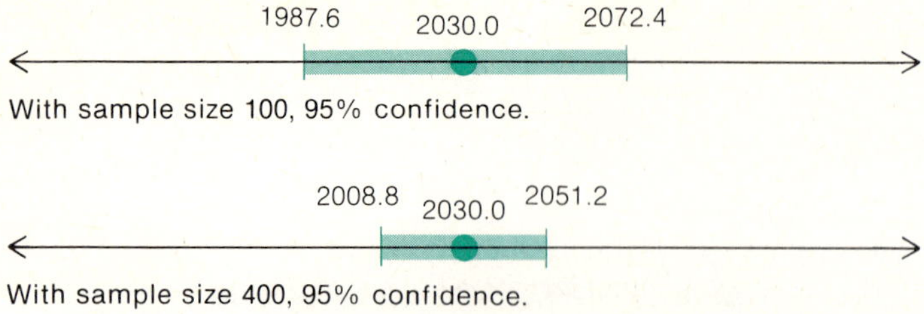

Figure 4.7

narrower (and thus better) 95% confidence interval would have been produced, because under these circumstances

$$\sigma_{\bar{x}} = \frac{\sigma}{\sqrt{n}} \approx \frac{s}{\sqrt{n}} = \frac{216.0}{\sqrt{400}} = \frac{216.0}{20} = 10.8,$$

only *half* as large as the previous $\sigma_{\bar{x}}$. That is, *at the larger sample size* 400 *the obtainable sample means* $\bar{x}$ *are much less dispersed* (much more similar to each other and thus more predictable) than at the smaller sample size 100. We must add and subtract (from the midpoint $\bar{x} = 2030.0$) the quantity

$$1.96\sigma_{\bar{x}} \approx 1.96(10.8) = 21.168$$

obtaining the endpoint values

$$\bar{x} - 1.96\sigma_{\bar{x}} = 2030.0 - 21.168 = 2008.832 \quad \text{and}$$

$$\bar{x} + 1.96\sigma_{\bar{x}} = 2030.0 + 21.168 = 2051.168,$$

which, when rounded off outward to widen the interval, give a 95% confidence interval estimate

$$2008.8 < \mu < 2051.2,$$

an interval only half as wide (and thus much more satisfactory) than the one in Example 4 at the smaller sample size. We have "bought" a better interval result by working with a more ambitious-sized sample. (Again, see Figure 4.7.)

6. To estimate the true mean salary (μ) of all employees in a certain industry, a statistician takes a random sample of size $n = 625$, obtaining a sample mean $\bar{x} = \$14,470$ and sample standard deviation \$1230. What is the 99% confidence interval estimate of μ?

We note that at the sample size $n = 625 > 30$, the sample means are approximately normally distributed with standard deviation

$$\sigma_{\bar{x}} = \frac{\sigma}{\sqrt{n}} \approx \frac{s}{\sqrt{n}} = \frac{1,230}{\sqrt{625}} = \frac{1,230}{25} = 49.2.$$

Then for a 99% confidence interval we should add and subtract (from the midpoint 14,470) the quantity

$$2.58\sigma_{\bar{x}} = 2.58(49.2) = 126.936$$

obtaining

$$\bar{x} - 2.58\sigma_{\bar{x}} = 14{,}470 - 126.936 = 14{,}343.064 \quad \text{and}$$

$$\bar{x} + 2.58\sigma_{\bar{x}} = 14{,}470 + 126.936 = 14{,}596.936$$

so that (with the appropriate round-offs to widen the interval) the 99% confidence interval estimate of μ is:

$$\$14{,}343.0 < \mu < \$14{,}597.0.$$

We can feel 99% confident that the overall population mean, the true mean salary of all the employees throughout the population in question, lies between these endpoints, which are rounded to tenths of a dollar.

7. If the sample results $\bar{x} = \$14{,}470$ and $s = \$1{,}230$ (of Example 6) had come from a sample of size $n = 900$, then the sample means (at that sample size) again could be assumed to be approximately normally distributed, but this time with the somewhat *smaller* standard deviation

$$\sigma_{\bar{x}} = \frac{\sigma}{\sqrt{n}} \approx \frac{s}{\sqrt{n}} = \frac{1{,}230}{\sqrt{900}} = \frac{1{,}230}{30} = 41.0$$

so that to compute a 99% confidence interval estimate of μ we must add and subtract (from the midpoint 14,470) the quantity

$$2.58\sigma_{\bar{x}} = 2.58(41.0) = 105.78$$

obtaining

$$\bar{x} - 2.58\sigma_{\bar{x}} = 14{,}470 - 105.78 = 14{,}364.22 \quad \text{and}$$

$$\bar{x} + 2.58\sigma_{\bar{x}} = 14{,}470 + 105.78 = 14{,}575.78$$

so that the desired 99% confidence interval is:

$$\$14{,}364.2 < \mu < \$14{,}575.8.$$

You will see that this interval is somewhat narrower (and therefore somewhat more specific and informative) than the one obtained in Example 6—the result, of course, of having a larger sample this time. (The interval of Example 6 shows a spread of \$254, whereas the interval of this present example shows a spread of only \$211.60.)

Because of the Central Limit Theorem, which assures us that (given a sample size above 30) repeatedly obtained sample means would be approximately normally distributed, there are certain kinds of probabilities that we can compute, concerning the mean of a sample, even a sample that has not yet been taken, provided the necessary population parameters are known. The following examples illustrate the method.

EXAMPLES

8. In a certain large factory complex employing thousands of workers, the true mean weight of all the employees is known to be $\mu = 64.0$ kilograms with standard deviation $\sigma = 7.0$ kilograms. If 100 employees show up at random in the cafeteria (i.e. if in effect a random sample of size $n = 100$ is taken), what is the probability that their mean weight (sample mean $\bar{x}$) will exceed 65.0 kilograms?

To solve this problem, we first note that all such sample means $\bar{x}$ would be approximately normally distributed with standard deviation

$$\sigma_{\bar{x}} = \frac{\sigma}{\sqrt{n}} = \frac{7.0}{\sqrt{100}} = \frac{7.0}{10} = 0.7.$$

Also, the distribution of such sample means $\bar{x}$ would be centered about the true mean μ, and we are really finding an area under the appropriate normal curve, beyond the value $\bar{x} = 65.0$ (see Figure 4.8). Again, it is the *sample means* $\bar{x}$ (at size $n = 100$) that are normally distributed and the appropriate standard deviation for them is *not* the original $\sigma = 7.0$ for the x's (the employee weights) of the population, but rather $\sigma_{\bar{x}} = 0.7$.

To find the desired area under the tail of the normal curve, we change the value $\bar{x} = 65.0$ into a z-score in the ordinary manner:

$$z = \frac{\bar{x} - \mu}{\sigma_{\bar{x}}} = \frac{65.0 - 64.0}{0.7} = \frac{1.0}{0.7} \approx 1.43$$

and because (from the normal area table) the area from μ *out to* $z = 1.43$ is

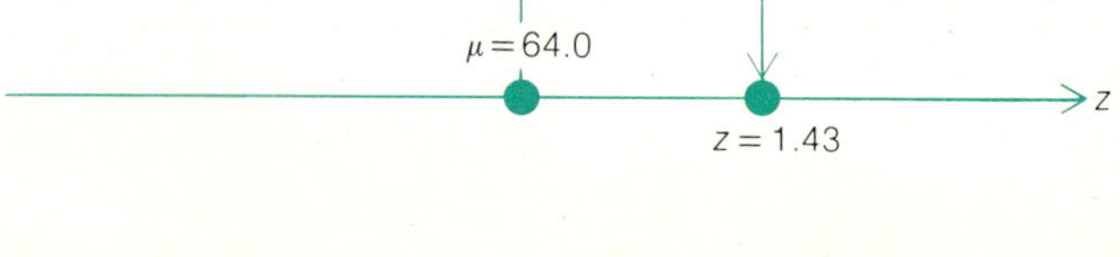

Figure 4.8

0.4236, the desired area (shaded in Figure 4.8) *beyond* $z=1.43$ is

$0.5000-0.4236=0.0764.$

That is, when 100 employees are assembled at random, the probability that their sample mean weight will exceed 65.0 kilograms is about 0.0764. If such random samples (at size $n=100$) were taken over and over again indefinitely, we would expect the *sample means* to exceed 65.0 kilograms about 7.64% of the time. (It is important to understand that *this does not mean that* 7.64% *of the employees weigh more than* 65 *kilograms*: the variable in use here is $\bar{x}$ (sample means at size 100), *not* the actual weights x of the original population.)

Note that because $\sigma_{\bar{x}}$ is quite small as a measure of the dispersion of the $\bar{x}$'s (only one-tenth as much as the original population parameter σ), the probability that the sample mean on any occasion *misses* μ by being 1 kilogram (or more) larger than μ, is rather small. The moral of the tale is that because of the nature of the distribution of the $\bar{x}$'s, *they are highly predictable*: more predictable even than the original weights x (which may or may not even be normally distributed).

9. In the situation of Example 8, with the employees having mean weight $\mu=64.0$ kilograms with standard deviation $\sigma=7.0$ kilograms, suppose 400 employees (instead of 100) assemble in the cafeteria at random. What is the probability that *their* sample mean weight will exceed 65.0 kilograms?

This time we have a sample of size $n=400$, so that the standard deviation of the sample means $\bar{x}$ would be even smaller than before:

$$\sigma_{\bar{x}}=\frac{\sigma}{\sqrt{n}}=\frac{7.0}{\sqrt{400}}=\frac{7.0}{20}=0.35.$$

And the value $\bar{x}=65.0$ as a z-score will be

$$z=\frac{\bar{x}-\mu}{\sigma_{\bar{x}}}=\frac{65.0-64.0}{0.35}$$

$$=\frac{1}{0.35}\approx 2.86.$$

From the normal table, the area *out to* this z-score is 0.4971, so that (referring again to the sort of picture as in Figure 4.8 but with a different z) the area *beyond* $\bar{x}=65.0$ (beyond $z=2.86$) is

$0.5000-0.4971=0.0029,$

indicating a very small probability. The moral of *this* story: The larger the sample size, at which sample means $\bar{x}$ are computed, the more predictable (because less dispersed) the $\bar{x}$'s are. At $n=400$, $\bar{x}$ simply *should not* be as far away from μ as 65.0 kilograms or beyond.

NEW TERMS IN THIS SECTION

point estimate	sampling distribution of the mean
interval estimate	standard error of the mean
Central Limit Theorem	confidence interval

EXERCISES 4-A

1. To estimate the true mean (μ) age of all the people using the public library in a certain town, a sociologist takes a random sample of 100 library users, finding a sample mean of $\bar{x} = 29.0$ years with standard deviation $s = 8.0$ years. Find (expressing endpoints in tenths):
   (a) the 95% confidence interval estimate of μ
   (b) the 99% confidence interval estimate of μ.

2. In Exercise 4-A.1, suppose the results $\bar{x} = 29.0$ and $s = 8.0$ had been obtained from a random sample of size $n = 400$. Find (and compare with your results in Exercise 4-A.1):
   (a) the 95% confidence interval estimate of μ
   (b) the 99% confidence interval estimate of μ.

3. An economist, wishing to estimate the true mean annual income of all families residing in a certain region, takes a random sample of 625 families, obtaining a sample mean of $\bar{x} = \$14{,}568$ and sample standard deviation $s = \$2{,}875$. Find (expressing endpoints in tenths):
   (a) the 95% confidence interval for μ
   (b) the 99% confidence interval for μ.

4. In Exercise 4-A.3, suppose the economist had obtained the results $\bar{x} = \$14{,}568$ and $s = \$2{,}875$ from a random sample of only 36 families. Find (and compare with your results in Exercise 4-A.3):
   (a) the 95% confidence interval for μ
   (b) the 99% confidence interval for μ.

5. To estimate the true mean length (in terms of numbers of words) of all the novels published in the United States during a certain decade, an editor takes a random sample of size $n = 200$ novels, obtaining a sample mean $\bar{x} = 81{,}000$ words, with $s = 3{,}600$ words. Find (expressing endpoints rounded to appropriate hundreds, that is, integers ending in two zeros—remember, round off in such a way as to *widen* the interval each time):
   (a) the 95% confidence interval for μ
   (b) the 99% confidence interval for μ.

6. Repeat Exercise 4-A.5 (assuming the same sample mean and standard deviation) with a sample size of
   (a) $n = 40$
   (b) $n = 500$.

7. A psychologist wishes to estimate the true mean score μ (obtainable for all the examinees who might be considered) for people taking a test that measures attitudes toward certain social

issues. A random sample is taken, yielding a sample mean score $\bar{x} = 230.0$ with $s = 87.5$. Find (expressing endpoints in tenths) the 95% confidence interval for μ if the sample size is $n =$

(a) 49 (b) 81 (c) 100 (d) 300 (e) 400
(f) 625 (g) 900 (h) 1,600 (i) 6,400 (j) 14,400

8. A forestry researcher wants to estimate the true mean height of all the trees in a certain wooded region, and takes a random sample of size $n = 100$ trees, obtaining a sample mean of 22.45 meters. Find the 95% confidence interval for μ (expressing endpoints to two decimal places) if the standard deviation as estimated from the sample is

(a) 7.00 meters (b) 6.50 meters (c) 6.00 meters (d) 5.50 meters
(e) 5.00 meters (f) 4.50 meters (g) 4.00 meters (h) 3.50 meters
(i) 3.00 meters (j) 2.50 meters (k) 2.00 meters

9. A business analyst wants to estimate the "average" (true mean μ) amount of money spent by individual shoppers entering a supermarket during certain hours of the day. The analyst takes a random sample of 250 shoppers and finds their sample mean purchase to be $\bar{x} = \$18.70$ with $s = \$3.90$. Find (expressing endpoints to two decimal points, that is, cents):

(a) the 95% confidence interval estimate of μ
(b) the 99% confidence interval estimate of μ.

10. For comparison, repeat Exercise 4-A.9 with the assumption that the same $\bar{x}$ and s were obtained from each of the following sample sizes, and compare your answers:

(a) 225 (b) 200 (c) 175 (d) 150
(e) 125 (f) 100 (g) 75 (h) 50

11. To estimate the true mean number of hours of use which television sets will give before needing repairs, an engineer samples 300 sets at random, finding that these 300 sets average 4,800 hours, with sample standard deviation $s = 450$ hours. Find (expressing endpoints in whole-number amounts of hours with round-off as appropriate for confidence intervals):

(a) the 95% confidence interval estimate of μ
(b) the 99% confidence interval estimate of μ.

12. Repeat Exercise 11 (assuming again $\bar{x} = 4{,}800$ and $s = 450$) at a sample size of

(a) 325 (b) 350 (c) 375 (d) 400 (e) 425
(f) 450 (g) 475 (h) 500 (i) 1,000 (j) 10,000

13. In a certain large industrial plant the true mean age of all the employees is $\mu = 39.5$ years, and the true standard deviation is $\sigma = 7.2$ years. A committee of 36 employees is to be chosen at random. Find the probability that when the committee is chosen, the mean age ($\bar{x}$) of its members will be

(a) between 39.5 and 40.7 years (b) between 38.9 and 40.1 years
(c) between 39.8 and 41.3 years (d) between 37.0 and 42.0 years
(e) between 36.0 and 43.0 years (f) between 36.5 and 39.0 years
(g) more than 41.5 years (h) more than 42.0 years
(i) more than 37.5 years (j) less than 41.4 years
(k) less than 37.0 years (l) between 39.0 and 40.0 years

14. For a large group of subjects undergoing psychological testing, the true mean score on an attitude-measuring test is $\mu = 320$ with standard deviation $\sigma = 40$. A random sample of these examinees is to be taken. Find the probability that the sample mean $\bar{x}$ will exceed 323 if the sample size is

(a) 36	(b) 49	(c) 64	(d) 81	(e) 100
(f) 225	(g) 400	(h) 625	(i) 900	(j) 1,600

PONDERABLES 4-A

1. Discuss, intuitively and mathematically, the results of Exercise 4-A.8.

2. Do the same for Exercise 4-A.14.

3. To estimate the true mean gas mileage of a car, a motorist takes a sample by computing the mileage on a number of occasions, finding the sample standard deviation to be $s = 2.80$ miles per gallon, and finding the 95% confidence interval for μ (the true mean mileage for the car's overall performance) to be

$$16.716 < \mu < 18.284.$$

Find
(a) the sample mean that must have been used and
(b) the sample size.

4. A medical lab technician, to estimate the true mean amount of a drug that patients must receive to register improvement from a certain disease, takes a random sample of 100 patients, finding the 99% confidence interval to be

$$235.872 < \mu < 244.128$$

(endpoints in milligrams). Find
(a) the sample mean that must have been used and
(b) the sample standard deviation that must have been used.

5. Using the normal table, find what z-value (instead of 1.96 or 2.58) one would use to compute a confidence interval estimate of μ at the level of confidence
(a) 98% (b) 97% (c) 96% (d) 94% (e) 93% (f) 92%

6. Using your results from the previous problem: a child psychologist wants to estimate the true mean age (in months) at which children become toilet-trained during the day. The researcher takes a random sample of size 100 and gets $\bar{x} = 38.0$ months, with $s = 8.0$ months. Find (expressing endpoints in tenths):
(a) the 99% confidence interval estimate of μ
(b) the 98% confidence interval estimate of μ
(c) the 97% confidence interval estimate of μ
(d) the 96% confidence interval estimate of μ

(e) the 95% confidence interval estimate of μ
(f) the 94% confidence interval estimate of μ
(g) the 93% confidence interval estimate of μ
(h) the 92% confidence interval estimate of μ

7. How does one intuitively explain the nature of the answers in the previous problem?

8. Wishing to estimate the true mean (μ) amount of time (in seconds) required for laboratory mice to run through a maze, an experimental psychologist randomly samples 65 mice, testing one at a time, and finding their times to be:

48	46	52	41	53
47	54	50	42	49
53	45	40	58	44
44	49	56	50	43
40	41	50	56	48
47	48	55	55	60
40	45	57	47	57
52	57	52	46	44
55	43	43	58	40
39	46	59	57	51
51	42	48	42	46
50	59	41	48	38
62	54	57	52	43

Compute
(a) the sample mean $\bar{x}$
(b) the sample standard deviation s (to the nearest tenth)
(c) the 99% confidence interval estimate of μ (round $\sigma_{\bar{x}}$ to tenths and express endpoints in tenths).

9. Suppose we wish to estimate the true mean lifetime (in hours) of a certain brand of toaster, computing a 95% confidence interval. σ has been estimated to be 25 hours. Suppose, further, that we want the interval estimate to run from 5 hours below the sample mean to 5 hours above it. (That is, we want the interval to be 10 hours wide.) How large a sample must be taken? (*Hint*: The radius of the interval is $\frac{1.96\sigma}{\sqrt{n}}$ and must equal 5.)

B SMALL-SAMPLE ESTIMATION

In the previous section we examined the idea of constructing a confidence interval estimate of a population mean μ. Our computations were based on the fact that for samples of fixed size $n > 30$, *sample means* $\bar{x}$ *tend to be normally distributed* with

mean $\mu_{\bar{x}} = \mu$ and

standard deviation $\sigma_{\bar{x}} = \dfrac{\sigma}{\sqrt{n}}$

(where μ and σ are the parameters of the given population). This is the Central Limit Theorem.

Because of this tendency of the sample means $\bar{x}$ to be normally distributed, we can construct a 95% or 99% confidence interval by computing a single sample mean $\bar{x}$ and constructing the intervals as:

$$(95\%) \quad \bar{x} - 1.96\frac{\sigma}{\sqrt{n}} < \mu < \bar{x} + 1.96\frac{\sigma}{\sqrt{n}}$$

$$(99\%) \quad \bar{x} - 2.58\frac{\sigma}{\sqrt{n}} < \mu < \bar{x} + 2.58\frac{\sigma}{\sqrt{n}}.$$

But—remember—this technique is only trustworthy *when the sample size* (n) *exceeds* 30. Conceivably, a sample this large may sometimes be difficult or impossible to come by. For example, we may wish to estimate the true mean weight of a certain ancient coin when only 23 of them have ever been found—we cannot very well just pick up the telephone and say, "Send up 400 of those coins." We would have to make do with our sample of size $n = 23$.

Obviously the question arises, then: What do we do if the sample size does *not* exceed 30? Can such a small sample lead to any kind of confidence interval estimation of μ?

The answer is that a **small sample** ($n \leq 30$) can indeed be used, but *not* with the usual normal curve approach, especially when the population standard deviation σ is unknown. We do *not* use the idea of subtracting and adding 1.96 standard deviations or 2.58 standard deviations, because these numbers 1.96 and 2.58 come from the normal table—but with $n \leq 30$, sample means $\bar{x}$ are *not* guaranteed to be normally distributed.

What, then, do we do?

The solution to this problem was first supplied in 1908, when William S. Gosset published a paper showing that when the sample size n is 30 or less, and when the population itself is not very different from normally distributed (as we will assume throughout this section), sample means $\bar{x}$ tend to be distributed *not* as suggested by a normal curve, but by a similar-looking curve called a t-curve. (See Figure 4.9.) There is a different t-curve for each sample size from 30 downward. Each t-curve resembles a normal curve in general shape and appearance, but Figure 4.9 shows the difference: Typically, a t-curve reaches a less high "peak" than a normal curve and has tails tapering off less quickly than a normal curve. Intuitively, the t-curve is a picture of a distribution more dispersed and less dramatically "bunched up" around the mean than what a normal curve represents.

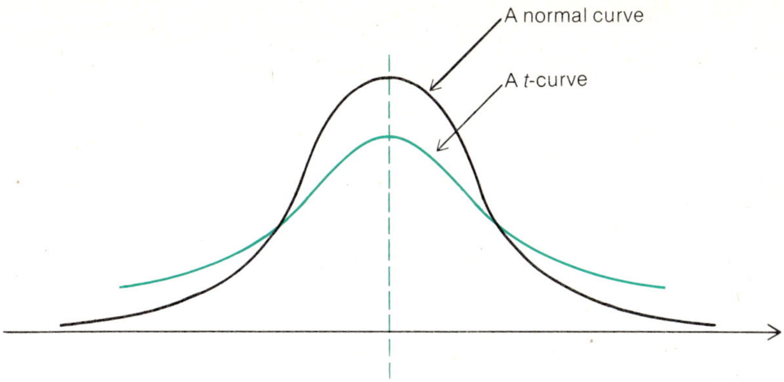

Figure 4.9

Tradition has it that when Mr. Gosset published his tract about the *t*-curves, he had to work around the difficulty that he was employed by the Guinness Brewing Company in Dublin, which did not allow its employees to publish individual research. He therefore published his research under the pen-name "Student," and we still use the term **Student's *t*-distribution** for the type of distribution that a *t*-curve pictures.

Just as with a normal curve, a *t*-curve can be used by referring to a table: The *t* Distribution, in the back of the book. Basically, the idea is this: with a small sample size, we construct a confidence interval for μ in the same way as before, but *in place of* 1.96 *or* 2.58 *we must use a t-value* (from the table) in the computations. Also, where we previously used the value $\frac{s}{\sqrt{n}}$ as a *necessary* stand-in for $\frac{\sigma}{\sqrt{n}}$, the standard deviation needed, *when using a t-curve we always use* $\frac{s}{\sqrt{n}}$ (with no regrets) as the appropriate standard deviation, because the *t*-curve is designed that way. Thus for $n \leq 30$, a confidence interval for μ takes the form

$$\bar{x} - (t\text{-value}) \cdot \frac{s}{\sqrt{n}} < \mu < \bar{x} + (t\text{-value}) \cdot \frac{s}{\sqrt{n}},$$

where the *t*-value (used in lieu of 1.96 or 2.58) depends on the sample size and on the confidence level desired.

Let us see how the appropriate *t*-value is obtained from the table.

You will recall that for large sample sizes, when a 95% confidence interval is desired, the *z*-value 1.96 from the normal table is used, because 2.5% (that is, 0.0250) of the area under a normal curve lies beyond $z = 1.96$, and another 2.5% lies beyond $z = -1.96$, under the tails of the curve, so that 95% of the area lies between $z = -1.96$ and $z = 1.96$. Similarly, in a small-sample situation, we wish to use a *t*-value beyond which the area is 0.0250 *on each side.* The

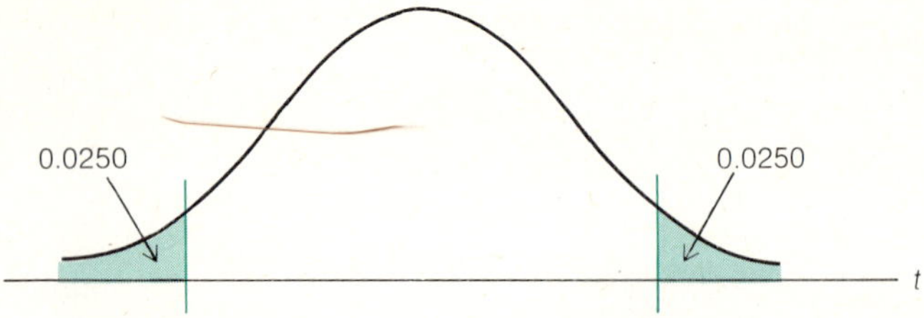

Figure 4.10

necessary t, therefore, is found in the table by first agreeing to use the column headed

$t_{.025}$

where the subscript .025 means that 0.0250 (or 2.5%) of the area lies under *each* tail of the curve, for a total of 0.0500 = 5% on both sides combined. (See Figure 4.10.) Thus it is important to realize that the subscript refers to the under-the-tail area *on one side only.*

Now that we know to use the column $t_{.025}$ for a 95% confidence situation, the next question is: which number in the column $t_{.025}$ do we use? To answer this, we need only choose the appropriate d.f. value in the left column (where **d.f.** means **degrees of freedom,** a mathematical term the precise meaning of which need not be examined here), and this part is easy:

d.f. = (sample size) − 1.

So, for example, if the sample size is $n = 27$, then we note that

d.f. = 27 − 1 = 26

and we find the needed t-value in the table:

d.f. $\quad t_{.025}$
↓ $\quad$ ↓
26 → 2.056.

This value $t_{.025} = 2.056$ is used in place of $z = 1.96$ in the construction of the 95% confidence interval.

(The notation *"inf."* just below d.f. = 29 means that starting with d.f. = 30 [that is, starting with $n = 31$], from there to infinity, so to speak, the t-values coincide in effect with familiar z-values, such as 1.960 in the $t_{.025}$ column.)

EXAMPLES

1. A psychologist wishes to find a 95% confidence interval estimate for the true mean time μ needed for a certain unusual stimulus response. The response has been observed in the laboratory only $n = 25$ times, and for those 25 times

the sample mean is $\bar{x}=3.35$ seconds with a sample standard deviation $s=0.05$ seconds.

The t-value needed for construction of a confidence interval with this small sample size $n=25$ is found in the table as follows. First,

$$\text{d.f.} = (\text{sample size}) - 1$$

$$= 25 - 1 = 24$$

and (as previously explained) we must use the column $t_{.025}$ to get 95% confidence:

$$\begin{array}{ll} \text{d.f.} & t_{.025} \\ \downarrow & \downarrow \\ 24 \rightarrow & 2.064 \end{array}$$

so that we use $t_{.025}=2.064$ just as we would have used $z=1.96$ if the sample size had been large. The quantity to be added and subtracted from the midpoint $\bar{x}=3.35$ is

$$t_{.025} \cdot \frac{s}{\sqrt{n}} = 2.064 \cdot \frac{0.05}{\sqrt{25}} = 2.064 \cdot \frac{0.05}{5}$$

$$= 2.064(0.01) = 0.02064.$$

Thus the confidence interval desired will have

left endpoint: $\bar{x}-0.02064=3.35-0.02064=3.32936$

right endpoint: $\bar{x}+0.02064=3.35+0.02064=3.37064$

so that (before round-off) we have

$$3.32936 < \mu < 3.37064.$$

If we, say, round these endpoints to three places (in such a way as to widen the interval), the final result is

$$3.329 < \mu < 3.371.$$

This interval is, of course, *wider* (and vaguer) than what would be obtained at a larger sample size (especially a large enough sample size to use the smaller $z=1.96$ instead of $t_{.025}=2.064$).

2. A liquor retailer wishes to compute a 95% confidence interval estimate of the true mean amount (in quarts) of wine sold at a store between 6 p.m. and 7 p.m. from day to day. Records are kept for 20 random days, showing a sample mean of $\bar{x}=32.0$ quarts and sample standard deviation of $s=4.0$ quarts.

To compute a 95% confidence interval one must first note that

$$\text{d.f.} = n - 1 = 20 - 1 = 19,$$

so that the desired t-value from Table III is 2.093:

$$\begin{array}{ll} \text{d.f.} & t_{.025} \\ \downarrow & \downarrow \\ 19 \rightarrow & 2.093. \end{array}$$

And the quantity to be subtracted and added to get the endpoints is

$$t_{.025} \cdot \frac{s}{\sqrt{n}} = 2.093 \cdot \frac{4.0}{\sqrt{20}} \approx 2.093 \cdot \frac{4.0}{4.472}$$

$$\approx 1.872$$

so that the endpoints for the interval are:

$$\text{(left)} \quad \bar{x} - 1.872 = 32.0 - 1.872 = 30.128$$

$$\text{(right)} \quad \bar{x} + 1.872 = 32.0 + 1.872 = 33.872$$

and if we round these off to appropriate tenths:

$$30.1 < \mu < 33.9.$$

If we desire a 99% confidence interval from a small sample, the reasoning is much the same as before: to have only 1% (=0.0100) of the area under the tails of the curve (both sides combined), we must have half of that, or 0.0050, under the tail of the curve on *each* side, so that in the table the appropriate column to use is

$$t_{.005}.$$

EXAMPLES

3. In the situation of Example 1, in which a psychologist wants to estimate the true mean time required for a certain stimulus response, let us suppose again that a sample of size $n = 25$ has yielded a sample mean of $\bar{x} = 3.35$ seconds with sample standard deviation $s = 0.05$ seconds. But this time we will construct a 99% confidence interval.

   First (as in Example 1),

   $$\text{d.f.} = n - 1 = 25 - 1 = 24.$$

The required t-value (to use in place of $z = 2.58$ because of the smallness of the sample) turns out to be 2.797 in the table:

$$\begin{array}{ll} \text{d.f.} & t_{.005} \\ \downarrow & \downarrow \\ 24 \rightarrow & 2.797. \end{array}$$

And the quantity to be added and subtracted is

$$t_{.005} \cdot \frac{s}{\sqrt{n}} = 2.797 \cdot \frac{0.05}{\sqrt{25}} = 2.797 \cdot \frac{0.05}{5}$$

$$= 2.797 \cdot (0.01) = 0.02797$$

so that the interval endpoints are:

$$\text{(left)}\ \bar{x} - 0.02797 = 3.35 - 0.02797 = 3.32203$$

$$\text{(right)}\ \bar{x} + 0.02797 = 3.35 + 0.02797 = 3.37797$$

and if we express the endpoints in thousandths with the appropriate round-off, the desired 99% confidence interval is

$$3.322 < \mu < 3.378.$$

(Notice that this 99% confidence interval is about 0.007 seconds farther out on each end than the previous 95% confidence interval of Example 1. As always, a 99% confidence interval must be wider than a 95% interval to accommodate 99% confidence.)

4. In the situation of Example 2, in which a liquor retailer wants an estimate of the true mean amount (in quarts) of wine sold during certain hours, with a sample ($n = 20$) yielding $\bar{x} = 32.0$ quarts, $s = 4.0$ quarts, suppose we compute a 99% confidence interval.

As before,

$$\text{d.f.} = n - 1 = 20 - 1 = 19$$

and we look up the appropriate t-value in the table:

$$\begin{array}{ll} \text{d.f.} & t_{.005} \\ \downarrow & \downarrow \\ 19 \rightarrow & 2.861 \end{array}$$

so that the quantity to be added and subtracted is

$$t_{.005} \cdot \frac{s}{\sqrt{n}} = 2.861 \cdot \frac{4.0}{\sqrt{20}} \approx 2.861 \cdot \frac{4.0}{4.472} \approx 2.559$$

and the interval endpoints are:

(left) $\bar{x}-2.559=32.0-2.559=29.441$

(right) $\bar{x}+2.559=32.0+2.559=34.559$

where rounding off to tenths in such a direction as to widen the interval would give

$29.4<\mu<34.6.$

Again, this result is a wider interval (by an extension of about 0.7 quarts on each side) than the previous 95% interval estimate. As usual, if we want to be 99% sure instead of only 95% sure, we must pay for the privilege by accepting a somewhat wider and vaguer interval estimate.

We have seen that, although the use of such values as $z=1.96$ or $z=2.58$ requires normality assumptions and sample sizes in excess of 30, even a small sample of size 30 or less can be used to generate an interval estimate of a population mean. By using table values in place of such z-values as 1.96 or 2.58, and otherwise following the usual kind of procedure, we can, even with a small sample, have our cup of t.

NEW TERMS IN THIS SECTION

small sample
Student's *t*-distribution
d.f. (degrees of freedom)

EXERCISES 4-B

1. A museum staff researcher wishes to estimate the true mean weight (μ) of a certain kind of rare fossil. Surveying the available ones in various museum collections, she finds only 25 of them available, and from this sample the mean is $\bar{x}=28.4$ grams with sample standard deviation $s=2.7$ grams. Find (expressing your endpoints in tenths):
(a) the 95% confidence interval estimate of μ
(b) the 99% confidence interval estimate of μ.

2. Suppose you wish to estimate your car's mean gasoline mileage μ on the basis of a random sample of 16 mileage computations, from which you have obtained a sample mean $\bar{x}=17.3$ miles per gallon and sample standard deviation $s=2.5$ miles per gallon. Find (expressing your endpoints in tenths):
(a) the 95% confidence interval estimate of μ
(b) the 99% confidence interval estimate of μ.

3. A quality control inspector wants to estimate the true mean time required for an operation on an assembly-line to be performed, but does not have time to take a large sample. Watching the operation performed 9 times chosen at random, the inspector finds a sample mean of $\bar{x} = 12.5$ minutes with sample standard deviation $s = 2.4$ minutes. Expressing endpoints in tenths, find:
   (a) the 95% confidence interval for μ
   (b) the 99% confidence interval for μ.

4. To analyze prices in the supermarkets in a certain area, a statistician samples 20 supermarkets, in each one tallying up the cost of a certain fixed list of grocery items. The results are: $\bar{x} = \$35.70$, $s = \$1.40$. Find (expressing endpoints in hundredths, that is, cents):
   (a) the 95% confidence interval for μ, the mean cost of the items in all the area supermarkets
   (b) the 99% confidence interval for μ.

5. A researcher wishes to estimate the true mean amount (in milliliters) of precipitate obtained by a certain chemical process. The procedure is time-consuming and expensive; thus the researcher samples it only by trying it 15 times, producing results of $\bar{x} = 216.3$ milliliters and $s = 7.1$ milliliters. Expressing endpoints in tenths, find:
   (a) the 95% confidence interval estimate of μ
   (b) the 99% confidence interval estimate of μ.

6. To estimate the true mean lifetime (in hours of continuous use) of a certain brand of high-intensity light bulb, an analyst randomly samples 28 of the bulbs, obtaining a sample mean of $\bar{x} = 448.4$ hours with sample standard deviation $s = 12.8$ hours. Expressing endpoints in tenths, find:
   (a) the 95% confidence interval for μ
   (b) the 99% confidence interval for μ.

7. To estimate the true mean response time (from initial call to arrival at the scene) of a city emergency medical crew, a researcher randomly observes the team responding to 12 calls, with the result that the sample mean is $\bar{x} = 8.6$ minutes with sample standard deviation $s = 2.2$ minutes. Expressing endpoints in hundredths, find:
   (a) the 95% confidence interval estimate of μ
   (b) the 99% confidence interval estimate of μ.

8. Suppose someone wishes to estimate the true mean weight of all the first-grade children in a certain region, and does so with a small random sample that has yielded a sample mean of $\bar{x} = 20.9$ kilograms and standard deviation $s = 2.4$ kilograms. Expressing endpoints in hundredths, find the 95% confidence interval if the sample size used was

(a) 30	(b) 29	(c) 28	(d) 27	(e) 26
(f) 25	(g) 24	(h) 23	(i) 22	(j) 21
(k) 20	(l) 19	(m) 18	(n) 17	(o) 16
(p) 15	(q) 14	(r) 13	(s) 12	(t) 11

9. Repeat Exercise 4-B.8 for 99% confidence intervals, and compare your answers with those

of Exercise 4-B.8 (and with each other). Again, $\bar{x}=20.9$, $s=2.4$, and sample size:

(a) 30	(b) 29	(c) 28	(d) 27	(e) 26
(f) 25	(g) 24	(h) 23	(i) 22	(j) 21
(k) 20	(l) 19	(m) 18	(n) 17	(o) 16
(p) 15	(q) 14	(r) 13	(s) 12	(t) 11

PONDERABLES 4-B

1. What confidence level is being achieved in building a small-sample confidence interval if, in the t-table, we use
   (a) column $t_{.100}$? (b) column $t_{.050}$? (c) column $t_{.010}$?

2. A motorist wants to estimate the true mean driving time (μ) required to drive between the motorist's home and a vacation camp, and takes a sample by driving the route 16 times over a period of time, obtaining a sample mean of $\bar{x}=78.0$ minutes with sample standard deviation $s=10.8$ minutes. Expressing endpoints in tenths, find:
   (a) the 90% confidence interval estimate of μ
   (b) the 98% confidence interval estimate of μ
   (c) the 80% confidence interval estimate of μ.

3. In any horizontal row in the t-table (say, for example, across from d.f. = 19: the entries 1.328, 1.729, 2.093, 2.539, 2.861), what intuitively do the relative sizes of the entries mean, in terms of constructing confidence intervals?

4. Across from d.f. = "inf." in the t-table (corresponding to sample sizes $n \geq 31$ and therefore d.f. ≥ 30) what does one notice about the t-values (in particular, in the columns $t_{.025}$ and $t_{.005}$)?

5. A sociologist wishes to estimate the true mean number of miles the adult people in a given region have traveled while changing places of residence during the past 20 years. A random sample of 25 people is taken. Compute the 95% confidence interval estimate of μ (expressing endpoints in hundredths) given the sociologist's raw sample data:

48	94	86	177	19
103	79	216	33	56
91	324	75	81	132
68	210	53	114	98
74	131	198	24	41

C INTERVAL ESTIMATES OF A PROPORTION

We have examined the process of estimating a population *mean* μ by taking a random sample, computing the sample mean ($\bar{x}$) as a point estimate of μ, and constructing an *interval* estimate of μ around the midpoint $\bar{x}$. The interval estimate will be characterized by a certain level of confidence (usually 95% or 99%), depending on how it is constructed.

Now we turn to a very similar procedure, in which we want to estimate not a population mean μ, but a **population proportion** (denoted π).

A population proportion π is simply the proportion or fraction of a population that has a given characteristic.

For example, suppose we are interested in left-handedness, and want to know what *proportion* (or fraction, or percentage) of a given population is left-handed. It may turn out that (say) 14% of the given population in question is left-handed, and we would write $\pi = 0.14$.

Or we might be interested in knowing what proportion of the voting-age people in a given area are Democrats. It may turn out that 68% of them are, in which case we would write $\pi = 0.68$.

Or we might be interested in knowing what proportion of a population of 100,000 oak trees are diseased. It may turn out that 7,400 of them are diseased, and we would express this proportion as

$$\pi = \frac{7{,}400}{100{,}000} = 0.074.$$

In general, if there are N items in a population and if m of them have a certain property, then the proportion of the population having that property is

$$\pi = \frac{m}{N}.$$

But the problem with populations, as we have seen before, is that typically they are very *large*, and it becomes difficult or impossible to know (by direct observation of the entire population) what a true population proportion π is.

Accordingly, the value of a population proportion π is often *estimated* (just as a population mean is), by sampling.

To estimate the *parameter* π (which exists for the whole population), we naturally employ a sample *statistic*, and this time the appropriate statistic is called a **sample proportion,** denoted p. (Again, we denote population parameters by such Greek letters as π, and sample statistics by such Roman letters as p.)

If a sample of size n is taken, and if k of the sample items turn out to have a certain property, then the sample proportion of items having that property is

$$p = \frac{k}{n}.$$

For example, suppose we want to know what true population proportion (π) of all the factory workers in a certain state are union members. Now because there are (let us say) many, many thousands of factory workers, we can best proceed by *sampling* the population. Suppose we take a random sample of size 1,000 and find that 790 of those are union members. Then the *sample proportion* is

$$p = \frac{790}{1{,}000} = 0.79.$$

(That is, 79% of the workers *sampled* are union members.)

This sample proportion p becomes the *point estimate* of the true overall population proportion π, so that

$$\pi \approx 0.79.$$

But as we have seen in the case of estimating a population mean μ by a point estimate $\bar{x}$, point estimates alone leave a lot to be desired. In the present example, we have point-estimated π (the true population proportion of union members) to be 0.79, but suppose we are asked questions about our confidence in the accuracy of the estimate. Someone may well ask, "But could the true proportion π be as far away from your point estimate 79% as, say, 75%, or 83%?" We admit, "Well, it *could* be. . . ." Next someone asks, "But then how *likely* is it that π is as far away as 75% or 83%?" At this point we begin to realize that with nothing but a simple point estimate on hand, we cannot hope to answer such questions. Put simply, a point estimate does not come equipped with any built-in measure of how *good* an estimate it is.

But, as we have seen, an *interval* estimate *does*. Thus what is really needed is an interval estimate of π.

And statistical theory supports our quest for an interval estimate, because the behavior of sample proportions (like the behavior of sample means) is *predictable*. Specifically:

If random samples of size n are taken repeatedly from a population such that some proportion π of the population has a certain property, then as long as $n\pi$ and $n(1-\pi)$ are at least 5, *the sample proportions p are approximately normally distributed* with

mean $\mu_p = \pi$ and

standard deviation $\sigma_p = \sqrt{\frac{\pi(1-\pi)}{n}}$.

That is, sample proportions tend to be normally distributed around the true population proportion π as a mean, and with the indicated standard deviation to measure their dispersion. (See Figure 4.11.)

This tendency toward normal distribution signifies that, just as we relied on similar behavior of sample means to build confidence intervals for μ, we may build confidence intervals for π in much the same way. The routine procedures involved in confidence interval building for μ still hold, in part: the z-score associated with a 95% confidence interval is still 1.96, and the z-score employed for a 99% interval is still 2.58.

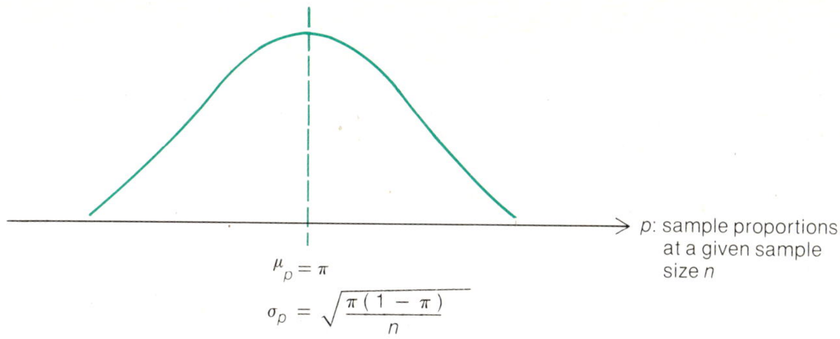

Figure 4.11

Just as we used a sample statistic, the sample mean $\bar{x}$, as the midpoint for a μ-interval, we similarly use the sample statistic p (the sample proportion obtained from a random sample) as the midpoint of a π-interval. Remember, p by itself is the *point* estimate for π.

The difference is in the standard deviation used. For estimating π by a confidence interval, the standard deviation which is called for, according to statistical theory, is

$$\sigma_p = \sqrt{\frac{\pi(1-\pi)}{n}}.$$

But the alert reader will perceive, in this, a problem. The problem is that to compute σ_p by this formula we would have to know the value of π, the population proportion—the very thing we are trying to estimate! If we *knew* π, we would not be trying to estimate it.

We get around this difficulty, however, in a natural manner. When we are estimating a population proportion π and have first computed a sample proportion p as a point estimate of π, then to approximate the needed standard deviation σ_p *we use the sample proportion p in the formula in place of* π. When this is done, we may signal the fact that we are *approximating* σ_p, by placing a "hat" (ˆ) over σ_p:

$$\hat{\sigma}_p = \sqrt{\frac{p(1-p)}{n}}.$$

The procedure can then go forward as illustrated by the following examples, which you should study carefully, and which should remind you of building confidence intervals for μ as in days of yore.

EXAMPLES

1. To return to the situation (discussed earlier in this section) in which we want to estimate the true proportion (π) of factory workers in a certain state who are union members: suppose, again, we have taken a random sample of 1,000 workers, finding that 790 are union members.

This sample result gives us, first, the *point* estimate of π: the sample proportion

$$p = \frac{790}{1,000} = 0.79.$$

That is, 79% of the *sample* are union workers, and if we have to state a single-number estimate of π, we would give 0.79 as our best guess: $\pi \approx 0.79$.

But we can do much better, by constructing (say) a 95% confidence interval estimate of π. It is safe to proceed with this, since $n\pi$ and $n(1-\pi)$, at least as suggested by $\pi \approx 0.79$, must both exceed 5 considerably (check this out). In a manner similar to the procedure for constructing a μ-interval, we simply *add and subtract* 1.96 standard deviations from the mean, where in this case "mean" means the point estimate p, and "standard deviation" means σ_p, or in practice its approximation $\hat{\sigma}_p$:

Left endpoint: $p - 1.96\hat{\sigma}_p$.

Right endpoint: $p + 1.96\hat{\sigma}_p$.

We already have $p = 0.79$, but we still need to compute $\hat{\sigma}_p$:

$$\begin{aligned}\hat{\sigma}_p &= \sqrt{\frac{p(1-p)}{n}} = \sqrt{\frac{0.79(1-0.79)}{1,000}} \\ &= \sqrt{\frac{0.79(0.21)}{1,000}} = \sqrt{\frac{0.1659}{1,000}} = \sqrt{0.0001659} \\ &\approx \sqrt{0.\overrightarrow{00}\overrightarrow{01}\overrightarrow{66}} \approx 0.01288 \approx 0.013\end{aligned}$$

(which we accept to three decimal places, one more than such values of the variable p as 0.79).

Thus the quantity $1.96\hat{\sigma}_p$ which is to be added and subtracted is

$$1.96(0.013) = 0.02548$$

so that the desired interval endpoints are found by computing

$$\text{(left)}\ p - 1.96\hat{\sigma}_p = 0.79 - 0.02548 = 0.76452$$

$$\text{(right)}\ p + 1.96\hat{\sigma}_p = 0.79 + 0.02548 = 0.81548.$$

We round these off in such a direction as to *widen* the interval (not to endanger its built-in 95% confidence level), so that our finished 95% confidence interval estimate of π is

$$0.76 < \pi < 0.82.$$

That is, we may state that the true proportion of factory workers who are union members is between 76% and 82%, and we have about 95% confidence in this estimate.

2. In the same situation as in Example 1, if we had desired a 99% confidence interval estimate of π, we would have used the z-score 2.58 instead of 1.96; all other details, including $\hat{\sigma}_p \approx 0.013$ would remain the same. Thus the quantity $2.58\hat{\sigma}_p$ (to be added and subtracted) would be

$$2.58\hat{\sigma}_p = 2.58(0.013) = 0.03354$$

so that we would obtain the endpoints

$$\text{(left)}\ p - 2.58\hat{\sigma}_p = 0.79 - 0.03354 = 0.75646$$

$$\text{(right)}\ p + 2.58\hat{\sigma}_p = 0.79 + 0.03354 = 0.82354$$

and (rounding off to widen the interval) the desired 99% confidence interval for π is

$$0.75 < \pi < 0.83.$$

Comparing this result with the 0.76-to-0.82 result of Example 1, you will see that we have had to yield to a widening of the interval, one percentage point on each side—this is the price that we have paid for wanting to be 99% confident that we have enclosed π in an interval, instead of only 95% confident.

In general, then:

A 95% confidence interval estimate of π is

$$p - 1.96\sqrt{\frac{p(1-p)}{n}} < \pi < p + 1.96\sqrt{\frac{p(1-p)}{n}}$$

and a 99% confidence interval estimate of π is

$$p - 2.58\sqrt{\frac{p(1-p)}{n}} < \pi < p + 2.58\sqrt{\frac{p(1-p)}{n}}.$$

EXAMPLES

3. In attempting to determine what proportion (π) of the voters in a certain locale plan to vote for a given candidate, campaign analysts take a random sample of $n = 500$ voters, and 285 say that they plan to vote for the candidate in question.

(They may be lying, of course, but that is a chance which must be taken.) Thus the *point* estimate of π is the sample proportion

$$p=\frac{285}{500}=0.57.$$

That is, $\pi \approx 0.57$. Suppose, further, the analysts want to construct a 95% confidence interval estimate of π. (To proceed safely we must have some assurance that both $n\pi$ and $n(1-\pi)$ exceed 5. Using $\pi \approx 0.57$ gives $n\pi \approx 285$ and $n(1-\pi) \approx 215$, both considerably greater than 5.) It will be necessary to add and subtract (from the midpoint $p=0.57$) the quantity $1.96\hat{\sigma}_p$, where

$$\begin{aligned}\hat{\sigma}_p &= \sqrt{\frac{p(1-p)}{n}}=\sqrt{\frac{0.57(1-0.57)}{500}}=\sqrt{\frac{0.57(0.43)}{500}}\\ &=\sqrt{\frac{0.2451}{500}}=\sqrt{0.0004902}\approx\sqrt{0.\overrightarrow{00}\overrightarrow{04}\overrightarrow{90}}\\ &\approx 0.02214 \approx 0.022.\end{aligned}$$

Thus $1.96\hat{\sigma}_p$ is

$$1.96(0.022)=0.04312$$

so that the endpoints are found by computing

$$p-1.96\hat{\sigma}_p=0.57-0.04312=0.52688$$

$$p+1.96\hat{\sigma}_p=0.57+0.04312=0.61312$$

and, with the appropriate "outward" round-off, the desired 95% confidence interval estimate is

$$0.52<\pi<0.62.$$

4. Suppose, in Example 3, the sample had been of size $n=5{,}000$, with 2,850 of those sampled saying that they would vote for the candidate. Then the sample proportion would be

$$p=\frac{2{,}850}{5{,}000}=0.57,$$

the same *point estimate* as before. *But*: This time the standard deviation required is smaller than before, because n is larger:

$$\begin{aligned}\hat{\sigma}_p &= \sqrt{\frac{p(1-p)}{n}}=\sqrt{\frac{0.57(1-0.57)}{5{,}000}}=\sqrt{\frac{0.57(0.43)}{5{,}000}}\\ &=\sqrt{\frac{0.2451}{5{,}000}}=\sqrt{0.00004902}\approx\sqrt{0.\overrightarrow{00}\overrightarrow{00}\overrightarrow{49}}=0.007.\end{aligned}$$

The quantity $1.96\hat{\sigma}_p$ (to be added and subtracted) is

$$1.96(0.007) = 0.01372$$

so that the endpoints are derived by computing

$$p - 1.96\hat{\sigma}_p = 0.57 - 0.01372 = 0.55628$$

$$p + 1.96\hat{\sigma}_p = 0.57 + 0.01372 = 0.58372$$

and the 95% confidence interval is

$$0.55 < \pi < 0.59.$$

Comparing this with the result of Example 3, we see that the larger sample size has yielded, naturally, a noticeably narrower (and more specific, more informative) interval: This is what we "buy" by investing in a more ambitious sample size. (And do not underestimate the possible importance of those few percentage-point differences—they could be critically important to the political candidate! And to the public.)

In a situation where the true population proportion π is *known*, certain probabilities can be computed from the fact that sample proportions can be expected to be approximately normally distributed. Consider the following examples carefully.

EXAMPLES

5. It is known that 40% of the employees in a large corporation have less than 10 years' seniority. (That is, it is known that $\pi = 0.40$.) A task force of 384 employees is to be chosen at random. What is the probability that the sample proportion (p) of chosen employees having less than 10 years' seniority will exceed 0.45?

We note that such sample proportions p are approximately normally distributed with mean

$$\pi = 0.40$$

and standard deviation

$$\sigma_p = \sqrt{\frac{\pi(1-\pi)}{n}} = \sqrt{\frac{0.40(1-0.40)}{384}}$$

$$= \sqrt{\frac{0.40(0.60)}{384}} = \sqrt{\frac{0.24}{384}} = \sqrt{0.000625} = 0.025.$$

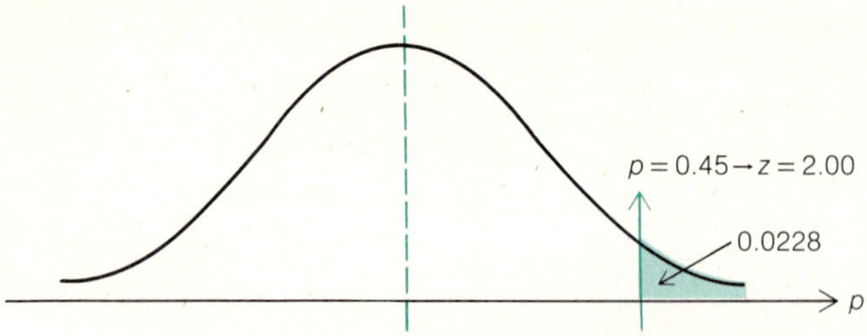

Figure 4.12

(Note that the true value of π—*not* the value 0.45—has been used in computing σ_p because in this case π is *known* and is not being estimated.)

Thus we wish to find the area, under the appropriate normal curve, *beyond* the value $p = 0.45$ (see Figure 4.12); that is, the area beyond

$$z = \frac{0.45 - 0.40}{0.025} = \frac{0.050}{0.025} = 2.00.$$

From the normal table, the required area (which is the probability that we seek) is

$$0.5000 - 0.4772 = 0.0228.$$

That is, if such 384-person task forces were repeatedly chosen at random, one could expect them to contain more than 45% people with less than 10 years' seniority only about 2.28% of the time.

6. It is known that 30% of the cars owned by people working in a city near the state line are registered out of state. If a random sample of 2,100 cars is taken in that location, what is the probability that the sample proportion (p) of out-of-state registrations will be between 0.27 and 0.33?

Again, we take such sample proportions p to be approximately normally distributed with mean $\pi = 0.30$ and standard deviation

$$\sigma_p = \sqrt{\frac{\pi(1-\pi)}{n}} = \sqrt{\frac{0.30(1-0.30)}{2{,}100}}$$
$$= \sqrt{\frac{0.30(0.70)}{2{,}100}} = \sqrt{\frac{0.21}{2{,}100}} = \sqrt{0.0001} = 0.01.$$

We are looking for the area under the appropriate normal curve between $p = 0.27$ and $p = 0.33$; that is, between

$$z = \frac{0.27 - 0.30}{0.01} = \frac{-0.03}{0.01} = -3.00 \quad \text{and}$$

$$z = \frac{0.33 - 0.30}{0.01} = \frac{0.03}{0.01} = 3.00.$$

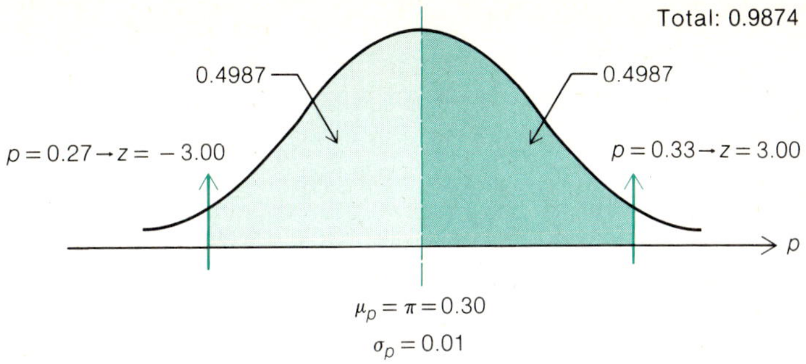

Figure 4.13

(See Figure 4.13.) Out to $z = 3.00$ from the mean, the area (in the normal table) is 0.4987, and out to $z = -3.00$ from the mean the area is of course again 0.4987. Thus the desired probability is

$$0.4987 + 0.4987 = 0.9874.$$

That is, there is a strong 98.74% chance that in such a sample the proportion of out-of-state registrations would be between 0.27 and 0.33.

NEW TERMS IN THIS SECTION

population proportion
sample proportion

EXERCISES 4-C

1. To estimate the true proportion of cans of beans that come off a processing belt with crooked or torn labels, a quality control researcher takes a random sample of $n = 1{,}900$ cans, finding that 361 cans have defective labels. What is the point estimate of the true proportion π? Find (expressing your endpoints to three decimal places)
   (a) the 95% confidence interval estimate of π
   (b) the 99% confidence interval estimate of π.

2. In the situation of Exercise 4-C.1, what would be the 95% confidence interval if the sample size were
   (a) 1,700, with 323 defectives?
   (b) 1,300, with 247 defectives?
   (c) 1,100, with 209 defectives?
   (d) 800, with 152 defectives?
   (e) 600, with 114 defectives?
   (f) 400, with 76 defectives?
   (g) 200, with 38 defectives?
   (h) 100, with 19 defectives?

3. What would the 99% confidence interval be, each time, in Example 4-C.2? (Use the same $\hat{\sigma}_p$ as before, in each case, but multiply it by $z = 2.58$ instead of $z = 1.96$.)

4. What do you observe in Exercises 4-C.1, 4-C.2, and 4-C.3 as the sample size decreases? Why does it happen?

5. To estimate the true proportion (π) of cases in which an experimental antibiotic drug is effective, a researcher tests the drug on a random sample of 225 patients, finding that 171 of the tests prove effective. What is the point estimate of π? Expressing endpoints to three decimal places, find

(a) the 95% confidence interval for π

(b) the 99% confidence interval for π.

6. In Exercise 4-C.5, what would be the 95% confidence interval if the sample size had been

(a) 250, with 190 effective tests?
(b) 275, with 209 effective tests?
(c) 300, with 228 effective tests?
(d) 325, with 247 effective tests?
(e) 350, with 266 effective tests?
(f) 375, with 285 effective tests?
(g) 400, with 304 effective tests?
(h) 425, with 323 effective tests?
(i) 450, with 342 effective tests?
(j) 475, with 361 effective tests?
(k) 500, with 380 effective tests?
(l) 1,000, with 760 effective tests?
(m) 10,000, with 7,600 effective tests?

7. What do you observe in Exercises 4-C.5 and 4-C.6 as the sample size increases? Why?

8. A sociologist wishes to estimate the true proportion π of childless marriages in a certain region. Randomly sampling 400 couples in the area, the researcher finds that 104 of these couples are childless. What is the point estimate of π? Expressing endpoints to three decimal places, find:

(a) the 95% confidence interval estimate of π

(b) the 99% confidence interval estimate of π.

9. What would have been the 99% confidence interval in each case in Exercise 4-C.8 if the sample size had been

(a) 500, with 130 childless couples?
(b) 600, with 156 childless couples?
(c) 700, with 182 childless couples?
(d) 800, with 208 childless couples?
(e) 1,000, with 260 childless couples?
(f) 1,500, with 390 childless couples?

10. To estimate the true proportion of the population of school children (in a given area) having certain reading difficulties, a group of psychologists take a random sample among the children. What is the 95% confidence interval estimate of π if

(a) 130 children out of 1,000 sampled have reading difficulties?
(b) 104 children out of 800 sampled have reading difficulties?
(c) 78 children out of 600 sampled have reading difficulties?
(d) 52 children out of 400 sampled have reading difficulties?
(e) 26 children out of 200 sampled have reading difficulties?
(f) 13 children out of 100 sampled have reading difficulties?

11. It is known that the true proportion of tenured instructors at a certain large university is $\pi = 0.36$. A task force of 100 faculty members is to be chosen at random. Find the probability that, in such a sample, the sample proportion of faculty members who are tenured will be greater than 0.40.

12. In Exercise 4-C.11, find the probability, as before, that the sample proportion will exceed 0.40 if the sample size is
(a) 64 (b) 400.

13. It is known that the true proportion of registered voters (in a certain city) who are Democrats is 0.75. We will regard a group of voters showing up at the polls at random (in a given time span) as the taking of a random sample. Find the probability that the sample proportion of Democrats will be between 0.73 and 0.77 if the sample size is
(a) 100 (b) 200 (c) 300 (d) 400
(e) 500 (f) 600 (g) 700 (h) 800

PONDERABLES 4-C

1. A team of market researchers has taken a random sample to estimate the true proportion of people (in a given region) who regularly watch a certain television program. They compute a 95% confidence interval: $0.624 < \pi < 0.656$. Before round-off, this interval appeared as:

$$0.62432 < \pi < 0.65568.$$

Find:
(a) the sample proportion p
(b) the sample size n that was used.

2. In computing a confidence interval estimate of a population proportion π, what specifically would happen if, in the sample, we obtained $p = 1.00$? What would $\hat{\sigma}_p$ be then? Explain.

3. Similarly, what if we get $p = 0.00$?

4. It is known that 49% of the adult people in a certain state own cars that are less than 5 years old. If random samples of size 5,100 were taken repeatedly, the sample proportions p would exceed——only about 7.64% of the time.

SUGGESTED READINGS

Concepts of Statistical Inference, by William C. Guenther (McGraw-Hill Book Company, 1965). Chapter 5 provides a close look at the algebra and logic of confidence interval estimation.

Introduction to Statistical Methods, by Basil P. Korin (Winthrop Publishers, Inc., 1977). This book offers, in Chapter 11, good discussion of various aspects of confidence interval estimation.

Statistics: A Beginning, by Roy A. Kuebler and Harry Smith, Jr. (John Wiley & Sons, Inc., 1976). Chapter 5 provides useful discussion of the algebraic details underlying the structure of confidence intervals.

5 Hypothesis Testing

The cautious seldom err.—Confucius

A HYPOTHESES ABOUT MEANS

In this section we move into the mainstream of inferential statistics. The concept of **hypothesis testing** will be discussed, in various forms, throughout the remainder of the book.

A *hypothesis*, for our purposes, is a *claim* or an *assertion* that someone puts forward, usually for the purpose of having it accepted and believed by other people.

For example, someone may make an assertion—state a **hypothesis—concerning the mean of a population.** Someone, for instance, may claim: "The mean weight of all the adults in this state is $\mu = 68.5$ kilograms." Or someone may claim: "The mean weight of all the adults in this state is *at least* 68.5 kilograms. That is, $\mu \geq 68.5$ kilograms."

Clearly, such assertions may or may not be true. The people making them may have sound evidence behind them, or may simply be guessing, or in fact may be lying outright. Our interest, then, is in *testing* such hypotheses, to clarify their possible truth or falsity.

How do we test a hypothesis? When someone does say, "The mean weight of adults in this state is $\mu = 68.5$ kilograms," how do we go about testing this claim—and possibly even *rejecting* it?

The answer is a natural one: we test a hypothesis by *sampling* the population that the hypothesis speaks of, and determining the extent to which our sample results tend to support or refute the claim. We take, of course, *random* samples, and (when possible) samples that are reasonably large. (Very small samples may not be as trustworthy, or at least may require special handling.)

Suppose we wish, for example, to test the hypothesis that in this state the mean adult weight is $\mu = 68.5$ kilograms.

First, it is common in statistics to refer to such a hypothesis—a hypothesis to be tested and perhaps rejected—as the **null hypothesis,** denoted H_0. Thus our null hypothesis is

$$H_0: \mu = 68.5.$$

Now, if we should end up **rejecting** the null hypothesis, then whatever there is left to believe, logically, is called the **alternative hypothesis,** denoted H_1. In this case, if we reject the null hypothesis $\mu = 68.5$, then we can only say that μ is *not* equal to 68.5, but rather some different value. That is, the alternative hypothesis here is

$$H_1: \mu \neq 68.5.$$

It is important to know from the beginning that, faced with a null hypothesis H_0 to test, we can end up doing only one of two things:

1. Reject the null hypothesis H_0, or
2. Fail to reject the null hypothesis H_0.

For instance, if H_0 is the assertion that $\mu = 68.5$, we may either *reject* this claim (and conclude that $\mu \neq 68.5$) or admit that we *cannot reject* the claim on the basis of the information available. Even when we end up saying that we cannot reject the null hypothesis, we are *not* really saying that we have proved it true—we are merely admitting that we have failed to reject it. "Only this and nothing more," as Poe says.

Now, suppose that to test the null hypothesis H_0: $\mu = 68.5$ we have taken a large random sample of the adult people in the state and have come up with a *sample* mean of

$$\bar{x} = 68.5.$$

There is certainly no way to argue for rejection of the original claim that $\mu = 68.5$ if our very own sample has yielded the value 68.5! Remember, the result $\bar{x} = 68.5$ does not *prove* that $\mu = 68.5$ (because $\bar{x}$ comes only from a sample), but it certainly fails to *dis*prove it.

But what if the sample mean differs from the claimed (hypothesized) value $\mu = 68.5$? For instance, what if our sample gives

$$\bar{x} = 68.3,$$

a value undershooting the claimed μ by 0.2 kilograms?

Well, maybe that difference of 0.2 kilograms is important enough to complain about, and maybe not—after all, the people who made the claim H_0 ($\mu = 68.5$) could say, "Your 68.3 is close enough to 68.5 to attribute the difference to *change fluctuation* of the sort that always occurs in sampling. We still say that 68.5 is the right value."

But now, what if our sample yields

$\bar{x} = 67.9$? or 64.3? or 61.4?

Or what if it yields

$\bar{x} = 69.1$? *or* 72.2? *or* 75.6?

Clearly (even though we admittedly are looking *only* at a sample) it must be true with a sample that is large and random, that the sample mean $\bar{x}$ should be *somewhere* in the vicinity of 68.5—*if* the 68.5 is right. The farther away the sample mean $\bar{x}$ is from the hypothesized 68.5, the more we grow suspicious—the more we begin to suspect that the claimed 68.5 is *not* right.

But how far away is *too* far away?

That is, how much does our sample mean $\bar{x}$ have to differ from the claimed $\mu = 68.5$ before we start saying that that claim ought to be *rejected*? Is $\bar{x} = 68.3$ close enough to 68.5 to fail to reject H_0? Is $\bar{x} = 72.2$ far enough away from 68.5 to reject H_0?

These are questions that take some specific analysis to answer. To put it simply: You can't tell by looking. The value 68.3 *looks* close but may not be close enough. The value 72.2 *looks* considerably removed from the claimed μ, but *may* be close enough. How, then, do we proceed?

First, we note that sample means have a nice kind of mathematical predictability about them. We have already looked, in Chapter 4, at the technique of building confidence intervals on the basis of the fact that

for random samples of size $n > 30$, sample means $\bar{x}$ tend to be *normally distributed* with mean

$$\mu_{\bar{x}} = \mu$$

and standard deviation

$$\sigma_{\bar{x}} = \frac{\sigma}{\sqrt{n}}$$

where μ and σ are the parameters of the given population from which the samples are taken. We can certainly make use of this fact—the Central Limit Theorem—here.

Suppose, again, that someone has made the claim

$$H_0\colon \mu = 68.5 \text{ kilograms}$$

and we wish to test this hypothesis to see whether or not we can *reject* it. Suppose, further, we take a random sample of size $n = 100$ people, finding their sample mean weight to be $\bar{x} = 69.1$ kilograms, with a sample standard deviation of $s = 7.0$ kilograms. (We will have to use $s = 7.0$ as an estimator of the population standard deviation σ, which is unknown.)

Now, let us say that in the event we do end up rejecting the null hypothesis H_0, *we want to be at least* 95% *confident that we are correct in doing so*, and otherwise will not reject H_0. That is, we are willing to accept no more than a 5% risk of rejecting the null hypothesis when we should *not* have: rejecting it and being wrong. In this situation, we are

testing the null hypothesis at the **level of significance** $\alpha = 0.05$.

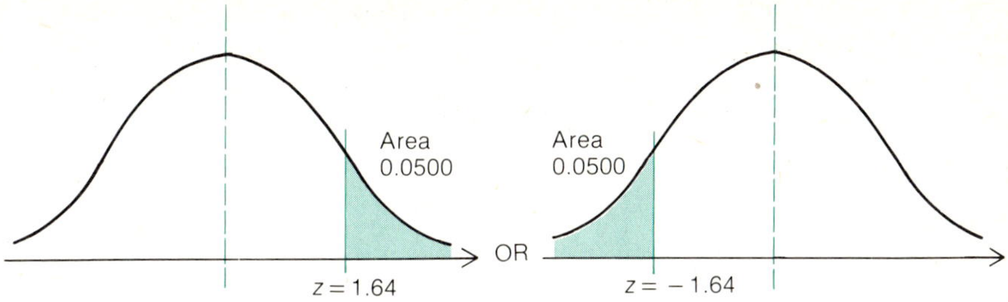

Figure 5.4 Colored rejection region (with $\alpha = 0.05$) existing only on one side or the other.

As before, the sample mean $\bar{x}$ must be changed to a z-score and compared with some cutoff or critical z-value beyond which rejection would be in order. The difference here is that for *one*-sided tests, the cutoff z-values are *not* 1.96 or 2.58. Let us suppose first that we want to work at the level $\alpha = 0.05$. Then the pertinent fact is that to have a rejection region under the normal curve with area $\alpha = 0.0500$ *all on one side* (instead of being split, 0.0250 on each side), we must start with

$$z = 1.64 \text{ (or } -1.64)$$

as may be verified from the normal table (see Figure 5.4). In other words, 5% of the area under the normal curve is beyond $z = 1.64$ (or, on the left, beyond $z = -1.64$), so that to test a null hypothesis by a one-sided test at the level of significance $\alpha = 0.05$, we compare the computed z-score with 1.64 or -1.64, whichever turns out to be appropriate.

If we wish, however, to work at the level of significance $\alpha = 0.01$ with a one-sided test, then the critical z-value (with which to compare computed z-scores) is $z = 2.33$ or $z = -2.33$, whichever turns out to be appropriate, because on either side the "area beyond," under the tail of the curve, is 0.0100. (See Figure 5.5.)

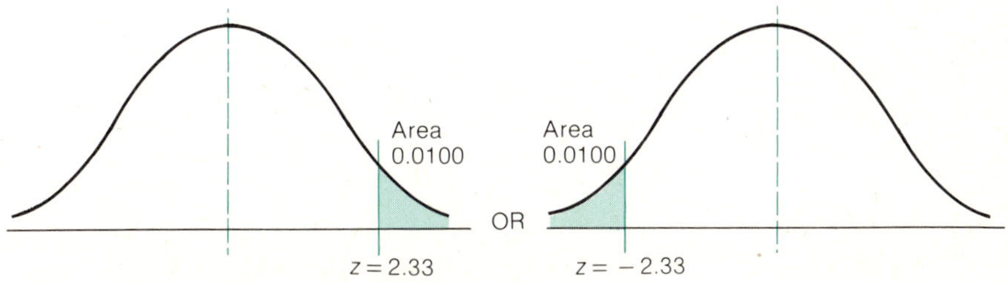

Figure 5.5 One-side rejection region at the level $\alpha = 0.01$.

EXAMPLES

7. An economist claims that the true mean family income in a certain region is *at most* \$12,500 a year. That is, the stated null hypothesis is

$$H_0: \mu \leq 12{,}500.$$

(The *alternative* hypothesis, which one would accept if rejecting H_0, is: $\mu >$ 12,500. That is, if we reject the claim that the true mean is 12,500 or less, we assert that the true mean is really more than 12,500.) To test this hypothesis at the level $\alpha = 0.05$, a sociological researcher takes a random sample of $n = 100$ families, obtaining a sample mean of $\bar{x} = \$12{,}550$ with sample standard deviation $s = \$300$. What is the conclusion?

The test must be *one-sided* because of the nature of the null hypothesis. If a sample yielded a sample mean $\bar{x}$ *smaller* than 12,500, then it would be compatible with the statement $\mu \leq 12{,}500$ and would present no possible argument against the null hypothesis. We will reject H_0 only if $\bar{x} = 12{,}550$ turns out to be *significantly more* than the claimed upper limit 12,500. (If the null hypothesis had simply said $\mu = 12{,}500$, we would use a *two*-sided test, because varying too far on *either* side would lead to rejection.)

We change $\bar{x} = 12{,}550$ into a z-score:

$$z = \frac{\bar{x} - \mu}{\frac{\sigma}{\sqrt{n}}} \approx \frac{\bar{x} - \mu}{\frac{s}{\sqrt{n}}} = \frac{12{,}550 - 12{,}500}{\frac{300}{\sqrt{100}}}$$

$$= \frac{50}{\frac{300}{10}} = \frac{50}{30} \approx 1.67.$$

And at the level of significance $\alpha = 0.05$ we must compare this result with the critical cutoff z-value 1.64 (see Figure 5.6), finding

$$z = 1.67 > 1.64$$

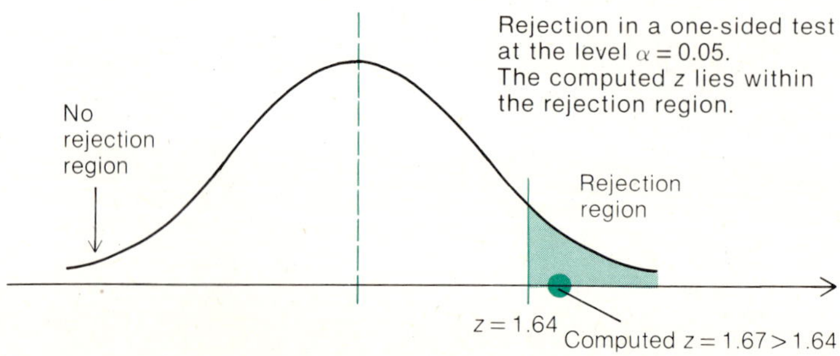

Figure 5.6

so that we *reject* the null hypothesis (taking a 5% chance of being wrong) and conclude that actually the true mean μ is *greater* than \$12,500.

8. An economist claims that the true mean prices of houses in a certain large suburban area is *at least* \$50,000. That is, the null hypothesis is

H_0: $\mu \geq 50{,}000$.

To test this hypothesis (at, say, the level 0.01) we must run a *one-sided* test, because we will reject H_0 only if our sample mean $\bar{x}$ is significantly *less* than \$50,000. (If $\bar{x}$ turns out *more* than \$50,000 then we have no way with the sample evidence to argue against the claim that μ is \$50,000 *or more*.)

Suppose we take a random sample of $n=64$ houses, obtaining a sample mean price $\bar{x}=\$49{,}500$ with sample standard deviation $s=\$1{,}600$. We change $\bar{x}$ into a z-score:

$$z=\frac{\bar{x}-\mu}{\frac{\sigma}{\sqrt{n}}}\approx\frac{\bar{x}-\mu}{\frac{s}{\sqrt{n}}}=\frac{49{,}500-50{,}000}{\frac{1{,}600}{\sqrt{64}}}$$

$$=\frac{-500}{\frac{1{,}600}{8}}=-\frac{500}{200}=-2.50.$$

We must compare this result (at the level of significance $\alpha=0.01$) with the critical z-value -2.33 (see Figure 5.7), noting that

$-2.50<-2.33$

(that is, -2.50 *is* out beyond -2.33 to the left) so that we do *reject* the null hypothesis H_0 and conclude that in reality the alternative hypothesis is acceptable:

H_1: $\mu<50{,}000$.

Of course, we accept a 1% probability (risk) of being wrong. In this example, the rejection region lies only on the left, but for some hypotheses, as in the previous example, the rejection region may lie only on the right.

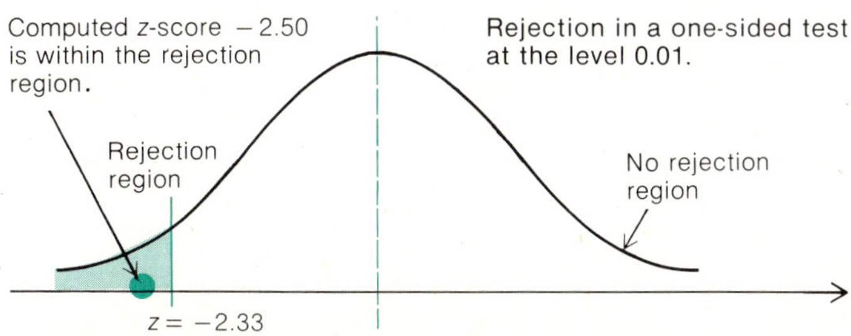

Figure 5.7

NEW TERMS IN THIS SECTION

hypothesis testing
hypothesis concerning a population mean
null hypothesis
rejection of null hypothesis
alternative hypothesis
level of significance
two-sided test
Type I error
Type II error
one-sided test

EXERCISES 5-A

1. It is claimed that the true mean price for new cars in a certain year was $5,100 (no more, no less). To test this hypothesis (at the level of significance $\alpha = 0.05$) an economist takes a random sample of 81 of that year's new car prices, for which the sample mean is $5,180 and the sample standard deviation is $580. Should the test be one-sided or two-sided? What is the result? Is the null hypothesis $\mu = \$5,100$ rejected at the level 0.05?

2. Suppose, in Exercise 5-A.1, the results $\bar{x} = \$5,180$ and $s = \$580$ had come from a sample of size $n = 225$ instead of $n = 81$. What would the result of the hypothesis test have been? Again, use the level of significance $\alpha = 0.05$. Intuitively and mathematically, why the difference?

3. It is claimed that the true mean height of all the adults in a certain region is $\mu = 163.0$ centimeters. A random sample of size 400 is taken, with sample standard deviation $s = 5.0$ centimeters. At the level of significance $\alpha = 0.01$, test the null hypothesis $\mu = 163.0$ (one-sided or two-sided test?) if the sample mean obtained was

(a) 163.1 centimeters (b) 163.2 centimeters
(c) 163.3 centimeters (d) 163.4 centimeters
(e) 163.5 centimeters (f) 163.6 centimeters
(g) 163.7 centimeters

4. In Exercise 5-A.3's progression of $\bar{x}$ values, if the level of significance had been $\alpha = 0.05$, at what point would $\bar{x}$ have been far enough away from the claimed $\mu = 163.0$ for a rejection of the hypothesis?

5. It is hypothesized that a certain chemical reaction takes, on the average, 24.0 seconds to run (no more, no less). To test this hypothesis $\mu = 24.0$ at the level of significance 0.05, a researcher runs the reaction 36 times, obtaining a sample mean $\bar{x} = 23.5$ seconds with sample standard deviation $s = 1.4$ seconds. What is the conclusion?

6. What if the test in Exercise 5-A.5 had been done at the level 0.01?

7. An engineering firm claims that on the average it takes an experimental rocket engine 18.35 seconds (no more, no less) to warm up to a certain level. A scientist wishes to test this hypothesis and sample-runs the engine 50 times, getting a sample mean of 18.34 seconds with sample standard deviation 0.03 seconds. What is the result of the test if the chosen level of significance is

(a) $\alpha = 0.01$? (b) $\alpha = 0.05$?

8. In the situation in Exercise 5-A.7, at the level of significance 0.01, what if the results $\bar{x} = 18.34$ and $s = 0.03$ had come from a sample size of 60 trial runs?

9. A council on nutrition claims that on the average a first-grade child derives at least 500 milligrams of vitamin C daily from the normal diet. A researcher wishes to test this null hypothesis (H_0: $\mu \geq 500$) at the level of significance $\alpha = 0.05$ (applying a one-sided or two-sided test?) and samples 100 children at random, obtaining a sample mean of $\bar{x} = 490$ milligrams and sample standard deviation $s = 120$ milligrams. What is the result of the test?

10. In Exercise 5-A.9, what would the results have been if the sample statistics $\bar{x} = 490$ and $s = 120$ had been obtained from a sample of size
(a) 125? (b) 150? (c) 175? (d) 200? (e) 225? (f) 250?
(g) 275? (h) 300? (i) 325? (j) 350? (k) 375? (l) 400?

11. An orange grower advertises the claim that a competitor's oranges are at most 8.5 centimeters in diameter. To settle the question, an outsider randomly samples the competitor's oranges, obtaining a sample mean of 8.7 centimeters and sample standard deviation 1.2 centimeters, from a sample of size 100. If the hypothesis H_0 ($\mu \leq 8.5$) is tested at the level of significance $\alpha = 0.01$ (one-sided or two-sided test?), what are the results?

12. What if the test in Exercise 5-A.11 had been run at the level of significance $\alpha = 0.05$?

13. In Exercise 5-A.11 at the level of significance 0.01, what would have been the result of the hypothesis test if the results $\bar{x} = 8.7$ and $s = 1.2$ had been obtained from a random sample of size
(a) 125? (b) 150? (c) 175? (d) 180? (e) 190? (f) 200?

14. It is claimed that on the average a laboratory mouse can run through a test maze in less than 40.0 seconds. A psychologist samples 400 mice, running each one through the maze, and obtains a sample mean time of 41.4 seconds with sample standard deviation 7.0 seconds. At the level of significance $\alpha = 0.05$, what is the conclusion, concerning the null hypothesis H_0 ($\mu <$ 40.0)?

15. In Exercise 5-A.14, what if the sample size (all else remaining the same) had been
(a) 350? (b) 300? (c) 250? (d) 200?
(e) 150? (f) 100? (g) 75? (h) 65?

16. A company that cans and distributes beans makes the claim that the true mean amount of beans in each of its cans labeled "8 ounces" is *at least* 8.00 ounces. To test this hypothesis (H_0: $\mu \geq 8.00$) a consumer protection agency randomly samples 250 cans of the company's beans, finding a sample mean of $\bar{x} = 7.98$ ounces with sample standard deviation $s = 0.15$ ounces. At the level of significance $\alpha = 0.01$, what is the result of the test? Is the difference between 7.98 ounces and 8.00 ounces significant at the level 0.01?

17. In Exercise 5-A.16, what if the results $\bar{x} = 7.98$ ounces and $s = 0.15$ ounces had come from a sample of size
(a) 260? (b) 270? (c) 280? (d) 290? (e) 300? 310?

18. What would the sample results of Exercises 5-A.16 and 5-A.17 have meant at the level of significance $\alpha = 0.05$?

19. A soup manufacturing plant claims that on the average its soup vats contain no more than 17.5 milligrams of rodent hairs. A health inspector, wishing to test this hypothesis at the level of significance $\alpha = 0.01$, samples 49 vats of soup, finding that the vats sampled have, on the average, $\bar{x} = 17.6$ milligrams of rodent hairs, with sample standard deviation $s = 0.28$ milligrams. What is the conclusion?

20. In Exercise 5-A.19, what if the health inspector had obtained the results $\bar{x} = 17.6$ and $s = 0.28$ from sampling
(a) 45 vats? (b) 44 vats? (c) 43 vats? (d) 42 vats?

21. The distributor of a machine that makes precision steel bolts says that the true mean length of the bolts made by the machine is $\mu = 40.00$ millimeters, no more, no less. A quality control inspector tests this hypothesis at the level of significance 0.01 by sampling 32 bolts, obtaining from them a sample mean $\bar{x} = 39.99$ millimeters with sample standard deviation 0.03 millimeters. What is the result of the test?

22. In Exercise 5-A.21, suppose the sample size had been
(a) 35 (b) 40 (c) 45 (d) 50 (e) 55 (f) 60

PONDERABLES 5-A

1. A medical researcher wishes to test the claim that a certain drug takes measurable effect in at most 4 minutes after it is administered. The researcher samples 100 patients at random, administering the drug and finding the sample mean time to be 4 minutes, 6 seconds. At the level of significance $\alpha = 0.05$, what is the conclusion if the sample standard deviation was
(a) 0.64 minutes? (b) 0.62 minutes? (c) 0.60 minutes?

2. It is hypothesized that the true mean age of people who attend lectures at a certain lecture hall is $\mu = 39.0$ years. Suppose that to test this hypothesis a sociologist takes a random sample of people attending and finds $\bar{x} = 38.0$ years with $s = 10.2$ years. What is the smallest sample size for which this sample information would lead to a rejection of the null hypothesis at the level of significance $\alpha = 0.05$?

3. (*Small-sample hypothesis testing.*) You should notice that, when using z-scores such as 1.96 and 2.58 as cutoff scores for testing hypotheses concerning a mean μ, you assume that sample means $\bar{x}$ are normally distributed, and this assumption is safe only when the sample size n exceeds 30. If $n \leq 30$, and if the original population is not terribly far from being normally distributed, we may still test hypotheses about μ by using t-values in place of the accustomed z-values. The appropriate t-values are found by referring to this table:

	Level $\alpha = 0.05$	Level $\alpha = 0.01$
Two-sided test	Use $t_{.025}$	Use $t_{.005}$
One-sided test	Use $t_{.05}$	Use $t_{.01}$

For example, to run a two-sided hypothesis test at the level of significance $\alpha = 0.05$ with a sample of size 28 (so that $d.f. = 27$), we look up

$$t_{.025} = 2.052$$

in the t-table and use that value (or -2.052) as a cutoff score to determine where rejections start (instead of using $z = 1.96$ or -1.96). That is, we compute, from the sample mean, not a z-value but a t-value:

$$t = \frac{\bar{x} - \mu}{\frac{s}{\sqrt{n}}}$$

and compare it with the table t-value to determine whether to reject the hypothesis H_0. Consider, then, the following:

(a) It is claimed that a certain extinct bird (for which remains are very rare and only found in certain museum collections) had a mean wingspread of at most 80.0 centimeters (H_0: $\mu \leq 80.0$). A researcher examines 25 sets of remains of these birds and finds the sample mean wingspread to be 81.2 centimeters with sample standard deviation $s = 2.5$ centimeters. At the level 0.01, what is the conclusion concerning the null hypothesis?

(b) In part (a), what if the sample mean had been 81.3 centimeters (all other details unchanged)?

(c) It is claimed that a certain rare Babylonian coin weighed, on the average, 8.0 grams. A researcher manages to find 21 of these coins in special collections, and tests the hypothesis (H_0: $\mu = 8.0$) by computing the sample mean and sample standard deviation, which turn out to be $\bar{x} = 7.6$ grams and $s = 0.9$ grams. At the level of significance 0.05, what is the result?

(d) In part (c), what if the sample size had been 24 (all other details unchanged)?

B DIFFERENCES BETWEEN MEANS

Suppose we have two distinct populations with means μ_1 and μ_2, respectively, and we want to investigate the **difference between means:** $\mu_1 - \mu_2$.

Specifically, $\mu_1 - \mu_2$ might be zero: the populations could have the *same* mean ($\mu_1 = \mu_2$). On the other hand, one mean might be considerably different from the other.

For example, if one population is the set of all lifetimes of Brand *A* lightbulbs (measured in hours), and if the other population is the set of all lifetimes of Brand *B* lightbulbs (also measured in hours), then let us say that μ_1 is the mean of the first population and μ_2 is the mean of the second population. That is, "on the average" Brand *A* lightbulbs burn for μ_1 hours and Brand *B* lightbulbs burn for μ_2 hours.

Now, what if $\mu_1 = \mu_2$? Then the two brands have the same "average" (mean) burning time, and in terms of life expectancy we would regard Brand *A* bulbs and Brand *B* bulbs as being of the same quality. We could just buy a lightbulb indiscriminately, not caring whether the bulb was Brand *A* or Brand *B*. In effect, the truth of the statement $\mu_1 = \mu_2$ would (in terms of life expectancy) make *one* population out of all the bulbs: all the Brand *A* and all the Brand *B* bulbs could be lumped together in one big bin, and the whole pile of bulbs would simply have mean life expectancy μ_1 $(= \mu_2)$.

But clearly μ_1 may *not* be the same as μ_2. It is entirely possible that the mean life expectancy of Brand *A* may be quite different from the mean life expectancy of Brand *B*. One brand, in terms of overall longevity, may be better than the other.

Obviously, it may be desirable to *know* whether $\mu_1 = \mu_2$ or not, or at least to be able to say with a certain amount of confidence that $\mu_1 \neq \mu_2$.

So: hypothesis testing again.

In this two-sample situation, the *null hypothesis* that is stated, and perhaps rejected or perhaps not, is

$$H_0: \mu_1 = \mu_2.$$

The null hypothesis states: The two population means are equal—there is really no difference between them. The *alternative* hypothesis (which we are left accepting if we reject H_0) is simply

$$H_1: \mu_1 \neq \mu_2.$$

The alternative hypothesis states: The two population means are different.

Now of course we test such a hypothesis as H_0 $(\mu_1 = \mu_2)$ by *sampling* each of the two populations at random. In the example of the lightbulbs, we would try a number of Brand *A* bulbs, computing their *sample* mean lifetime (call it $\bar{x}_1$), and try a number of Brand *B* bulbs, computing *their* sample mean lifetime (call it $\bar{x}_2$).

Intuitively, it should be clear that *the more difference there is* between the sample means $\bar{x}_1$ and $\bar{x}_2$, the more suspicious we become that the *population* means which they approximate (μ_1 and μ_2) are different, too. That is, the greater the difference between $\bar{x}_1$ and $\bar{x}_2$, the more likely we are to reject the null hypothesis H_0.

But how much difference is *too* much difference? If $\bar{x}_1 = 780$ and $\bar{x}_2 = 785$, is that enough difference to be significant—that is, enough difference to suggest that $\mu_1 \neq \mu_2$? If $\bar{x}_1 = 780$ and $\bar{x}_2 = 810$, is *that* difference significant? Or can we attribute such differences to mere chance fluctuation? As usual, in statistics, you can't tell by looking. A definite test procedure is called for, and "significant" must mean "significant at a given level."

To see exactly how all of this works, we have to look at the question of **the sampling distribution of the difference of two sample means,** and it turns out to have a familiar flavor to it.

Put simply, if such experiments (sampling from Brand *A*, then sampling from Brand *B*) were done repeatedly, and if the difference $\bar{x}_1 - \bar{x}_2$ were computed each time, such differences would (at large sample sizes) tend to be *normally distributed.* Specifically:

Given a population with mean μ_1 and standard deviation σ_1, and given a second population with mean μ_2 and standard deviation σ_2: If an experiment consists of taking a random sample of size $n_1 > 30$ from the first population (producing a sample mean $\bar{x}_1$) and taking also a random sample of size $n_2 > 30$ from the second population (producing a sample mean $\bar{x}_2$), then if the experiment were done repeatedly, the differences $\bar{x}_1 - \bar{x}_2$ would tend to be normally distributed, with mean and standard deviation

$$\mu_{\bar{x}_1-\bar{x}_2} = \mu_1 - \mu_2$$

$$\sigma_{\bar{x}_1-\bar{x}_2} = \sqrt{\frac{\sigma_1^2}{n_1} + \frac{\sigma_2^2}{n_2}}.$$

See Figure 5.8.

Of course, this tendency toward normal distribution makes the behavior of such differences $\bar{x}_1 - \bar{x}_2$ highly predictable; it also makes it possible to test the hypothesis H_0 ($\mu_1 = \mu_2$) at specific levels of significance (such as 0.05 or 0.01).

Before we turn to examples of the procedure, let us note that, unfortunately, the two population variances σ_1^2 and σ_2^2 are usually unknown, so that we are obliged to substitute the *sample* variances s_1^2 and s_2^2 (computed from the respective samples):

$$\sigma_{\bar{x}_1-\bar{x}_2} \approx \sqrt{\frac{s_1^2}{n_1} + \frac{s_2^2}{n_2}}.$$

This latter expression is sometimes denoted $\hat{\sigma}_{\bar{x}_1-\bar{x}_2}$, where the "hat" (^) indicates that we have dealt in approximation (substituting the sample variances s_1^2 and s_2^2 for the population variances σ_1^2 and σ_2^2).

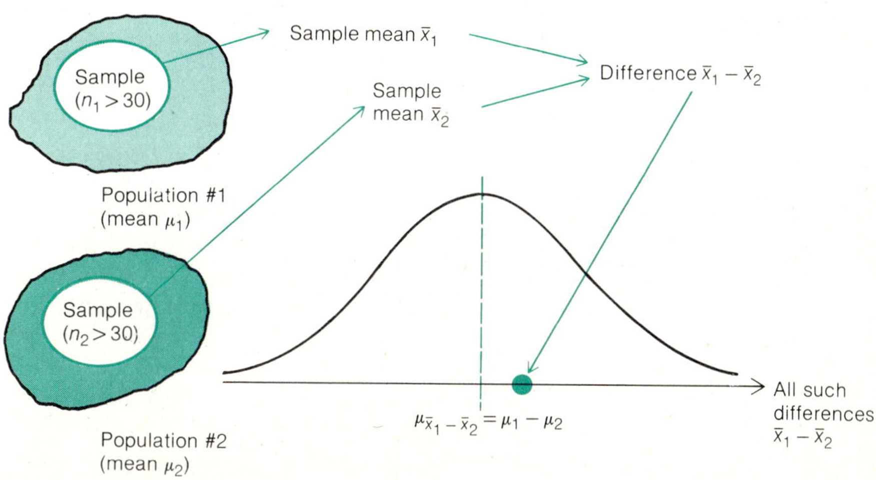

Figure 5.8

To test a null hypothesis H_0 of the form $\mu_1 = \mu_2$, then, we go through the following procedure.

From random samples, each of size exceeding 30, we obtain the sample means $\bar{x}_1$ (from the population with mean μ_1) and $\bar{x}_2$ (from the population with mean μ_2) and consider their difference $\bar{x}_1 - \bar{x}_2$. We convert this difference into a z-score by the formula

$$z = \frac{\bar{x}_1 - \bar{x}_2}{\hat{\sigma}_{\bar{x}_1 - \bar{x}_2}} = \frac{\bar{x}_1 - \bar{x}_2}{\sqrt{\dfrac{s_1^2}{n_1} + \dfrac{s_2^2}{n_2}}}$$

and reject the null hypothesis H_0 ($\mu_1 = \mu_2$) only if z goes out beyond 1.96 or -1.96 (at the level of significance $\alpha = 0.05$) or out beyond 2.58 or -2.58 (at the level of significance $\alpha = 0.01$). The idea behind this formula for the z-score is that *if* the null hypothesis H_0 were true (that is, *if* $\mu_1 = \mu_2$) then $\mu_1 - \mu_2$ would be 0, and one would naturally convert $\bar{x}_1 - \bar{x}_2$ (which is an approximation of $\mu_1 - \mu_2$) into a z-score by computing

$$z = \frac{(\bar{x}_1 - \bar{x}_2) - (\mu_1 - \mu_2)}{\sigma_{\bar{x}_1 - \bar{x}_2}}$$

but $\mu_1 - \mu_2 = 0$ *if* the null hypothesis is true, so that the formula reduces to the original form given. *If* the null hypothesis is true, then $\bar{x}_1 - \bar{x}_2$ should be small and z should be small, so that obtaining a large z out beyond ± 1.96 or ± 2.58 arouses suspicion that the null hypothesis is *not* true. We are dealing here with a *two-sided* test, because the null hypothesis $\mu_1 = \mu_2$ would be rejected by our getting z too far out in either direction. There are also one-sided tests of this type, and they will be examined soon. The following examples should illustrate the type of two-sided tests described here.

EXAMPLES

1. To test the null hypothesis that the true population mean lifetime (μ_1) of Brand *A* lightbulbs is the same as the true population mean lifetime (μ_2) of Brand *B* lightbulbs, a statistician first randomly samples 50 Brand *A* lightbulbs, testing them to see how long they will burn continuously (and obtaining these sample results: mean $\bar{x}_1 = 790.0$ hours, standard deviation $s_1 = 20.0$ hours); then randomly sampling 75 Brand *B* lightbulbs (obtaining sample results $\bar{x}_2 = 780.0$ hours, $s_2 = 25.0$ hours). At the level of significance $\alpha = 0.05$, what is the result? That is, can the null hypothesis H_0 ($\mu_1 = \mu_2$) be rejected at the level 0.05?

   We note first that the sample sizes are $n_1 = 50$ and $n_2 = 75$, both exceeding 30. Thus, in general, the obtainable sample differences $\bar{x}_1 - \bar{x}_2$ will be approximately normally distributed, and we may proceed by determining their standard deviation:

$$\hat{\sigma}_{\bar{x}_1-\bar{x}_2}=\sqrt{\frac{s_1^2}{n_1}+\frac{s_2^2}{n_2}}$$

$$=\sqrt{\frac{20^2}{50}+\frac{25^2}{75}}=\sqrt{\frac{400}{50}+\frac{625}{75}}$$

$$=\sqrt{8+\frac{25}{3}}=\sqrt{\frac{49}{3}}\approx\sqrt{16.3}\approx 4.04.$$

Now we compute the test statistic z:

$$z=\frac{\bar{x}_1-\bar{x}_2}{\hat{\sigma}_{\bar{x}_1-\bar{x}_2}}=\frac{790-780}{4.04}$$

$$=\frac{10}{4.04}\approx 2.48$$

and, because $2.48>1.96$, we *reject* (at the level of significance $\alpha=0.05$) the hypothesis that $\mu_1=\mu_2$. We conclude, on the contrary, that the population means μ_1 and μ_2 are not equal, the bulbs are not of identical quality. Our *sample* means $\bar{x}_1=790$ and $\bar{x}_2=780$, in other words, are **significantly different** at the level 0.05: too different still to let us believe that μ_1 and μ_2 are equal.

Notice that this result $z=2.48$ is not quite strong enough to warrant a rejection of H_0 at the level of significance $\alpha=0.01$, which would have required $z>2.58$. We may reject H_0, on the basis of the sample data, with 95% confidence but not quite with 99% confidence. (The difference between the sample means is not significant at the 0.01 level.)

2. Suppose, in the previous situation, the sample results had been:

Sample of Brand *A* ($n_1=50$): $\bar{x}_1=790.0$, $s_1=20.0$.

Sample of Brand *B* ($n_2=75$): $\bar{x}_2=783.0$, $s_2=25.0$.

That is, everything is the same as in Example 1 except that the Brand *B* mean is 783, a bit closer to Brand *A* this time. Again we test (at the level 0.05) the null hypothesis H_0 ($\mu_1=\mu_2$) by computing

$$z=\frac{\bar{x}_1-\bar{x}_2}{\sqrt{\frac{s_1^2}{n_1}+\frac{s_2^2}{n_2}}}=\frac{790-783}{\sqrt{\frac{20^2}{50}+\frac{25^2}{75}}}$$

$$=\frac{7}{4.04}\approx 1.73$$

(because the standard deviation in the denominator is the same as in Example 1). And because $z=1.73$ is *not* beyond the critical value 1.96, this time we do *not* reject the null hypothesis H_0 at the level 0.05. That is, the sample means 790 and 783 are not significantly different at the level 0.05—not different enough to give us a "95% sure" rejection of the claim $\mu_1=\mu_2$. (We have not *proved* the statement $\mu_1=\mu_2$; we have simply failed to reject it at the specified level of significance 0.05.)

3. Suppose, again in the lightbulb situation of the previous two examples, the sample results (from Example 2)

$$\bar{x}_1=790.0, \quad s_1=20.0$$

$$\bar{x}_2=783.0, \quad s_2=25.0$$

had come from samples of size $n_1=100$ (Brand *A*) and $n_2=100$ (Brand *B*). To test the null hypothesis H_0 $(\mu_1=\mu_2)$ under these circumstances at the level of significance $\alpha=0.05$, we would compute

$$z=\frac{\bar{x}_1-\bar{x}_2}{\sqrt{\frac{s_1^2}{n_1}+\frac{s_2^2}{n_2}}}=\frac{790-783}{\sqrt{\frac{20^2}{100}+\frac{25^2}{100}}}$$

$$=\frac{7}{\sqrt{\frac{400}{100}+\frac{625}{100}}}=\frac{7}{\sqrt{4+\frac{25}{4}}}$$

$$=\frac{7}{\sqrt{\frac{41}{4}}}=\frac{7}{\frac{\sqrt{41}}{2}}=\frac{7\cdot 2}{\sqrt{41}}$$

$$\approx\frac{14}{6.403}\approx 2.19>1.96$$

so that now (with sample means differing by 7 as in Example 2) we *do* reject H_0 at the level 0.05, even though we did not do so in Example 2. The difference has been the sample sizes: *If* it were true that $\mu_1=\mu_2$, then at the *larger* sample sizes one would be less likely to get sample means $\bar{x}_1$ and $\bar{x}_2$ differing by as much as 7.

Now, in contrast to a null hypothesis like $\mu_1=\mu_2$, suppose someone has made a claim that *compares* μ_1 with μ_2. For instance, someone (let us say Jones) has a vested interest in Brand *A* lightbulbs as opposed to Brand *B*, and puts forward the claim that Brand *A* bulbs burn *longer* (on the overall average) than Brand *B*. That is, Jones puts forth the null hypothesis

$$H_0: \mu_1>\mu_2.$$

(The alternative hypothesis, if Jones is wrong, is H_1: $\mu_1\leq\mu_2$.)

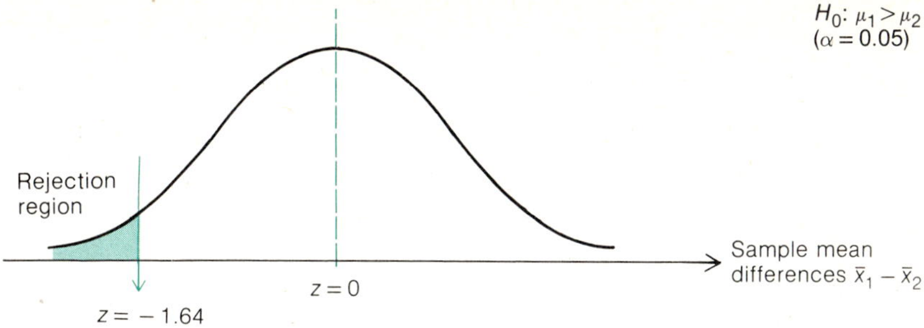

Figure 5.9

We may test this null hypothesis by computing z from the difference $\bar{x}_1 - \bar{x}_2$ as before, but this time because of the structure of the hypothesis H_0, the test is *one-sided*. Our suspicion of H_0 "runs only in one direction." After all, suppose $\bar{x}_1$ turns out to be *larger* than $\bar{x}_2$—then how can we hope to argue with the statement $\mu_1 > \mu_2$? A sample result $\bar{x}_1 > \bar{x}_2$ tends to support H_0, not reject it. But when $\bar{x}_1$ turns out to be *smaller* than $\bar{x}_2$ (that is, when $\bar{x}_1 - \bar{x}_2$ is negative), then we get suspicious about the statement $\mu_1 > \mu_2$. Because in this case only $\bar{x}_1 - \bar{x}_2$ being negative raises doubts, there is a rejection region *only on the left,* starting (as has been the case before with such tests) at $z = -1.64$ if the level of significance is 0.05. (See Figure 5.9.)

On the other hand, suppose someone (say Smith) with a vested interest in Brand *B* lightbulbs claims that Brand *A* bulbs burn *less* long on the average than Brand *B* bulbs. That is, Smith puts forth the null hypothesis:

H_0: $\mu_1 < \mu_2$.

(The alternative hypothesis, if Smith is wrong, is H_1: $\mu_1 \geq \mu_2$.) Then again, to test H_0, we would require a *one*-sided test. Our suspicion runs again only in one direction, because if the sample means turn out such that $\bar{x}_1 < \bar{x}_2$, we have no argument against the claim $\mu_1 < \mu_2$, but if the sample means turn out such that $\bar{x}_1 > \bar{x}_2$ (that is, $\bar{x}_1 - \bar{x}_2 > 0$), then we may suspect the statement $\mu_1 < \mu_2$ of being false.

Because only the circumstance $\bar{x}_1 - \bar{x}_2 > 0$ raises our suspicions, there is a rejection region *only on the right,* starting at $z = 1.64$ if the level of significance is 0.05. (See Figure 5.10.)

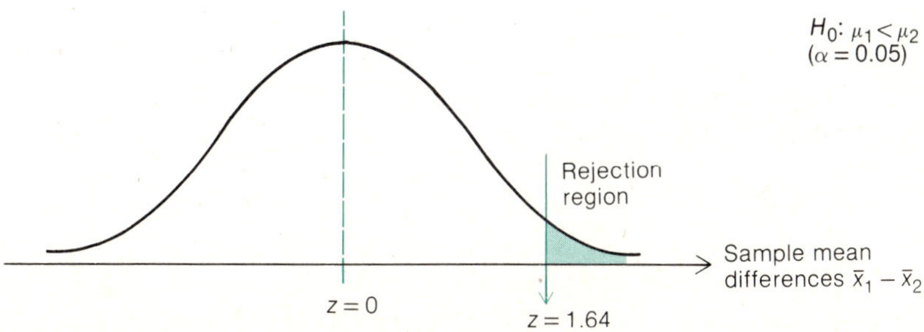

Figure 5.10

In either case, if the level of significance for the one-sided test had been $\alpha = 0.01$ instead of 0.05, then (as before) we would regard the appropriate rejection region as starting at $z = -2.33$ or $z = 2.33$.

Still pursuing the idea of the Brand *A* and Brand *B* bulb comparisons, the following examples should be illuminating.

EXAMPLES

4. Suppose Jones manufactures Brand *A* lightbulbs and Smith manufactures Brand *B*. Jones claims that

$$H_0: \mu_1 > \mu_2.$$

That is, on the overall average, Brand *A* bulbs burn longer than Brand *B* bulbs. An independent researcher decides to test this hypothesis at the level of significance $\alpha = 0.05$ and randomly samples 200 Brand *A* bulbs and 200 Brand *B* bulbs, obtaining these sample results:

(Brand *A*) $\bar{x}_1 = 791.0$ hours, $s_1 = 30.0$ hours.

(Brand *B*) $\bar{x}_2 = 796.0$ hours, $s_2 = 28.0$ hours.

Because $\bar{x}_1 < \bar{x}_2$, it is *conceivable* that the null hypothesis $\mu_1 > \mu_2$ can be rejected on the basis of the samples. The necessary *z*-score is:

$$z = \frac{\bar{x}_1 - \bar{x}_2}{\sqrt{\frac{s_1^2}{n_1} + \frac{s_2^2}{n_2}}} = \frac{791 - 796}{\sqrt{\frac{30^2}{200} + \frac{28^2}{200}}}$$

$$= \frac{-5}{\sqrt{\frac{900}{200} + \frac{784}{200}}} = -\frac{5}{\sqrt{\frac{1{,}684}{200}}}$$

$$= -\frac{5}{\sqrt{8.42}} \approx -\frac{5}{2.90} \approx -1.72.$$

And because, under the left tail of the curve (see Figure 5.9 again) -1.72 *is* out beyond $z = -1.64$, the null hypothesis H_0 is *rejected* at the level of significance $\alpha = 0.05$. Thus with 95% confidence, the independent researcher can state that Jones's advertising claim $\mu_1 > \mu_2$ is false. Note, however, that that statement could not be made with 99% confidence, because to do so would require a rejection of H_0 at the level of significance $\alpha = 0.01$, which is only obtained for a computed *z*-score out beyond -2.33.

5. Suppose the same sample results as shown in Example 4 had been obtained by sampling only 100 Brand *A* bulbs and 100 Brand *B* bulbs (instead of 200 of each). Then the *z*-score would have been

$$z=\frac{791-796}{\sqrt{\frac{30^2}{100}+\frac{28^2}{100}}}=\frac{-5}{\sqrt{\frac{900}{100}+\frac{784}{100}}}$$

$$=\frac{-5}{\sqrt{\frac{1,684}{100}}}=\frac{-5}{\sqrt{16.84}}$$

$$\approx\frac{-5}{\sqrt{16.8}}\approx-\frac{5}{4.10}\approx-1.22,$$

which is *not* out beyond -1.64 to the left; we would *not* reject the null hypothesis $\mu_1>\mu_2$ at the level 0.05 this time. The difference between $\bar{x}_1=791$ and $\bar{x}_2=796$ does not turn out to be significant at these smaller sample sizes. That is, we do not have enough evidence to dispute Jones's advertising claim this time, because the samples (unlike those of Example 4) are too small to establish that the observed difference (between 791 and 796) is statistically significant.

6. Again letting μ_1 be the overall population mean lifetime of Brand *A* bulbs and μ_2 the mean for Brand *B* bulbs, suppose Smith (the manufacturer of Brand *B* bulbs, Jones's rival) makes the claim

H_0: $\mu_1<\mu_2$.

That is, the null hypothesis is that Brand *A* bulbs do not burn as long, on the overall average, as Brand *B* bulbs. Again, a one-sided test is appropriate; there will be a rejection region only on the *right*, because we grow suspicious only if $\bar{x}_1>\bar{x}_2$ (that is, only if $\bar{x}_1-\bar{x}_2>0$).

To test the hypothesis (at the level 0.05), suppose a researcher samples 50 Brand *A* bulbs and 40 Brand *B* bulbs, finding sample results as follows:

(Brand *A*) $\bar{x}_1=784.0$ hours, $s_1=10.0$ hours.

(Brand *B*) $\bar{x}_2=781.0$ hours, $s_2=8.0$ hours.

Now, it is *conceivable* that H_0 (the statement $\mu_1<\mu_2$) can be rejected because sampling has turned up the result $\bar{x}_1>\bar{x}_2$, opposite to the sort of thing that the null hypothesis claims. But as always, one cannot tell by just looking. It is necessary to compute the z-score:

$$z=\frac{\bar{x}_1-\bar{x}_2}{\sqrt{\frac{s_1^2}{n_1}+\frac{s_2^2}{n_2}}}=\frac{784-781}{\sqrt{\frac{10^2}{50}+\frac{8^2}{40}}}$$

$$=\frac{784-781}{\sqrt{\frac{100}{50}+\frac{64}{40}}}=\frac{3}{\sqrt{2+\frac{8}{5}}}$$

$$=\frac{3}{\sqrt{3.6}}\approx\frac{3}{1.90}\approx 1.58<1.64.$$

The computed z-score 1.58 is not far enough out under the righthand tail of the curve (see Figure 5.10 again) to be in the rejection region, and we do not reject H_0 (the claim $\mu_1<\mu_2$) at the level of significance 0.05. Under the circumstances, the fact that $\bar{x}_1$ is three units larger than $\bar{x}_2$ is not statistically significant.

7. Suppose that in Example 6 the same sample results had been obtained from samples of size $n_1=100$ (for Brand A) and $n_2=80$ (for Brand B). That is, suppose we had doubled the size of each sample. Then the z-score would be

$$z=\frac{\bar{x}_1-\bar{x}_2}{\sqrt{\frac{s_1^2}{n_1}+\frac{s_2^2}{n_2}}}=\frac{784-781}{\sqrt{\frac{10^2}{100}+\frac{8^2}{80}}}$$

$$=\frac{3}{\sqrt{\frac{100}{100}+\frac{64}{80}}}=\frac{3}{\sqrt{1+\frac{4}{5}}}$$

$$=\frac{3}{\sqrt{1.8}}\approx\frac{3}{1.34}\approx 2.24>1.64$$

so that this time the z-score *is* significantly far out under the tail of the curve (where it probably should *not* be if the null hypothesis were true) and we *reject* H_0 at the level 0.05. With 95% confidence we may state that Smith's advertising claim $\mu_1<\mu_2$ is false. The difference of 3 between $\bar{x}_1=784$ and $\bar{x}_2=781$ does turn out to be significant under these circumstances, at the larger sample sizes.

Note that we could *not*, however, reject H_0 at the level 0.01, because such a rejection would call for z to exceed 2.33, but we have $2.24<2.33$. Thus, although we may deny Smith's claim with 95% confidence, we do not quite have strong enough evidence to deny it with 99% confidence. Our confidence in the correctness of rejecting H_0 lies between 95% and 99%.

Remember that these techniques (using a normal curve) are valid only when both sample sizes n_1 and n_2 exceed 30, because otherwise the normality assumption involved may not be justified. If either sample size fails to exceed 30, then a special "small sample" technique can be invoked, which is described in Ponderables Problem 4 at the end of Section 5-B.

Also, we may wish to compare *more* than two sample means. In that case we employ a technique called the *F*-test, which will be examined in Chapter 6.

NEW TERMS IN THIS SECTION

hypotheses concerning the difference between two population means
sampling distribution of the difference between two sample means
statistically significant difference (between sample means)
***t*-test**

EXERCISES 5-B

1. A consumer-interest agency claims that two brands of flashlight battery (Neverdie and Everglow) are of equal quality, in the sense that their overall mean lifetimes (in continuous use) are of equal length. (The null hypothesis is H_0: $\mu_1 = \mu_2$.) Another agency decides to test this claim at the level of significance $\alpha = 0.05$ by sampling 36 batteries of each brand, obtaining sample results as follows:

(Neverdie) $\bar{x}_1 = 42.0$ hours, $s_1 = 4.0$ hours.

(Everglow) $\bar{x}_2 = 43.0$ hours, $s_2 = 3.0$ hours.

What is the computed z for $\bar{x}_1 - \bar{x}_2$? What is the conclusion?

2. In Exercise 5-B.1, repeat the test assuming that the same sample results had been obtained from samples of size $n_1 = n_2$ that equal

(a) 40 (b) 50 (c) 60 (d) 70 (e) 75
(f) 80 (g) 85 (h) 90 (i) 95 (j) 100

3. To test (at the level of significance $\alpha = 0.01$) the null hypothesis that 5 p.m. traffic on the Gauss Turnpike and 5 p.m. traffic on the Gosset Turnpike are equally fast, a statistician using radar techniques clocks the speed in miles per hour (mph) of 80 cars chosen at random on the Gauss Turnpike and 100 cars chosen at random on the Gosset Turnpike, gathering the following sample data:

(Gauss) $\bar{x}_1 = 61.0$ mph, $s_1 = 4.0$ mph.

(Gosset) $\bar{x}_2 = 59.0$ mph, $s_2 = 5.0$ mph.

Compute the z-score from $\bar{x}_1 - \bar{x}_2$. What is the conclusion, concerning the null hypothesis that the overall average speeds are equal ($\mu_1 = \mu_2$)?

4. In Exercise 5-B.3, what if the sample sizes (with all other details remaining the same) had been

(a) $n_1 = 70$, $n_2 = 90$? (b) $n_1 = 70$, $n_2 = 80$? (c) $n_1 = 70$, $n_2 = 75$?

5. To test (at the level 0.05) the null hypothesis that two routes to work (on the overall average) involve the same amount of driving time ($\mu_1 = \mu_2$), a commuter clocks herself driving to work on each route 40 times, finding these sample statistics:

(Route 1) $\bar{x}_1 = 48.0$ minutes, $s_1 = 3.5$ minutes.

(Route 2) $\bar{x}_2 = 47.0$ minutes, $s_2 = 5.5$ minutes.

What is the conclusion?

6. What would have been the result of the test in Exercise 5-B.5 if the commuter (assuming all other details unchanged) had driven each route

(a) 60 times? (b) 80 times? (c) 100 times? (d) 120 times?
(e) 140 times? (f) 160 times? (g) 165 times?

7. A meat distributing company claims that its ground meat contains, on the average overall, less fat per kilogram than a competitor's. A consumer agency wishes to test this hypothesis (H_0: $\mu_1 < \mu_2$) at the level 0.05 and randomly examines forty 1-kilogram packages of each distributor's ground meat, with the following results:

(First distributor) $\bar{x}_1 = 172.0$ grams, $s_1 = 22.0$ grams.

(Competitor) $\bar{x}_2 = 168.0$ grams, $s_2 = 21.0$ grams.

What is the result of the test?

8. In Exercise 5-B.7, with the same sample results assumed, what if the sample size had been (for each distributor)

(a) 50 packages? (b) 75 packages? (c) 100 packages?
(d) 125 packages? (e) 150 packages? (f) 160 packages?

9. A news writer claims that the average (population mean) house price in Theobald, Nebraska, is higher than the average house price in Chicago. An economist, wishing to test this claim (H_0: $\mu_1 > \mu_2$) at the level of significance $\alpha = 0.05$, samples 100 houses at random in Theobald and 200 in Chicago, finding these sample statistics:

(Theobald) $\bar{x}_1 = \$44{,}800$, $s_1 = \$2{,}400$.

(Chicago) $\bar{x}_2 = \$45{,}400$, $s_2 = \$2{,}800$.

What is the result of the test? Are the sample means significantly different at the level 0.05?

10. In Exercise 5-B.9, what if the sample sizes had been $n_1 = 100$ for Theobald but only $n_2 = 70$ for Chicago (all other details remaining the same)?

11. A sociologist claims that, on the average, families moving from the West to the East have more household goods (by weight, in kilograms say) than families moving from the East to the West. To test this hypothesis (H_0: $\mu_1 > \mu_2$) at the level 0.01, a statistician randomly samples 80 families of each type, determining the weight of their household goods, with the following results:

(West to East) $\bar{x}_1 = 3{,}650$ kilograms, $s_1 = 270$ kilograms.

(*East to West*) $\bar{x}_2 = 3{,}750$ kilograms, $s_2 = 260$ kilograms

What is the conclusion?

12. In Exercise 5-B.11, what if the given sample results had been obtained by sampling at sizes $n_1 = n_2 = 50$ instead of 80?

13. An anthropologist claims that, on the average, members of a certain tribe on the Amazon (call it Tribe *A*) are shorter than members of a second tribe (*B*). To test this null hypothesis (H_0: $\mu_1 < \mu_2$) at the level of significance $\alpha = 0.01$, a researcher samples 60 Tribe *A* members and 50 Tribe *B* members, finding the statistics relating to their heights to be:

(Tribe *A*) $\bar{x}_1 = 178.0$ centimeters, $s_1 = 10.0$ centimeters.

(Tribe *B*) $\bar{x}_2 = 171.0$ centimeters, $s_2 = 11.0$ centimeters.

What is the result of the hypothesis test? Are the sample means significantly different at the level 0.01?

14. In Exercise 5-B.13, with all other details assumed unchanged, what if the sample means had been

(a) $\bar{x}_1 = 177.0$ centimeters and $\bar{x}_2 = 171.0$ centimeters?
(b) $\bar{x}_1 = 176.0$ centimeters and $\bar{x}_2 = 171.0$ centimeters?
(c) $\bar{x}_1 = 175.0$ centimeters and $\bar{x}_2 = 171.0$ centimeters?

15. The manufacturer of a pain-relief drug (*A*) claims, in advertising, that drug *A* takes less time to achieve definite results than does a competitor's drug (*B*). To test the null hypothesis (how is it stated?) at the level of significance 0.01, a researcher randomly samples 100 users of each drug, recording the times required for the drug to achieve the specified results. The sample results are:

(*A*) $\bar{x}_1 = 21.5$ minutes, $s_1 = 3.5$ minutes.

(*B*) $\bar{x}_2 = 19.6$ minutes, $s_2 = 3.8$ minutes.

What is the conclusion?

16. In Exercise 5-B.15 (with all other details assumed unchanged), what if the sample means had been
(a) 21.5 and 19.7? (b) 21.5 and 19.8? (c) 21.5 and 19.9? (d) 21.5 and 20.0?
(e) 21.5 and 20.1? (f) 21.5 and 20.2? (g) 21.5 and 20.3? (h) 21.5 and 20.4?

17. An orange grower claims that his oranges are, on the average, larger in diameter than those of a competitor. To test the hypothesis (how is it stated?) at the level $\alpha = 0.05$, a consumer-interest researcher samples, at random, 160 of the first grower's oranges and 200 of the competitor's oranges, measuring the diameters and computing sample statistics as follows:

(First grower) $\bar{x}_1 = 11.2$ centimeters, $s_1 = 2.0$ centimeters.

(Competitor) $\bar{x}_2 = 11.3$ centimeters, $s_2 = 1.8$ centimeters.

What are the results of the test?

18. In Exercise 5-B.17, with all other details assumed unchanged, what if the sample means had been
(a) 11.2 and 11.4? (b) 11.2 and 11.5? (c) 11.2 and 11.6?

19. A consumer report claims that two makes of automobile get, on the overall average, the same gasoline mileage in highway driving. To test this null hypothesis at the level $\alpha = 0.01$, a researcher samples 35 of each of the two makes of car, obtaining sample mileage figures as follows:

(Make *A*) $\bar{x}_1 = 21.5$ miles per gallon, $s_1 = 1.7$ miles per gallon.

(Make *B*) $\bar{x}_2 = 20.8$ miles per gallon, $s_2 = 1.9$ miles per gallon.

What is the conclusion from the sample data?

20. All other details in Exercise 5-B.19 unchanged, what if the sample sizes ($n_1 = n_2$) had been
(a) 40? (b) 45? (c) 50? (d) 55? (e) 60? (f) 65?
(g) 70? (h) 75? (i) 80? (j) 85? (k) 90?

PONDERABLES 5-B

1. A researcher wishes to test the null hypothesis (at the level $\alpha = 0.05$) that Brand *X* and Brand *Y* television sets each operate on the average for the same amount of time before needing major repairs. To test the hypothesis, the researcher samples both brands, taking (as it happens) samples of equal size, $n_1 = n_2$, and obtaining these results:

(Brand *X*) $\bar{x}_1 = 7{,}250$ hours, $s_1 = 30$ hours.

(Brand *Y*) $\bar{x}_2 = 7{,}240$ hours, $s_2 = 40$ hours.

What is the smallest sample size $n_1 = n_2$ at which this difference between $\bar{x}_1$ and $\bar{x}_2$ would be significant (i.e., would lead to the rejection of the hypothesis $\mu_1 = \mu_2$) at the level 0.05?

2. What effect do the sizes of s_1 and s_2 have, in the hypothesis testing described in this section, on the likelihood of rejecting a hypothesis of the form $\mu_1 = \mu_2$? Intuitively, how is this explained?

3. A periodical claims that first-grade children in Mode City, Iowa, are, on the average, taller than the first-grade children in Dispersion, Kansas. To test this hypothesis at the level 0.05, suppose you have randomly sampled children in each place, with raw data (heights in centimeters) as follows:

Mode City, Iowa

91	88	96	94	90	97	109	94
95	93	103	90	96	107	86	91
98	92	89	92	91	88	95	89
95	89	87	89	104	96	102	85
101	92	91	100	99	87	103	86

Dispersion, Kansas

99	89	90	99	94	98	88	106
94	105	103	94	106	110	98	92
93	112	89	98	91	107	92	93
92	95	95	92	102	91	95	94
97	90	99	100	104	96	101	97

(a) Compute $\bar{x}_1$ and $\bar{x}_2$.
(b) Compute s_1^2 and s_2^2.
(c) Compute z from $\bar{x}_1-\bar{x}_2$.
(d) Interpret the result, in terms of testing the original hypothesis.
(e) What if the test had been done at the level $\alpha=0.01$?

4. (*Small-sample technique.*) Recall that testing a hypothesis concerning the difference of two means $\mu_1-\mu_2$ by computing the statistic

$$z=\frac{\bar{x}_1-\bar{x}_2}{\sqrt{\dfrac{\sigma_1^2}{n_1}+\dfrac{\sigma_2^2}{n_2}}}\approx\frac{\bar{x}_1-\bar{x}_2}{\sqrt{\dfrac{s_1^2}{n_1}+\dfrac{s_2^2}{n_2}}}$$

requires assuming that the sample mean differences $\bar{x}_1-\bar{x}_2$ are normally distributed, and *that* requires assuming that both sample sizes n_1 and n_2 are greater than 30. The question certainly arises, then: What if one or the other sample size (n_1 or n_2 or both) does *not* exceed 30? If one or both samples are small in this sense, then we are *not* guaranteed that sample mean differences $\bar{x}_1-\bar{x}_2$ are normally distributed. In that case, computing a z-score may be inappropriate, and we must instead proceed by computing the statistic

$$t=\frac{\bar{x}_1-\bar{x}_2}{\sqrt{\dfrac{(n_1-1)s_1^2+(n_2-1)s_2^2}{n_1+n_2-2}}\sqrt{\dfrac{1}{n_1}+\dfrac{1}{n_2}}}$$

which we compare with the appropriate t-value from the t-distribution table:

	$\alpha=0.05$	$\alpha=0.01$
One-sided	$t_{.050}$	$t_{.010}$
Two-sided	$t_{.025}$	$t_{.005}$

where the number of degrees of freedom in the table is

$$\text{d.f.} = n_1 + n_2 - 2.$$

The approach is similar to the ordinary z-statistic approach of this section, in that the table t-value (say $t_{.025}$ for a two-sided test at the level 0.05) determines where rejections start: any *computed* t falling farther out (under the tail of the t-curve) than the table-value, signals a rejection. Note that, although we use s_1^2 and s_2^2 generally of necessity in place of σ_1^2 and σ_2^2 when doing a z-type test, we use s_1^2 and s_2^2 with no regrets when doing a **t-test,** the sample variances being entirely appropriate. (It must be assumed, however, in t-testing with small samples, that the parent populations from which the samples are taken are at least approximately normally distributed with at least approximately equal variances, and that the two samples are independent of each other.)

Consider the following example.

To test, at the level 0.05, the null hypothesis that two different rocket engines burn a given fuel package, on the overall average, in the same amount of time (H_0: $\mu_1 = \mu_2$), an engineer observes 10 burnings of Engine A and 15 burnings of Engine B, noting the time in seconds that is required each time to burn the fuel package. Thus $n_1 = 10$ and $n_2 = 15$, sample sizes smaller than 30. (Larger samples are for the moment unavailable, hence a z-type test cannot be employed.) The sample statistics are:

(Engine A) $\bar{x}_1 = 45.5$ seconds, $s_1 = 0.08$ seconds.

(Engine B) $\bar{x}_2 = 45.4$ seconds, $s_2 = 0.07$ seconds.

The t-statistic needed is:

$$t = \frac{45.5 - 45.4}{\sqrt{\dfrac{(10-1)(0.08)^2 + (15-1)(0.07)^2}{10+15-2}}\sqrt{\dfrac{1}{10}+\dfrac{1}{15}}}$$

$$= \frac{0.01}{\sqrt{\dfrac{9(0.0064) + 14(0.0049)}{23}}\sqrt{\dfrac{5}{30}}}$$

$$= \frac{0.1}{\sqrt{\dfrac{0.0576 + 0.0686}{23}}\sqrt{\dfrac{1}{6}}}$$

$$= \frac{0.1}{\sqrt{\dfrac{0.1262}{23(6)}}} \approx \frac{0.1}{\sqrt{0.000914}}$$

$$\approx \frac{0.1}{0.030} \approx 3.333.$$

To see whether this result amounts to a rejection, we note that the test is two-sided at the level

0.05, and we look up (with d.f. $= n_1 + n_2 - 2 = 23$) the critical t-value:

$t_{.025} = 2.069.$

Because $3.333 > 2.069$, we *reject* H_0 at the level 0.05. We feel 95% confident that the true means μ_1 and μ_2, for the engines' overall performance, are *not* equal. The difference between $\bar{x}_1 = 45.5$ and $\bar{x}_2 = 45.4$ turns out to be significant at the level 0.05. Even a tenth of a second can be significant! Note, also, that if we had been testing at the level 0.01, the comparison of $t = 3.333$ with the table value $t_{.005} = 2.807$ would still lead to a rejection, even at that level.

The following problems will give you practice in working with the t-test in small-sample situations.

5. A collector of antique coins claims that, on the average overall, a certain type (A) of rare Babylonian coin is the same weight as another type of rare Babylonian coin (B). Only a few coins of each type are available, and a researcher wishing to test (at the level 0.01) this hypothesis $\mu_1 = \mu_2$ examines 10 coins ($n_1 = 10$) of type A and 12 coins ($n_2 = 12$) of Type B, finding:

(A) $\bar{x}_1 = 4.65$ grams, $s_1 = 0.11$ grams.

(B) $\bar{x}_2 = 4.52$ grams, $s_2 = 0.13$ grams.

Running a t-test, what would the researcher conclude at the level of significance 0.01?

6. In the previous problem, what if the sample mean $\bar{x}_2$ had been 4.50 grams (all other details unchanged)?

7. A manufacturer of a certain plant food claims that, on the average, a given type of house plant grows taller in a given amount of time with the plant food than without it. To test this hypothesis (H_0: $\mu_1 > \mu_2$), a botanist samples 25 of the plants that have been given the plant food and 20 that have not, measuring their heights in centimeters and finding the following sample statistics:

(food) $\bar{x}_1 = 41.0$ centimeters, $s_1 = 7.0$ centimeters.

(no food) $\bar{x}_2 = 42.0$ centimeters, $s_2 = 6.0$ centimeters.

At the level of significance $\alpha = 0.05$, what is the result of the t-test?

8. In problem 7 (with all other details assumed unchanged), what if the sample mean $\bar{x}_2$ had been

(a) 42.4 centimeters? (b) 42.8 centimeters? (c) 43.2 centimeters?
(d) 43.6 centimeters? (e) 44.0 centimeters? (f) 44.4 centimeters?

9. A business analyst claims that on the average people spend less time standing in the checkout line at Topfeed Supermarket than at Bestgrub Supermarket. To test this hypothesis (H_0: $\mu_1 < \mu_2$) at the level of significance $\alpha = 0.01$, a statistician randomly samples 15 shoppers at

Topfeed and 14 at Bestgrub, timing each one in the checkout line and coming up with these results:

(Topfeed) $\bar{x}_1 = 7.5$ minutes, $s_1 = 2.0$ minutes.

(Bestgrub) $\bar{x}_2 = 6.0$ minutes, $s_2 = 1.8$ minutes.

What is the result of a *t*-test?

10. In problem 9, what if the second sample mean $\bar{x}_2$ (all other details assumed unchanged) had been
(a) 5.9 minutes? (b) 5.8 minutes? (c) 5.7 minutes?

C HYPOTHESES ABOUT PROPORTIONS

We have seen in previous sections that it is possible, through sampling and proper analysis, to test hypotheses concerning the *means* of populations. Now we will look at the question of testing a **hypothesis concerning the proportion (π) of a population** that has a given characteristic.

For example, someone may claim that: "Three-fifths of the people in New York State eat Sterko Cereal." That is, the null hypothesis may be: H_0: $\pi = 0.60$.

Or someone may claim (with a vested interest in the cereal): "*At least* three-fifths of the people in New York State eat Sterko Cereal." That is, the null hypothesis H_0 may be: $\pi \geq 0.60$.

Or someone may claim (with a vested interest in the cereal's competitors): "No more than one-tenth of the people in New York State eat Sterko Cereal." That is, the null hypothesis H_0 may state: $\pi \leq 0.10$.

To test any of these hypotheses (you should notice by now that the first one requires a two-sided test, whereas the last two require one-sided tests), we must rely on the fact that, as long as $n\pi$ and $n(1-\pi)$ both exceed 5,

sample proportions p are approximately normally distributed with mean $\mu_p = \pi$ (the true population proportion) and standard deviation

$$\sigma_p = \sqrt{\frac{\pi(1-\pi)}{n}},$$

where n is the sample size.

The procedure, then, is very similar to that of testing hypotheses about means, where we changed a sample mean $\bar{x}$ into a *z*-score and checked it against the appropriate critical *z*-value. Here, similarly, we

change the sample proportion p into a *z*-score by the formula

$$z = \frac{p-\pi}{\sqrt{\frac{\pi(1-\pi)}{n}}}$$

(where π is the claimed population proportion) and check it against the appropriate critical or cutoff z-value:

1.96 (or −1.96) for a two-sided test at level 0.05

2.58 (or −2.58) for a two-sided test at level 0.01

1.64 (or −1.64) for a one-sided test at level 0.05

2.33 (or −2.33) for a one-sided test at level 0.01.

As usual, the null hypothesis is rejected only when the computed z-score exceeds the critical table-value of z appropriate. Note that when we test a hypothesis about π, the logic is that we do so by computing the z-score (from p) *as if* the *null hypothesis were true*. Thus the standard deviation formula uses the *claimed* proportion π

$$\sigma_p = \sqrt{\frac{\pi(1-\pi)}{n}}$$

and *not* the sample proportion p.

EXAMPLES

1. The manufacturer of Roadking auto tires claims, in advertising, that at least 70% of the drivers in Los Angeles use Roadking tires. To test this hypothesis (H_0: $\pi \geq 0.70$), a consumer agency randomly samples 100 drivers, finding that 67 of them use Roadking tires: $p = \frac{67}{100} = 0.67$. At the level of significance $\alpha = 0.05$, what is the conclusion?

The hypothesis H_0 is tested by changing the sample proportion $p = 0.67$ into a z-score:

$$z = \frac{p-\pi}{\sqrt{\frac{\pi(1-\pi)}{n}}} = \frac{0.67-0.70}{\sqrt{\frac{0.70(1-0.70)}{100}}}$$

$$= \frac{-0.03}{\sqrt{\frac{0.21}{100}}} = -\frac{0.03}{\sqrt{0.0021}}$$

$$\approx -\frac{0.03}{0.04583} \approx -0.65.$$

And because −0.65 is not out beyond −1.64 on the left, the result $p = 0.67$ does *not* justify rejecting the null hypothesis. (See Figure 5.11.) The test is one-sided because our suspicion of H_0 runs in only one direction: we suspect that the

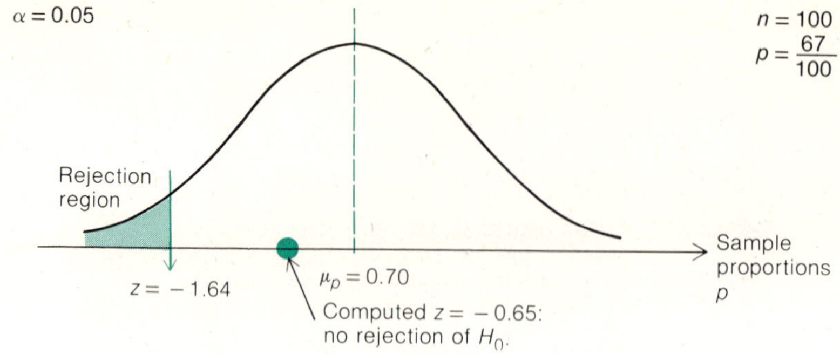

Figure 5.11

statement $\pi \geq 0.70$ *may* be shown false only if we get a sample proportion *less* than 0.70.

2. In Example 1, suppose the sample had been of size $n = 1{,}000$, with 670 of them using Roadking tires. Again the sample proportion turns out to be

$$p = 0.67 = \frac{670}{1{,}000}.$$

The z-score would be

$$z = \frac{p - \pi}{\sqrt{\dfrac{\pi(1-\pi)}{n}}} = \frac{0.67 - 0.70}{\sqrt{\dfrac{0.70(1-0.70)}{1{,}000}}}$$

$$= \frac{-0.03}{\sqrt{\dfrac{0.21}{1{,}000}}} = -\frac{0.03}{\sqrt{0.00021}}$$

$$\approx -\frac{0.03}{0.01449} \approx -2.07$$

and because -2.07 *is* out beyond -1.64 to the left, this time the sample proportion $p = 0.67$ means that the null hypothesis *is* rejected at the level 0.05. Comparing Example 1 with this example, the moral of the story is that *if* the null hypothesis $\pi \geq 0.70$ were true, it would be harder to get a sample proportion as small as 0.67 at a sample size of 1,000 than at a sample size of only 100, and at the larger sample size, the result $p = 0.67$ just should not be occurring if the statement $\pi \geq 0.70$ is true (see Figure 5.12).

3. A medical association claims that a certain experimental drug produces measurable improvement in asthma patients in 60% of the cases (no more, no less: hence any test must be a two-sided test). To test this hypothesis at the level 0.05, a researcher randomly samples 200 asthma patients who have used

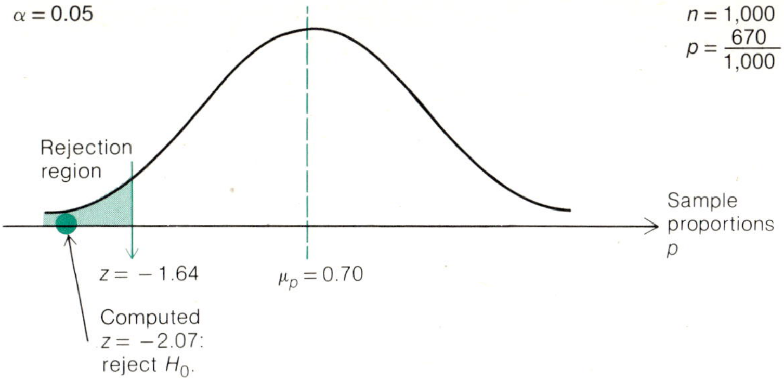

Figure 5.12

the drug, finding that 134 of them are improved. Thus the sample proportion is $p = \dfrac{134}{200} = 0.67$. What is the conlusion? That is, can the hypothesis H_0 $(\pi = 0.60)$ be rejected as stating the wrong population proportion?

The z-score needed is

$$z = \frac{p - \pi}{\sqrt{\dfrac{\pi(1-\pi)}{n}}} = \frac{0.67 - 0.60}{\sqrt{\dfrac{0.60(1-0.60)}{200}}}$$

$$= \frac{0.07}{\sqrt{\dfrac{0.60(0.40)}{200}}} = \frac{0.07}{\sqrt{0.0012}}$$

$$\approx \frac{0.07}{0.03464} \approx 2.02 > 1.96$$

so that the null hypothesis is rejected. We are at least 95% confident that the population proportion π has been misstated. (Note that a rejection at the level 0.01 would have required z to exceed 2.58, so that we are not 99% confident here that H_0 is false.) See Figure 5.13.

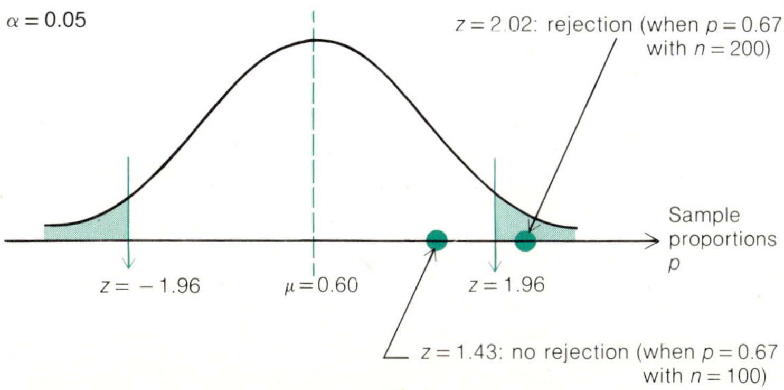

Figure 5.13

4. Suppose in Example 3 the sample size had only been 100, with 67 patients showing improvement $\left(\text{so that again } p=0.67=\frac{67}{100}\right)$. This time the z-score would be

$$z=\frac{0.67-0.60}{\sqrt{\frac{0.60(1-0.60)}{100}}}=\frac{0.07}{\sqrt{\frac{0.24}{100}}}$$

$$=\frac{0.07}{\sqrt{0.0024}}\approx\frac{0.07}{0.04899}\approx 1.43<1.96$$

so that we do *not* reject H_0. This time, the sample is too small for the result $p=0.67$ to be significantly different (at the level 0.05) from the claimed $\pi=0.60$. Again, see Figure 5.13.

5. A used car dealer claims that the used cars on a certain lot require substantial repair within a year of sale in less than 30% of the cases. A researcher wishes to test this hypothesis (H_0: $\pi<0.30$) at the level of significance $\alpha=0.01$, and samples 84 cases, finding that 26 of the cars needed substantial repair within a year. $\left(\text{Thus } p=\frac{26}{84}\approx 0.31.\right)$ What is the conclusion? (Note that this is a one-sided test, because the researcher would have no argument against the claim $\pi<0.30$ with a sample proportion *less* than 0.30.)

The z-score is

$$z=\frac{p-\pi}{\sqrt{\frac{\pi(1-\pi)}{n}}}=\frac{0.31-0.30}{\sqrt{\frac{0.30(1-0.30)}{84}}}$$

$$=\frac{0.01}{\sqrt{\frac{0.21}{84}}}=\frac{0.01}{\sqrt{0.0025}}$$

$$=\frac{0.01}{0.05}=0.20<2.33$$

and the null hypothesis is *not* rejected. The difference between the claimed upper limit 0.30 and the observed 0.31 is not significant at the level 0.01.

6. Suppose, in Example 5, the researcher had sampled 84 cars and found that 36 (instead of 26) of them had needed substantial repairs within a year. Then the

sample proportion would be $p = \frac{36}{84} \approx 0.43$, and the z-score would be

$$z = \frac{0.43 - 0.30}{\sqrt{\frac{0.30(1-0.30)}{84}}} = \frac{0.13}{\sqrt{\frac{0.21}{84}}}$$

$$= \frac{0.13}{\sqrt{0.0025}} = \frac{0.13}{0.05} = 2.60 > 2.33,$$

and the null hypothesis is rejected at the level 0.01. The sample result $p = 0.43$ is simply too far away from 0.30 for the claim $\pi \leq 0.30$ to be believed.

NEW TERMS IN THIS SECTION

hypothesis concerning a population proportion
***sampling distribution of the difference between two sample proportions**
***hypothesis concerning the difference between two population proportions**

EXERCISES 5-C

1. The manager of a large department store claims that 40% (no more, no less) of the people entering the store actually make a purchase. To test this hypothesis at the level 0.05, an analyst takes a random sample of 300 people entering the store, noting that 126 of them make purchases. What is the sample proportion? Is a one-sided or two-sided test appropriate? What is the conclusion?

2. In Exercise 5-C.1, what if the analyst had sampled
(a) 400 people, finding 168 purchasers?
(b) 800 people, finding 336 purchasers?
(c) 1,200 people, finding 504 purchasers?
(d) 1,600 people, finding 672 purchasers?
(e) 2,000 people, finding 840 purchasers?
(f) 2,400 people, finding 1,008 purchasers?

3. A television producer claims that at least 70% of the people in a given area regularly watch a certain program. To test this hypothesis at the level 0.05, a statistician samples 500 people at random and finds that 330 of them are regular watchers of the program in question. Should the test be one-sided or two-sided? What is the sample proportion? What is the result of the hypothesis test?

4. In Exercise 5-C.3, what if the survey had been of
(a) 450 people, with 297 watchers?
(b) 400 people, with 264 watchers?
(c) 300 people, with 198 watchers?

5. It is claimed that at most 20% of the cars on the road in a certain region have defective mufflers. A random sample of 1,600 cars turns up 336 cars with defective mufflers. At the level of significance $\alpha = 0.05$, what is the conclusion?

6. In Exercise 5-C.5, what if the random sample had been of size
(a) 2,000, with 420 defectives? (b) 2,400, with 504 defectives?
(c) 2,800, with 588 defectives? (d) 3,200, with 672 defectives?
(e) 3,600, with 756 defectives? (f) 4,000, with 840 defectives?
(g) 4,400, with 924 defectives?

7. A sociologist claims that 75% no more, no less) of the working people in a certain state commute from their homes to another town to work. To test this hypothesis (at the level 0.01) a researcher takes a random sample of 40 working people, finding that 24 of them are town-to-town commuters. What is the result of the test? If the conclusion surprises you at first glance, what accounts for its nature?

8. In Exercise 5-C.7, would the observed difference between p and the hypothesized π be significant at the level 0.05?

9. In Exercise 5-C.7 (again at the original $\alpha = 0.01$), what if the sample had been of size
(a) 45, with 27 commuters? (b) 50, with 30 commuters?
(c) 60, with 36 commuters?

10. A literary researcher claims that at least 50% of the college students in New England have read some of the works of H. P. Lovecraft. To test this hypothesis at the level 0.01, another researcher randomly samples 400 students, finding that 196 of them have read Lovecraft. What is the result of the test?

11. In Exercise 5-C.10, what if the sample had been of size
(a) 500, with 245 readers? (b) 1,000, with 490 readers?
(c) 2,000, with 980 readers? (d) 4,000, with 1,960 readers?
(e) 8,000, with 3,920 readers?

12. The directors of an extrasensory perception laboratory predict (as a null hypothesis H_0) that a certain "mildly sensitive" person to be tested would succeed in a mindreading experiment on no more than 30% of the test items if tested over and over again indefinitely. The person is tested on 500 items randomly and succeeds on 180 occasions. At the level of significance $\alpha = 0.01$, what can be said about the directors' prediction concerning the subject's overall potential?

13. In Exercise 5-C.12, what if the subject had been tested on
(a) 400 items, with 144 successes? (b) 350 items, with 126 successes?
(c) 300 items, with 108 successes?

14. A drug manufacturer claims that a certain drug designed for hay fever relief will be effective in at least three-fourths of the cases. A medical researcher wishes to test this hypothesis at the level 0.01 and randomly samples 600 users of the drug, finding that in 420 cases the drug is effective. What is the result of the test?

15. In Exercise 5-C.14, what if the number of successes had been
(a) 421? (b) 422? (c) 423?

16. An oil company claims that at least 80% of its service stations employ full-time general mechanics. To test this hypothesis, a researcher samples 200 of the company's stations at random, finding that 158 of them have such mechanics employed. At the level of significance $\alpha = 0.05$, what is the conclusion?

17. In Exercise 5-C.16, what if the number of sampled stations employing mechanics had been
(a) 156? (b) 154? (c) 152? (d) 150?

18. The manufacturer of a certain brand of ballpoint pen claims that the pen ink will dry up prematurely in no more than 10% of the pens. To test this claim, a consumer-protection group randomly tests 1,000 pens, finding that 128 of them dry up prematurely. At the level of significance 0.01, what is the conclusion? Express your sample proportion to three decimal places.

19. In Exercise 5-C.18, what if the number of sampled pens which dried up had been
(a) 127? (b) 126? (c) 125? (d) 124? (e) 123? (f) 122? (g) 121?

20. A city council claims that 85% (no more, no less) of the residents have household pets. To test this claim a researcher randomly samples 500 residents, finding that 445 of them do have household pets. At the level 0.05, what can be said?

21. In Exercise 5-C.20, what if the number of sampled persons having pets had been (express sample proportions to three decimal places)
(a) 444? (b) 443? (c) 442? (d) 441? (e) 440?

PONDERABLES 5-C

1. (*Differences between Proportions.*) It may be of interest at times to look at

a. a sample proportion p_1 taken from a sample (of size n_1) from one population (which has population proportion π_1) and
b. a sample proportion p_2 taken from a sample (of size n_2) from another population (which has population proportion π_2)

and ask whether $\pi_1 = \pi_2$ (the null hypothesis) or whether, on the other hand, H_0 can be rejected ($\pi_1 \neq \pi_2$), by virtue of the fact that the *sample* proportions p_1 and p_2 are too far apart. To proceed we need to know that as long as n_1p_1, $n_1(1-p_1)$, n_2p_2, and $n_2(1-p_2)$ are each at least 5, and as long as the two samples are independent, then

sample proportion *differences* $p_1 - p_2$ tend to be normally distributed (this normal distribution is called the **sampling distribution of the difference between two sample proportions**), with mean zero and standard deviation approximated by

$$\hat{\sigma}_{p_1-p_2} = \sqrt{\frac{p_1(1-p_1)}{n_1} + \frac{p_2(1-p_2)}{n_2}},$$

so that to test a difference $p_1 - p_2$ we may compute a z-score

$$z = \frac{p_1 - p_2}{\hat{\sigma}_{p_1 - p_2}}$$

and use the customary critical cutoff scores (for example, ±1.96 for a two-sided test at the level 0.05). Either a one-sided or two-sided test is possible, as in the case of testing hypotheses concerning the difference between two means.

For example, suppose a sample of 300 people in one town shows that 180 of them regularly watch a certain network news program, while a sample of 400 people in another town across the county shows that 280 of them watch the same program. Thus $p_1 = \frac{180}{300} = 0.60$ and $p_2 = \frac{280}{400} = 0.70$. Let us say that the null hypothesis H_0 here is that $\pi_1 = \pi_2$: the true **population proportions** watching the program are the same in both towns. To test this hypothesis at (say) the level 0.01, we compute

$$\begin{aligned}
z &= \frac{p_1 - p_2}{\sqrt{\dfrac{p_1(1-p_1)}{n_1} + \dfrac{p_2(1-p_2)}{n_2}}} \\
&= \frac{0.60 - 0.70}{\sqrt{\dfrac{0.60(1-0.60)}{300} + \dfrac{0.70(1-0.70)}{400}}} \\
&= \frac{-0.10}{\sqrt{\dfrac{0.24}{300} + \dfrac{0.21}{400}}} = -\frac{0.10}{\sqrt{0.0008 + 0.000525}} \\
&= -\frac{0.10}{\sqrt{0.001325}} \approx -\frac{0.10}{\sqrt{0.00132}} \\
&\approx -\frac{0.10}{0.0363} \approx -2.75
\end{aligned}$$

and because −2.75 is out beyond −2.58 to the left, we reject the null hypothesis H_0 and conclude that $\pi_1 \neq \pi_2$. That is, the difference between $p_1 = 0.60$ and $p_2 = 0.70$ is significant at the level 0.01. The true proportion of people in one town who watch the TV program in question is not, we feel "99% sure," the same as the proportion in the other town.

The following problems will provide practice with this kind of hypothesis test. (Statisticians sometimes may also want to consider the differences among several proportions instead of only two. This consideration leads to a province of statistics called *chi-square* analysis, which will be examined in Chapter 6.)

2. A market analyst wishes to test (at the level of significance $\alpha = 0.05$) the claim H_0 that the proportion of white-collar workers (in a certain region) who use Fang toothpaste is the same as the proportion of blue-collar workers who use it. Samples are taken. Of 500 white-collar workers sampled, 140 use the toothpaste; and of 600 blue-collar workers sampled, 150 use it. What are

the sample proportions, what is the z-score for $p_1 - p_2$, and what is the result of the hypothesis test?

3. In Ponderable 5-C.2, what if the numbers of sampled white-collar and blue-collar workers using Fang toothpaste had been
(a) 150 and 156 respectively? (b) 170 and 174 respectively?
(c) 160 and 156 respectively?

4. A sociological journal claims that the proportion of drug addicts treated by conventional methods, who return to their addiction, is greater on the East Coast than the proportion on the West Coast. (That is, H_0 is the statement $\pi_1 > \pi_2$.) A researcher, interested in testing this hypothesis at the level 0.05 (one-sided or two-sided test?), samples 500 East Coast cases of addiction, finding that 285 of them have reverted to addiction after treatment; the researcher then samples 400 West Coast cases, finding 244 reversions. What is the conclusion?

5. In problem 4, what if the number of reversions in the East and West Coast samples had respectively been
(a) 285 and 248? (b) 285 and 252?

6. A manufacturer claims that the proportion of radios needing repair within one year of purchase is smaller than the proportion of a competitor's radios requiring repair. To test this claim H_0 at the level 0.01, a researcher samples 300 of the first manufacturer's sets, finding that 87 radios needed repair within one year; and samples 400 of the competitor's sets, finding that 72 needed repair within one year. What is the result of the hypothesis test?

7. In problem 6, what if the numbers of radio sets needing repairs, in the manufacturer sample and competitor sample respectively, had been
(a) 84 and 72? (b) 81 and 72? (c) 78 and 72? (d) 75 and 72?

D HYPOTHESES ABOUT VARIANCES

Just as we may sample independently from two populations (with means μ_1 and μ_2 respectively) and compute two sample means $\bar{x}_1$ and $\bar{x}_2$ to test such hypotheses as $\mu_1 = \mu_2$, we may also sample from two populations (with standard deviations σ_1 and σ_2, respectively) and compute two sample **variances** s_1^2 and s_2^2 for the purpose of testing the null hypothesis

$$H_0: \sigma_1^2 = \sigma_2^2.$$

(Remember, variance is the square of standard deviation; we compute variance by following the same procedure as for computing standard deviation except that we stop short of taking square root.)

It may be claimed that the two populations in question are equally dispersed in the sense that they have the same variance, and we may wish to test this hypothesis, at some given level of significance as usual. After all, if the hypothesis is false, then an important difference has been pointed up: The populations are differently dispersed.

It is natural that we would appeal to the variances of the samples themselves and compare them in some fashion. But here the procedure is not the same as it is for means. In the case of means, statistical theory shows that differences $\bar{x}_1 - \bar{x}_2$ between sample means tend to be normally distributed, and for means we proceed on that basis. But sample *variances* do not give rise to such a normal-curve distribution, and when we test the hypothesis $\sigma_1^2 = \sigma_2^2$ we do so not on the basis of normal curves and *z*-scores, but rather on the basis of what is called an **F-ratio.** (The present use of the *F*-ratio is very limited; the more general form of the *F*-ratio will be described in Chapter 6.)

Here, then, is the procedure. Given two approximately normal populations, with variances σ_1^2 and σ_2^2, to test (for example, at the level of significance $\alpha = 0.05$) the null hypothesis $\sigma_1^2 = \sigma_2^2$, we first sample each population, computing the sample variance of each sample (the samples need not necessarily be the same size). That is, we compute s_1^2 for the first sample and s_2^2 for the second sample, in each case using the sample variance formula

$$s^2 = \frac{\sum (x - \bar{x})^2}{n - 1}.$$

Then we form the *F*-ratio, which is simply the ratio of the two sample variances

$$F = \frac{s_1^2}{s_2^2} \quad \text{or} \quad F = \frac{s_2^2}{s_1^2}$$

where the *larger* of the two sample variances is always placed on top as the numerator. Having computed the ratio *F*, we then look up a *critical value* for *F* in the table *F-Ratios*, $\alpha = 0.05$ in the back of the book (we also include an *F*-table for the level $\alpha = 0.01$ when needed). And we *reject* the null hypothesis $\sigma_1^2 = \sigma_2^2$ (concluding that on the contrary, the population variances are different) only if

$$F_{\text{computed}} > F_{.05},$$

that is, only when the computed *F*-ratio value exceeds the table value. And the table-value is found as follows.

Across the top of the *F*-table there is the designation **Degrees of Freedom for Numerator,** and this is determined by taking *one less than the sample size* of the sample used to form the numerator in the *F*-ratio. Similarly, down the left side of the table are numbers designated **d.f. for Denominator,** and this is determined by taking one less than the sample size of the denominator sample in the *F*-ratio. For example, suppose the sample sizes were 16 (numerator) and

20 (denominator). Then the degrees of freedom would be:

(numerator) $16-1=15$

(denominator) $20-1=19$

so that the desired critical value of F would be found by "crossing up" these values in the table:

$$\begin{array}{ll} & 15 \\ & \downarrow \\ 19 \rightarrow & 2.23 \end{array}$$

Thus under these circumstances (at these sample sizes) we would reject the null hypothesis $\sigma_1^2=\sigma_2^2$ at the level 0.05 only if the computed F-ratio value exceeded 2.23. The following examples should clarify the use of this hypothesis test.

EXAMPLES

1. An analyst wants to determine if the lifetimes of Brand A lightbulbs are differently dispersed from the lifetimes of Brand B lightbulbs. That is, he will test the null hypothesis $\sigma_A^2=\sigma_B^2$, where σ_A^2 and σ_B^2 are the population variances for *all* the Brand A and Brand B bulb lifetimes. The analyst will use the level of significance $\alpha=0.05$.

He takes a random sample of $n_1=25$ Brand A bulbs (observing their lifetimes in hours) and, using the sample variance formula, finds (from a sample mean $\bar{x}$ computed for the Brand A bulbs) a sample variance (let us say) of $s_A^2=31.4$. Then he takes a random sample of $n_2=24$ Brand B bulb lifetimes and (again using the same sample variance formula, but with the appropriate mean and sample size) finds a sample variance of $s_B^2=38.3$. To test the null hypothesis at the level 0.05, he computes the F-ratio, placing the larger sample variance (which happens to be s_B^2) on top:

$$F=\frac{s_B^2}{s_A^2}=\frac{38.3}{31.4}\approx 1.22.$$

Now, this computed F-ratio must be compared with a critical or cutoff value from the F table. For that purpose the degrees of freedom are:

(numerator) $n_1-1=25-1=24$

(denominator) $n_2-1=24-1=23$

so that the critical value is found by looking under 24 and across from 23:

$$\begin{array}{ll} & 24 \\ & \downarrow \\ 23 \rightarrow & 2.01 \end{array}$$

That is, $F_{.05}=2.01$. To reject the null hypothesis $\sigma_A^2=\sigma_B^2$ at the level 0.05, the computed F-ratio would have to exceed 2.01, but because $1.22<2.01$, the analyst does *not* reject the null hypothesis at the level 0.05. In other words, the sample variances $s_A^2=31.4$ and $s_B^2=38.3$ are not different enough (at these sample sizes) to make it possible to claim with 95% confidence that the true population variances σ_A^2 and σ_B^2 are different.

2. Suppose in Example 1 (still assuming sample sizes $n_1=25$ and $n_2=24$) the sample variances had been

$$s_A^2=21.4, \qquad s_B^2=43.8.$$

Then the computed F-ratio would be

$$F=\frac{s_B^2}{s_A^2}=\frac{43.8}{21.4}\approx 2.05.$$

As before, the critical value is $F_{.05}=2.01$. And because $2.05>2.01$, we would *reject* (at the level 0.05) the hypothesis $\sigma_A^2=\sigma_B^2$. That is, we may feel 95% confident from the sample data that the true population variances are *not* equal.

3. A group of sales representatives frequently having to drive from Boston to New York may select either one of two routes. They sample each route by driving it 10 times ($n_1=10$, $n_2=10$), recording the following driving times (in minutes):

Route 1: 281, 298, 312, 277, 308, 268, 289, 301, 258, 318

$$\left(\text{sample mean: } \bar{x}_1=\frac{2{,}910}{10}=291.0\right).$$

Route 2: 296, 294, 287, 305, 291, 285, 297, 284, 303, 298

$$\left(\text{sample mean: } \bar{x}_2=\frac{2{,}940}{10}=294.0\right).$$

They wish to test (at the level of significance $\alpha=0.01$) the null hypothesis that (in the sense of true population variances σ_1^2 and σ_2^2) the two routes provide, overall, driving times that are *equally dispersed:* $\sigma_1^2=\sigma_2^2$.

First, the sample variances must be computed:

Sample 1

x:	281	298	312	277	308	268	289	301	258	318
$x-\bar{x}_1$:	−10	7	21	−14	17	−23	−2	10	−33	27
$(x-\bar{x}_1)^2$:	100	49	441	196	289	529	4	100	1,089	729

$$\sum(x-\bar{x}_1)^2=3{,}526$$

$$s_1^2=\frac{\sum(x-\bar{x}_1)^2}{n_1-1}=\frac{3{,}526}{10-1}\approx 391.778$$

Sample 2

x:	296	294	287	305	291	285	297	284	303	298
$x-\bar{x}_2$:	2	0	−7	11	−3	−9	3	−10	9	4
$(x-\bar{x}_2)^2$:	4	0	49	121	9	81	9	100	81	16

$$\sum (x-\bar{x}_2)^2 = 470$$

$$s_2^2 = \frac{\sum (x-\bar{x}_2)^2}{n_2-1} = \frac{470}{10-1} \approx 52.222$$

Then (s_1^2 being the larger sample variance) the F-ratio is

$$F = \frac{s_1^2}{s_2^2} = \frac{391.778}{52.222} \approx 7.50.$$

To determine the critical value for F, against which the computed $F=7.50$ must be compared, we note the degrees of freedom:

(numerator) $n_1-1=10-1=9.$

(denominator) $n_2-1=10-1=9.$

And from the $F_{.01}$ table (*not* the $F_{.05}$ table; the level of significance is 0.01 this time) the critical value, found by looking under 9 and across from 9, is

$$F_{.01}=5.35.$$

Because $7.50>5.35$, the sales representatives would *reject*, at the level 0.01, the hypothesis $\sigma_1^2=\sigma_2^2$. Thus they may feel 99% confident, from the sample data observed, that the variances of the overall possible driving times for Route 1 and for Route 2 *are different*, and that the apparent difference showing up in the sample results $s_1^2=391.778$ and $s_2^2=52.222$ is statistically significant, and not merely an illusion generated by chance.

4. (*Interpolation.*) It can easily happen, when we look for a critical value $F_{.05}$ or $F_{.01}$ in the table, that the desired degrees of freedom values (determined by the sample sizes) may not actually be listed. For example, with a numerator sample of size $n=18$, the d.f. value would be $n-1=18-1=17$, but the available d.f. values in the table skip from 15 to 20.

In such an event, we may **interpolate in the F-table** to find (at least approximately) the desired F-value. For instance, suppose the sample sizes were $n_1=18$ and $n_2=20$ (with the larger sample variance turning out to be the one from the sample of size 18, so that that becomes the "numerator sample"). The d.f. values are:

(numerator) $n_1-1=18-1=17.$

(denominator) $n_2-1=20-1=19.$

The **denominator d.f.** is listed, but the **numerator d.f.** is not (in "Ponderables" problem 3 at the end of this section there is a discussion of what can be done if neither d.f. is listed). Not finding d.f. = 17 at the top of the table, we note that 17 is between the listed values 15 and 20. If we look up $F_{.05}$ (with the denominator d.f. = 19) at 15 and at 20, on both sides of the desired 17 as it were, we get:

$$\begin{array}{cc} 15 & 20 \\ \downarrow & \downarrow \\ 19 \rightarrow 2.23 & 19 \rightarrow 2.16 \end{array}$$

so that the desired F value should lie between the results 2.23 and 2.16 (which are 0.07 units apart). In fact, since 17 is *two-fifths* of the way between 15 and 20

we would expect the desired F value to lie approximately two-fifths of the way between 2.23 and 2.16. Two-fifths of the difference 0.07 is

$$\frac{2}{5}(0.07) = 0.4(0.07) = 0.028$$

and subtracting this from 2.23 (since the F values decrease from 2.23 to 2.16) we get

$$2.23 - 0.028 = 2.202 \approx 2.20$$

so that the desired critical value for F is approximately $F_{.05} = 2.20$. The following example also illustrates this procedure.

5. For an F-test at the level $\alpha = 0.01$, suppose the sample sizes were $n_1 = 41$ (numerator sample) and $n_2 = 38$ (denominator sample). Then the desired d.f. values would be

(numerator) $n_1 - 1 = 41 - 1 = 40.$

(denominator) $n_2 - 1 = 38 - 1 = 37.$

The problem this time is that the denominator d.f. does not appear in the table, which skips from 30 to 40. Again, interpolation is called for. If the denominator d.f. values 30 and 40 (on both sides of the needed 37) were used, the table results would be

$$\begin{array}{cc} 40 & 40 \\ \downarrow & \downarrow \\ 30 \rightarrow 2.30 & 40 \rightarrow 2.11 \end{array}$$

so that the actual $F_{.01}$-value desired should lie between 2.30 and 2.11. Note first that the difference between these values is

$$2.30-2.11=0.19.$$

Now, d.f. = 37 is *seven-tenths* of the way between 30 and 40, so that we take seven-tenths of the difference:

$$\frac{7}{10}(0.19)=0.7(0.19)=0.133$$

and subtract it from the larger F-value 2.30 (because the F values are decreasing as d.f. goes from 30 to 40):

$$2.30-0.133=2.167\approx 2.17.$$

The desired critical value $F_{.01}$ is (approximately) 2.17.

NEW TERMS IN THIS SECTION

hypothesis concerning population variances
***F*-ratio**
degrees of freedom for numerator
degrees of freedom for denominator
interpolation in the *F*-tables

EXERCISES 5-D

1. Two salespersons wish to determine whether their overall weekly incomes are differently dispersed. (They wish to test, at the level 0.05, the null hypothesis $\sigma_1^2=\sigma_2^2$, where the variances are for the overall incomes of the two people. Rejecting this hypothesis would imply that one salesperson's income is less consistent than the other's.) Each salesperson's weekly income is recorded at random (in dollars) for eleven weeks, with the following results:

Salesperson 1: 310 284 330 270 295 335 260
315 300 265 325

Salesperson 2: 255 290 312 282 300 345 270
305 263 330 260

Compute the sample variances and the F-ratio. What is the critical value $F_{.05}$? Is the null hypothesis rejected at the level 0.05?

2. A consumer protection analyst wishes to test (at the level 0.05) the hypothesis that the variance σ_1^2 of the retail prices charged (at various stores) in one city for a certain item, is the same as the variance σ_2^2 of the prices charged for the same item in another city. The analyst

randomly samples the prices at 13 stores in each city, finding the following prices:

First city:	$12.75	12.79	12.49	12.55	12.85	12.69	12.45
	12.75	12.59	12.79	12.85	12.57	12.98	
Second city:	$12.65	12.89	12.79	12.75	12.70	12.59	12.88
	12.69	12.72	12.55	12.78	12.59	12.91	

Compute the sample variances and the F-ratio; to what critical value $F_{.05}$ from the table is this compared, and with what result? Is the null hypothesis $\sigma_1^2 = \sigma_2^2$ rejected at the level 0.05?

3. To investigate whether the overall times required for two different experimental drugs to take effect on laboratory mice are differently dispersed (testing the hypothesis $\sigma_1^2 = \sigma_2^2$), an observer samples by administering each drug 10 times, obtaining the following results (in seconds):

Drug A:	42	51	47	44	54	50	49	41	53	49
Drug B:	48	59	30	28	68	39	24	74	40	70

A level of significance $\alpha = 0.01$ is chosen. What are the sample variances? The F-ratio? The critical value $F_{.01}$? Is the hypothesis rejected?

4. Wanting to determine whether the weights of the first-grade children in one city are differently dispersed from the weights of the first-grade children in another city (testing at level 0.05 the null hypothesis $\sigma_1^2 = \sigma_2^2$), a statistician randomly samples the weights (in kilograms), sampling 21 children in one city and 16 in the other:

(First city) 21, 24, 23, 20, 22, 20, 21, 17, 19, 22, 18,
23, 24, 24, 18, 21, 23, 19, 20, 25, 17

(Second city) 23, 20, 18, 22, 21, 20, 23, 21, 19, 19, 22,
24, 18, 24, 25, 17

Compute the sample variances and the F-ratio. What critical value $F_{.05}$ must the computed F exceed for a rejection of the null hypothesis? What happens?

5. A statistician wishes to test (at the level of significance $\alpha = 0.01$) the null hypothesis ($\sigma_1^2 = \sigma_2^2$) that the heights of workers in one industry are dispersed by the same amount as the heights of workers in another industry. (That is, the hypothesis is that in neither industry are workers' heights more or less consistent than in the other.) Sampling the two populations of workers (with sample sizes of 16 and 25 respectively), the statistician finds the following heights (in centimeters):

(First sample)	170	164	162	173	171	166	186	160	
	179	168	172	160	174	174	158	183	
(Second sample)	166	168	162	170	165	163	161	162	165
	164	161	166	170	169	168	160	165	163
	170	161	169	163	165	170	159		

What are the sample variances? Compute the *F*-ratio. To what critical value from the table must it be compared? Is the null hypothesis rejected at the level 0.01?

6. Using interpolation in the *F*-tables (see Examples 4 and 5 in Section 5-D), determine what critical table value for $F_{.05}$ would be used if the sample sizes were as follows (numerator sample size always being given first):

(a) 19 and 18	(b) 20 and 19	(c) 23 and 23	(d) 12 and 16
(e) 28 and 25	(f) 29 and 24	(g) 26 and 26	(h) 27 and 23
(i) 22 and 25	(j) 14 and 22	(k) 22 and 24	(l) 35 and 23
(m) 40 and 21	(n) 36 and 31	(o) 32 and 26	(p) 39 and 25
(q) 34 and 31	(r) 33 and 22	(s) 38 and 41	(t) 37 and 24
(u) 49 and 25	(v) 47 and 31	(w) 51 and 61	(x) 45 and 26
(y) 44 and 25	(z) 52 and 61		

7. Repeat Exercise 5-D.6 for values in the $F_{.01}$ table (instead of $F_{.05}$).

8. A psychologist wishes to test the hypothesis ($\sigma_1^2 = \sigma_2^2$, level 0.05) that the reading speeds (in words per minute) for one group of students are equally dispersed as the reading speeds of another group of students. Sampling the two groups, the psychologist observes the following sample data:

(Group 1)	140	131	168	156	150	137	152	157	144	134
	155	162	160	172	133	151	166	173	128	
(Group 2)	152	143	129	137	156	160	140	133	158	175
	126	130	148	160	135	170				

Compute the sample variances. Against what critical value $F_{.05}$ (by interpolation) is the computed *F* compared? What is *F*, and what is the conclusion?

9. Wishing to determine whether the mileages (in miles per gallon to the nearest tenth) of two cars are differently dispersed, and testing the null hypothesis $\sigma_1^2 = \sigma_2^2$ at the level 0.05, an automotive engineer samples the two cars' mileage figures at random, coming up with the following data:

(First car)	18.0	18.4	17.7	19.4	17.0	19.0	16.7	17.2
	19.1	16.5	17.3	17.4	18.7	19.6		
(Second car)	16.2	19.5	17.4	17.7	17.9	19.7	20.2	19.3
	16.8	18.4	16.1	16.9	19.6	17.3	17.0	

Compute the sample variances and *F*. What is the critical value $F_{.05}$ for the experiment? Can it be concluded that the two population variances (for overall performance of the cars) are different?

10. An investigator wants to determine whether the dispersion in size for one shipment of peaches is different from the dispersion in size for another shipment from a different source. Each shipment is randomly sampled, sizes being recorded as peach diameters measured in

centimeters:

(Shipment A)	7.3	8.9	10.0	7.8	8.3	8.5	9.4	7.6	8.4
	8.5	7.8	9.2	10.2	7.9	8.3	9.3	8.8	
(Shipment B)	7.2	8.9	7.7	9.9	7.5	10.5	9.6	9.8	7.3
	8.5	7.4	9.7	10.2	7.1	7.7	9.3	8.0	10.3

At the level of significance 0.05, is the null hypothesis $\sigma_1^2=\sigma_2^2$ rejected? What happens at the level 0.01?

PONDERABLES 5-D

1. What is the numerically smallest size that the ratio $F=\frac{s_1^2}{s_2^2}$ can have? Under what circumstances is this value attained?

2. If sample data are given in centimeters, in what units is each sample variance (s_1^2 or s_2^2) expressed? What units does the F ratio itself have? Why?

3. (*Double interpolation.*) In Examples 4 and 5 we looked at the problem of interpolating in the F-tables to find the critical values $F_{.05}$ or $F_{.01}$ when one sample size led to a d.f. value that was not listed. But suppose *both* d.f. values were not listed. Then, to be sure that we are using at least a good approximation to the right F-value, we must interpolate *doubly*. Study the following example to see how this is done.

Suppose the d.f. values needed are:

(numerator) d.f. = 37.

(denominator) d.f. = 34.

Note that the F-tables skip over these d.f. values both horizontally and vertically, skipping in fact from 30 to 40. If we are dealing, say, with the level of significance $\alpha=0.01$, we find in the table these entries, for d.f. values of 30 or 40:

	30	40
30	2.39	2.30
40	2.20	2.11

Horizontally, our numerator d.f. of 37 suggests being *seven-tenths* of the way between 2.39 and 2.30, or seven-tenths of the way between 2.20 and 2.11. We do both these simple interpolations:

$$0.7(2.39-2.30)=0.7(0.09)=0.063$$

$$\text{and}\quad 2.39-0.063=2.327$$

$$0.7(2.20-2.11)=0.7(0.09)=0.063$$

$$\text{and}\quad 2.20-0.063=2.137.$$

We should now think in terms of the difference between these two results:

$$\begin{array}{lll|l} & & 2.327 & \\ 30 & 2.39 & \downarrow 2.30 & \text{difference} \\ 40 & 2.20 & \uparrow 2.11 & 0.190 \\ & & 2.137 & \end{array}$$

Now because we have a denominator d.f. = 34, we go *four-tenths* of the way between these values 2.327 and 2.137, doing a new interpolation:

$$0.4(2.327 - 2.137) = 0.4(0.190)$$
$$= 0.076$$
and $\quad 2.327 - 0.076 = 2.251 \approx 2.25.$

This $F_{.01}$ value, between 2.39 and 2.11 (look again at the given table-values) is a reasonably good approximation to the critical value needed.

4. Using double interpolation as described previously, find $F_{.01}$, given that the d.f. values for numerator and denominator respectively are:
(a) 38 and 33 (b) 36 and 39 (c) 32 and 35 (d) 35 and 37
(e) 23 and 28 (f) 22 and 27 (g) 21 and 29 (h) 50 and 50
(i) 48 and 52 (j) 54 and 46

SUGGESTED READINGS

Statistics in the Real World: A Book of Examples, by Richard J. Larsen and Donna Fox Stroup (Macmillan Publishing Co., Inc., 1976). This book provides, in Chapters 2 and 3, intriguing examples of one- and two-sample t-tests, with reference to such phenomena as the "Transylvania effect."

The World of Mathematics, presented with commentaries and notes by James R. Newman (Simon & Schuster, 1956). Part VIII, Chapter 6 ("Mathematics of a Lady Tasting Tea," by Sir Ronald A. Fisher) provides an enjoyable discussion of hypothesis testing in general.

Everything You Always Wanted To Know About Elementary Statistics (but were afraid to ask), by Jerald G. Schutte (Prentice-Hall, Inc., 1977). Chapter 3 gives an interesting and simple treatment of one- and two-sample hypothesis testing.

6 And More Hypothesis Testing

Chance generally favors the prudent.—Joubert

A ANALYSIS OF VARIANCE

In Chapter 5, Section B, we looked at the problem of testing an hypothesis concerning the difference, or lack of difference, between two population means μ_1 and μ_2 by considering the difference between the sample means $(\bar{x}_1 - \bar{x}_2)$ representing them. In the case of large samples we may compute a z-score

$$z = \frac{\bar{x}_1 - \bar{x}_2}{\sqrt{\dfrac{\sigma_1^2}{n_1} + \dfrac{\sigma_2^2}{n_2}}} \approx \frac{\bar{x}_1 - \bar{x}_2}{\sqrt{\dfrac{s_1^2}{n_1} + \dfrac{s_2^2}{n_2}}}.$$

In the case of small samples (see Ponderable problems at the end of Section 5-B) we may compute a t-score

$$t = \frac{\bar{x}_1 - \bar{x}_2}{\sqrt{\dfrac{(n_1 - 1)s_1^2 + (n_2 - 1)s_2^2}{n_1 + n_2 - 2}}\sqrt{\dfrac{1}{n_1} + \dfrac{1}{n_2}}}$$

to determine if the observed difference between *sample* means $\bar{x}_1$ and $\bar{x}_2$ is *significant*—that is, to determine whether to reject a null hypothesis which claims that $\mu_1 = \mu_2$.

But what if there are *more* than two population means involved?

That is, what if there are several populations, with means μ_1, μ_2, μ_3, and so on and we have to approximate these respectively by sample means $\bar{x}_1$, $\bar{x}_2$, $\bar{x}_3$, and so on? (We will assume throughout that the samples involved are all the same size, because this is usually the procedure in the kind of analysis to be discussed here, and because it simplifies the computations.) The null hypothesis to be tested might be

$$\mu_1 = \mu_2 = \mu_3 = \ldots$$

and we could reject it only by arguing that the population means are *not all* equal.

At first glance, one might think that with several samples, we could simply run a z-test or a t-test on the sample means two at a time. For example, with five populations *A, B, C, D, and E,* with population means μ_A, μ_B, μ_C, μ_D, and μ_E that are approximated respectively by sample means $\bar{x}_A$, $\bar{x}_B$, $\bar{x}_C$, $\bar{x}_D$, and $\bar{x}_E$, one might suppose that we could just do a z (or t) comparison on

$\bar{x}_A$ and $\bar{x}_B$, $\bar{x}_A$ and $\bar{x}_C$, $\bar{x}_A$ and $\bar{x}_D$, $\bar{x}_A$ and $\bar{x}_E$,

$\bar{x}_B$ and $\bar{x}_C$, $\bar{x}_B$ and $\bar{x}_D$, $\bar{x}_B$ and $\bar{x}_E$, $\bar{x}_C$ and $\bar{x}_D$,

$\bar{x}_C$ and $\bar{x}_E$, $\bar{x}_D$ and $\bar{x}_E$.

Clearly, this much two-sample testing would be a lot of work—but that is not the most serious objection to this approach! The real objection is that if each one of those two-sample tests were done at (say) the level of significance $\alpha = 0.05$, so that in each case where we rejected a hypothesis $\mu_i = \mu_j$ we were 95% confident that it was proper to do so, then the level of confidence in the whole experiment (if we rejected all 10 hypotheses) could be as little as

$$(0.95)(0.95)(0.95)(0.95)(0.95)(0.95)(0.95)(0.95)(0.95)(0.95) \approx 0.60.$$

Now wait a minute! All of a sudden we have dropped to a possible low level of only 60% confidence? That is, if we conclude with a rejection we could be taking as much as a 40% risk of being wrong? Clearly, this will never do.

Instead, in a situation where we want to test the null hypothesis

$$\mu_1 = \mu_2 = \mu_3 = \ldots$$

we turn to an area of statistics known as **analysis of variance** (often abbreviated, in the literature of statistics, to **ANOVA**).

To illustrate this technique, we will go through an example step by step.

Suppose a commuter has three different routes (*A, B, C*) that he can drive to work, and consider μ_A, μ_B, and μ_C to be the true overall mean driving times required, in minutes. (That is, μ_A, for instance, is the mean time that could be computed, theoretically, "in the long run" by considering *all* the times that the commuter drives route *A* to work.) Let us state a null hypothesis:

$$\mu_A = \mu_B = \mu_C.$$

That is to say, "overall" or ultimately, all the potential mean driving times are the same: The routes are uniform in terms of quickness of getting to work by them.

To test this hypothesis (at a specified level of significance as usual, say $\alpha = 0.05$), suppose the commuter samples the routes by driving each route seven times, timing each one always in minutes, with the following results:

Route *A*: 47 53 43 50 41 56 53

Route *B*: 44 50 46 40 45 42 48

Route *C*: 54 44 55 50 46 53 55

The three sample means are readily computed:

$$\bar{x}_A = \frac{47+53+43+50+41+56+53}{7} = \frac{343}{7} = 49.0 \text{ minutes.}$$

$$\bar{x}_B = \frac{44+50+46+40+45+42+48}{7} = \frac{315}{7} = 45.0 \text{ minutes.}$$

$$\bar{x}_C = \frac{54+44+55+50+46+53+55}{7} = \frac{357}{7} = 51.0 \text{ minutes.}$$

Now, if we had found exactly the same value for all three sample means, we would have *no* way, on the basis of such evidence, to argue a rejection of the hypothesis $\mu_A = \mu_B = \mu_C$. As it is, though, the sample means observed are a bit different from each other. The whole question is: are they different enough to cast real doubt ("95% confidence" doubt) on the hypothesis $\mu_A = \mu_B = \mu_C$? That is, are the sample means *significantly* different from each other at the level 0.05?

To answer this question, we apply the techniques of analysis of variance, running what is commonly called an **F-test.** Study the following procedure carefully to get a feeling for the mechanics of it; a discussion of its rationale will follow.

First, we need to record the *total* from each sample; because the sample data are written horizontally, we will call these sums the **row totals:**

								Row totals
A:	47	53	43	50	41	56	53	$T_A = 343$
B:	44	50	46	40	45	42	48	$T_B = 315$
C:	54	44	55	50	46	53	55	$T_C = 357$
							Grand total:	$\sum x = 1{,}015$

Now, we will be *squaring* all the sample data, as well as the totals just shown, in order to compute three quantities called **sums of squares (SS);** these are called

SS_{total}, SS_{factor}, and SS_{error}.

These are computed by the following formulas:

$$SS_{\text{total}} = \sum (x^2) - \frac{(\sum x)^2}{n}$$

(where the summations are taken over all 21 numbers, and n is how many numbers there are altogether: $n = 21$).

$$SS_{\text{factor}} = \frac{T_A^2 + T_B^2 + T_C^2}{c} - \frac{(\sum x)^2}{n}$$

(where c is the number of items in *each* sample: $c = 7$). We refer to the choice of three routes A, B, and C as the **factor** in our analysis, so that by looking at the row-total squares, we are looking at an aspect of the factor. Because there are three routes, we sometimes refer to the three **levels of the factor.** (Because only one factor is involved, namely, the question of which route to drive to work, this type of statistical analysis is called "single-factor" or *one-way* ANOVA. It is also possible to do *two*-factor or "two-way" ANOVA, as shown in Ponderables at the end of this section.) The third sums-of-squares formula is:

$$SS_{\text{error}} = SS_{\text{total}} - SS_{\text{factor}}.$$

We actually do the computations now:

$$SS_{\text{total}} = \sum (x^2) - \frac{(\sum x)^2}{n}$$

$$= 47^2 + 53^2 + 43^2 + 50^2 + 41^2 + 56^2 + 53^2$$

$$+ 44^2 + 50^2 + 46^2 + 40^2 + 45^2 + 42^2 + 48^2$$

$$+ 54^2 + 44^2 + 55^2 + 50^2 + 46^2 + 53^2 + 55^2$$

$$- \frac{(1{,}015)^2}{21}$$

$$= 2{,}209 + 2{,}809 + 1{,}849 + 2{,}500 + 1{,}681 + 3{,}136 + 2{,}809$$

$$+ 1{,}936 + 2{,}500 + 2{,}116 + 1{,}600 + 2{,}025 + 1{,}764 + 2{,}304$$

$$+ 2{,}916 + 1{,}936 + 3{,}025 + 2{,}500 + 2{,}116 + 2{,}809 + 3{,}025$$

$$- \frac{(1{,}1015)^2}{21} = 49{,}565 - \frac{1{,}030{,}225}{21}$$

$$= 49{,}565 - 49{,}058.33 = 506.67.$$

Easier, now, is the computation of

$$SS_{\text{factor}} = \frac{T_A^2 + T_B^2 + T_C^2}{c} - \frac{(\sum x)^2}{n}$$

$$= \frac{343^2 + 315^2 + 357^2}{7} - \frac{(1{,}015)^2}{21}$$

$$= \frac{117{,}649 + 99{,}225 + 127{,}449}{7} - \frac{(1{,}015)^2}{21}$$

$$= \frac{344{,}323}{7} - \frac{1{,}030{,}225}{21} = 49{,}189 - 49{,}058.33$$

$$= 130.67.$$

And we subtract the two results to get

$$SS_{\text{error}} = SS_{\text{total}} - SS_{\text{factor}}$$

$$= 506.67 - 130.67 = 376.00.$$

We now arrange this information, and some other information yet to be derived, in an **ANOVA table,** as shown in Figure 6.1.

We fill in the rest of this table as follows. The **d.f. (degrees of freedom)** are found this way:

For FACTOR: $\text{d.f.}_{\text{factor}} = r - 1$

(where r is the number of samples: in this case, $r = 3$, so that $\text{d.f.}_{\text{factor}} = r - 1 = 2$).

For TOTAL: $\text{d.f.}_{\text{total}} = n - 1$

(where n is the total number of items overall: in this case, $n = 21$ so that $\text{d.f.}_{\text{total}} = n - 1 = 20$).

For ERROR: $\text{d.f.}_{\text{error}} = \text{d.f.}_{\text{total}} - \text{d.f.}_{\text{factor}}$

Source	Sum of squares (SS)	Degrees of freedom (df)	Mean square (MS)	*F* ratio (*F*)
Factor	130.67			
Error	376.00			
Total	506.67			

Figure 6.1 ANOVA table

Source	SS	df	MS	*F*
Factor	130.67	2		
Error	376.00	18		
Total	506.67	20		

Degrees of freedom:

- $r-1$
- $r(c-1)$ or df_{total} minus df_{factor}
- $n-1$

Figure 6.2

(in this case $20-2=18$). *Or*: $d.f._{error}$ can be computed by the formula

$$d.f._{error}=r(c-1)$$

(where again r is the number of samples and c is the size of each, so that $r(c-1)=3(7-1)=3(6)=18$).

These d.f. values can now be entered into the ANOVA table, as in Figure 6.2.

Next we fill in two quantities called **mean squares,** in the "factor" and "error" parts of the table: in each case, *MS* is the *SS* value divided by the corresponding d.f. value:

$$MS_{factor}=\frac{SS_{factor}}{d.f._{factor}}=\frac{SS_{factor}}{r-1}$$

$$=\frac{130.67}{2}=65.335$$

$$MS_{error}=\frac{SS_{error}}{d.f._{error}}=\frac{SS_{error}}{r(c-1)}$$

$$=\frac{376.00}{18}\approx 20.889.$$

These values may now be filled in, and we finally compute the **F-ratio** (see Figure 6.3) by simply dividing the last two results:

$$F=\frac{MS_{factor}}{MS_{error}}=\frac{65.335}{20.889}\approx 3.13.$$

Source	SS	df	MS	*F*
Factor	130.67	2	65.335	3.13
Error	376.00	18	20.889	
Total	506.67	20		

Figure 6.3

This F-ratio is our final result—but what does it mean, in terms of rejecting or not rejecting the original hypothesis $\mu_A = \mu_B = \mu_C$?

To answer this question we must turn to the table F-ratios ($\alpha = 0.05$) in the back of the book and look up the *critical* F-value against which our computed $F = 3.13$ is compared. In that table, you will notice that two different "degrees of freedom" are considered: a **numerator d.f.** listed horizontally across the top, and a **denominator d.f.** listed vertically at the left. The d.f. values to look at are simply those two which have already been used in the ANOVA table computations:

$$\text{d.f.(numerator)} = \text{d.f.}_{\text{factor}} = r - 1 = 2.$$

$$\text{d.f.(denominator)} = \text{d.f.}_{\text{error}} = r(c-1) = 18.$$

Then in the F-table we find:

........2
:   ↓
18 → 3.55

and we *reject the null hypothesis only if our computed F-value exceeds the table value.* (Note that there are F-tables both for level 0.05 and 0.01.) See Figure 6.4. In this case,

$$F_{\text{computed}} = 3.13 < 3.55$$

so that we do *not* reject the null hypothesis $\mu_A = \mu_B = \mu_C$ at the level of significance 0.05.

That is, our sample evidence does *not* provide an adequate argument (at the level 0.05) against the claim that the routes all take, in the long run, the same amount of time on the average to drive. (If the computed F had been, say, $F = 3.87 > 3.55$, we would have rejected that claim, and would have concluded with 95% confidence that the route means are *not* all the same.)

A few words now about the theory behind this technique.

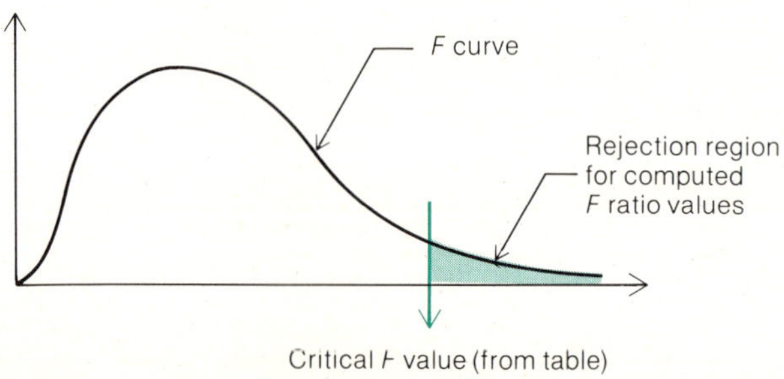

Figure 6.4

In working with the squares of the sample numbers, we are actually dealing with the *variability* of the data. For example, when we compute

$$SS_{\text{total}} = \sum (x^2) - \frac{(\sum x)^2}{n},$$

this can be shown to be equivalent to the form

$$\frac{\sum(x-\bar{x})^2}{n},$$

a *variance* formula measuring dispersion of the x's. (Hence the title of the technique: analysis of variance.) Likewise,

$$SS_{\text{factor}} = \frac{\sum (T^2)}{c} - \frac{(\sum x)^2}{n}$$

is a computation which measures variability, and because it is computed on the basis of the various *row totals*, the variability here being measured is *the variability due to the factor*: That is, the amount, loosely speaking, of variation due to the fact that the routes A, B, and C themselves may produce somewhat different driving times on the average. Because SS_{total} measures *all* the variability of the data, and because SS_{factor} measures the amount of "cross-sample" variability that is due to the routes A, B, and C, the difference

$$SS_{\text{error}} = SS_{\text{total}} - SS_{\text{factor}}$$

measures the *rest* of the variability: the amount of data variability not due to the route factor but attributable only to the fact that the samples *internally* have, each of them, some variability. (Expressing SS_{total} as $SS_{\text{factor}} + SS_{\text{error}}$ is often called **partitioning the squares**.) Thus the F-ratio is actually, in a sense,

$$\frac{\text{variability between samples}}{\text{variability within samples}}.$$

If such a null hypothesis as $\mu_A = \mu_B = \mu_C$ is true, then there should not be much "between-sample" variability, as compared with the amount of internal or "within-sample" variability, and the **F-ratio** should be small. Thus we reject such a hypothesis by finding F to be large: specifically, larger than whatever the F-table shows the critical value to be, larger than what can be attributed to chance alone.

Strictly speaking, in applying the F-test approach, we should make some assumptions about the populations involved (in our example, three populations: for each route, all possible driving times):

1. that each population is approximately normally distributed; and
2. that all the population variances (σ^2) are equal (this equality of variances is called *homoscedasticity* in the literature of statistics).

However, the analysis of variance is what we call a "robust" or "sturdy" statistical technique, in that its usefulness, applicability, and accuracy are not greatly impaired if the conditions are somewhat relaxed. That is, if the parent populations (from which the samples are drawn) are not quite normally distributed, or if the variances (say $\sigma_A^2, \sigma_B^2, \sigma_C^2$) are not exactly equal, then even though it would have been optimal to have these somewhat unsatisfied conditions exactly satisfied, we can still generally proceed with the F-test and expect reasonably accurate results.

The F-test is a widely applied statistical technique and is very useful to know about.

EXAMPLES

1. The claim is made that, on the average, sixth-grade students' math scores on final exams are the same at the four elementary schools in Gauss City. That is, the null hypothesis is a statement of the form $\mu_A = \mu_B = \mu_C = \mu_D$.

To test this hypothesis at the level 0.01, an educational researcher randomly samples math scores at each of the four schools, finding the following sample data:

School							Row totals	Sample means
A	67	81	74	92	76	84	474	$\bar{x}_A = 79.0$
B	88	83	98	85	97	95	546	$\bar{x}_B = 91.0$
C	62	72	80	58	75	67	414	$\bar{x}_C = 69.0$
D	74	79	65	70	80	76	444	$\bar{x}_D = 74.0$
						$\sum x =$	1,878	

We note that the sample means are somewhat different; but as usual in statistics, you can't tell by looking. Some specific statistical technique must be used to determine whether the dissimilarity of the sample means is significant, and the appropriate technique here is the F-test.

We compute the sums of squares needed, noting first that $n = 24$, $r = 4$, and $c = 6$:

$$SS_{\text{total}} = \sum (x^2) - \frac{(\sum x)^2}{n}$$

$$= 67^2 + 81^2 + 74^2 + 92^2 + 76^2 + 84^2$$

$$+ 88^2 + 83^2 + 98^2 + 85^2 + 97^2 + 95^2$$

$$+ 62^2 + 72^2 + 80^2 + 58^2 + 75^2 + 67^2$$

$$+ 74^2 + 79^2 + 65^2 + 70^2 + 80^2 + 76^2 - \frac{(1{,}878)^2}{24}$$

$$= 4{,}489 + 6{,}651 + 5{,}476 + 8{,}464 + 5{,}776 + 7{,}056$$
$$+ 7{,}744 + 6{,}889 + 9{,}604 + 7{,}225 + 9{,}409 + 9{,}025$$
$$+ 3{,}844 + 5{,}184 + 6{,}400 + 3{,}364 + 5{,}625 + 4{,}489$$
$$+ 5{,}476 + 6{,}241 + 4{,}225 + 4{,}900 + 6{,}400 + 5{,}776$$
$$- \frac{3{,}526{,}884}{24}$$
$$= 149{,}642 - 146{,}953.5 = 2{,}688.5.$$

$$SS_{\text{factor}} = \frac{\sum (T^2)}{c} - \frac{(\sum x)^2}{n}$$
$$= \frac{474^2 + 546^2 + 414^2 + 444^2}{6} - \frac{(1{,}878)^2}{24}$$
$$= \frac{224{,}676 + 298{,}116 + 171{,}396 + 197{,}136}{6} - \frac{(1{,}878)^2}{24}$$
$$= \frac{891{,}324}{6} - \frac{3{,}526{,}884}{24} = 148{,}554 - 146{,}953.5 = 1{,}600.5$$

And

$$SS_{\text{error}} = SS_{\text{total}} - SS_{\text{factor}} = 2{,}688.5 - 1{,}600.5$$
$$= 1{,}088.0.$$

We enter these results in the ANOVA table (see Figure 6.5) along with the d.f. values

$$\text{d.f.}_{\text{total}} = n - 1 = 24 - 1 = 23$$
$$\text{d.f.}_{\text{factor}} = r - 1 = 4 - 1 = 3$$
$$\text{d.f.}_{\text{error}} = r(c - 1) = 4(6 - 1) = 20.$$

Source	SS	df	MS	*F*
Factor	1,600.5	3	533.5	9.81
Error	1,088.0	20	54.4	
Total	2,688.5	23		

Figure 6.5

And we compute (and enter in the table)

$$MS_{\text{factor}}=\frac{SS_{\text{factor}}}{\text{d.f.}_{\text{factor}}}=\frac{1{,}600.5}{3}=533.5$$

$$MS_{\text{error}}=\frac{SS_{\text{error}}}{\text{d.f.}_{\text{error}}}=\frac{1{,}088.0}{20}=54.4$$

$$F=\frac{MS_{\text{factor}}}{MS_{\text{error}}}=\frac{533.5}{54.4}=9.81.$$

To tell whether this computed F-ratio is large enough to justify a rejection, at the level 0.01, of the null hypothesis $\mu_A=\mu_B=\mu_C=\mu_D$, we must look up the critical or cutoff F-value in the F-table for level 0.01, using the d.f. values

$$\text{d.f.}_{\text{numerator}}=\text{d.f.}_{\text{factor}}=3$$

$$\text{d.f.}_{\text{denominator}}=\text{d.f.}_{\text{error}}=20.$$

In the table we find:

.....3
↓
20 → 4.94

Thus any computed F-value larger than 4.94 is good enough for a rejection. Because $9.81>4.94$ (in fact, 9.81 is considerably in excess of 4.94), we resoundingly reject H_0 at the level 0.01. We are 99% confident, at the very least, in our conclusion that the four population means—the overall math score means for the four schools—are *not* all equal. The sample means have shown differences that, it turns out, are significant at the level 0.01. These differences are too great, under the circumstances, to allow us to believe that the μ's that they approximate are all equal.

It should be pointed out that this rejection indicates that not all the population means, all considered *together*, are equal; if we wish more specifically to compare, say, μ_A and μ_B only with each other, we should run a separate statistical test (probably a t-test) on only those two.

2. It is hypothesized that the overall average (mean) age of people using the special collections rooms at three large urban libraries, in three different cities, is uniform from library to library. To test this hypothesis (H_0: $\mu_A=\mu_B=\mu_C$) at the level 0.05, a researcher randomly samples the library patrons who use the special collections rooms at the three locations, recording the ages of the people sampled as follows:

									Row totals	Sample means
A:	20	32	26	18	35	25	40	52	248	$\bar{x}_A = 31.0$
B:	17	29	30	37	46	60	19	30	268	$\bar{x}_B = 33.5$
C:	33	44	21	58	42	24	27	35	284	$\bar{x}_C = 35.5$
								$\sum x =$	800	

The differences among the sample means $\bar{x}_A$, $\bar{x}_B$, $\bar{x}_C$ may or may not be significant at the level 0.05: it may still be that the population means which they represent are equal. (There is always some chance fluctuation in sampling, and it may be that alone which accounts for the observed differences; but a rejection would argue otherwise.) We proceed with the F-test.

$$SS_{\text{total}} = \sum (x^2) - \frac{(\sum x)^2}{n}$$

$$= 20^2+32^2+26^2+18^2+35^2+25^2+40^2+52^2$$
$$+17^2+29^2+30^2+37^2+46^2+60^2+19^2+30^2$$
$$+33^2+44^2+21^2+58^2+42^2+24^2+27^2+35^2$$
$$-\frac{800^2}{24}$$

$$= 400+1{,}024+676+324+1{,}225+625+1{,}600+2{,}704$$
$$+289+841+900+1{,}369+2{,}116+3{,}600+361+900$$
$$+1{,}089+1{,}936+441+3{,}364+1{,}764+576+729+1{,}225$$
$$-\frac{800^2}{24}$$

$$= 30{,}078 - \frac{640{,}000}{24} = 30{,}078 - 26{,}666.667$$

$$= 3{,}411.333.$$

$$SS_{\text{factor}} = \frac{\sum (T^2)}{c} - \frac{(\sum x)^2}{n}$$

$$= \frac{248^2+268^2+284^2}{8} - \frac{800^2}{24}$$

$$= \frac{61{,}504+71{,}824+80{,}656}{8} - \frac{640{,}000}{24}$$

$$= \frac{213{,}984}{8} - \frac{640{,}000}{24}$$

$$= 26{,}748 - 26{,}666.667 = 81.333.$$

Source	SS	df	MS	F
Factor	81.333	2	40.667	0.26
Error	3,330	21	158.571	
Total	3,411.333	23		

Figure 6.6

$$SS_{\text{error}} = SS_{\text{total}} - SS_{\text{factor}}$$
$$= 3{,}411.333 - 81.333 = 3{,}330.$$

We enter these results, and the d.f. values

$$\text{d.f.}_{\cdot\text{total}} = n - 1 = 24 - 1 = 23$$

$$\text{d.f.}_{\cdot\text{factor}} = r - 1 = 3 - 1 = 2$$

$$\text{d.f.}_{\cdot\text{error}} = r(c-1) = 3(8-1) = 21$$

into the ANOVA table (see Figure 6.6) and compute

$$MS_{\text{factor}} = \frac{SS_{\text{factor}}}{\text{d.f.}_{\cdot\text{factor}}} = \frac{81.333}{2} \approx 40.667$$

$$MS_{\text{error}} = \frac{SS_{\text{error}}}{\text{d.f.}_{\cdot\text{error}}} = \frac{3{,}330}{21} \approx 158.571$$

$$F = \frac{MS_{\text{factor}}}{MS_{\text{error}}} = \frac{40.667}{158.571} \approx 0.26.$$

Actually, when the computed value of F turns out to be less than 1, it may be concluded immediately that there is no rejection. But to follow through the usual procedure, finding the critical F-value in the $F(0.05)$ table with $\text{d.f.}_{\cdot\text{numerator}} = \text{d.f.}_{\cdot\text{factor}} = 2$ and with $\text{d.f.}_{\cdot\text{denominator}} = \text{d.f.}_{\cdot\text{error}} = 21$, we obtain $F = 3.47$. Because $0.26 < 3.47$, we do not reject, at the level 0.05, the null hypothesis $\mu_A = \mu_B = \mu_C$. The sample information does not provide an adequate argument against the claim that the population means (average ages) are all the same.

NEW TERMS IN THIS SECTION

analysis of variance (ANOVA)
***F*-test**
row totals
sums of squares: SS_{factor}, SS_{error}, SS_{total}
factor
levels of a factor
ANOVA table
degrees of freedom: $\text{d.f.}_{\cdot\text{factor}}$ (or $\text{d.f.}_{\cdot\text{numerator}}$), $\text{d.f.}_{\cdot\text{error}}$ (or $\text{d.f.}_{\cdot\text{denominator}}$), $\text{d.f.}_{\cdot\text{total}}$
mean squares: MS_{factor}, MS_{error}
***F*-ratio**
partitioning the squares
***two-way ANOVA**
***factorial design**
***interaction**

EXERCISES 6-A

1. It is claimed that when psychological examinees from three different areas are tested on a given item and timed (in seconds) on their completion of it, the three groups of examinees on the overall average will score the same. To test this null hypothesis (H_0: $\mu_A = \mu_B = \mu_C$) at the level of significance $\alpha = 0.05$, a psychologist randomly samples six from each group, with the following completion times:

Group A:	8	10	11	7	12	6
Group B:	11	15	9	7	10	14
Group C:	10	7	13	4	9	5

What are the sample means? Construct an ANOVA table. What is the computed F-ratio, and what is the conclusion?

2. The claim is made that four drugs (all designed to lower systolic blood pressure) all work, on the average, uniformly well compared to each other, in the sense that they lower blood pressure by the same amount. A medical researcher wishes to test this hypothesis at the level 0.01 and samples eight patients for each drug, recording the number of millimeters their blood pressure has been lowered:

Drug A:	7	9	6	4	8	3	10	5
Drug B:	14	10	11	9	13	8	12	11
Drug C:	6	7	8	10	9	5	7	8

What are the sample means? Are they significantly different? (That is, do we reject the null hypothesis $\mu_A = \mu_B = \mu_C$?)

3. The president of a company with four separate plants claims that on the average the distances which employees drive to work are the same at all four plants. To test this hypothesis at the level 0.05, a researcher randomly samples six employees at each plant, recording their commuting distances (in miles) as follows:

Plant A:	9	12	3	5	9	7
Plant B:	10	7	8	2	11	4
Plant C:	12	9	14	7	15	15
Plant D:	6	3	10	12	9	8

Compute the sample means. Are they significantly different, so that the hypothesis $\mu_A = \mu_B = \mu_C = \mu_D$ can be rejected?

4. In Exercise 6-A.3, suppose each person sampled had driven 1 mile less than the original number reported. Redo the F-test; what is the conclusion? Why does this happen?

5. It is claimed that the plums from three different distributors are, on the average, the same size. To test this hypothesis at the level 0.01, a researcher randomly samples 11 plums from each source, recording their diameters in centimeters:

Source A: 5 4 5 6 3 7 5 4 4 6 5

Source B: 4 5 6 4 4 7 6 5 4 4 5

Source C: 7 8 7 6 8 5 8 7 6 7 7

What are the sample means? What is the result of an F-test? What would you do if you owned an interest in the plums from source C?

6. A computer firm asserts that, for four types of computer, on the average the amounts of time per month which the computers spend "down" in need of repair are the same for all three types. To test the hypothesis at the level 0.05, a market analyst randomly samples computers of all four types, recording hours of downtime as follows:

Type A: 4 6 5 7 4

Type B: 3 7 4 9 6

Type C: 2 7 5 5 8

Type D: 4 7 7 5 4

What are the sample means? What can be concluded from an F-test?

7. A consumer agency claims that of four brands of flashlight batteries, on the average all will work for the same amount of time in continuous use before burning out. An analyst decides to test this claim at the level 0.01, sampling seven batteries of each brand and recording how many continuous hours they work:

Brand A: 7 6 7 8 9 8 9

Brand B: 10 11 10 13 9 9 12

Brand C: 6 4 5 7 5 4 8

Brand D: 7 4 5 7 6 7 4

Compute the sample means. What can be concluded from an F-test about the hypothesis $\mu_A = \mu_B = \mu_C = \mu_D$?

8. It is claimed that employees in four departments of a corporation have the same mean weight. To test this claim at the level 0.05, a statistician randomly samples six people in each department, recording their weights (in kilograms) as follows:

Department A: 72 60 78 68 67 63

Department B: 78 73 69 80 63 75

Department C: 65 74 70 67 79 77

Department D: 74 71 63 77 69 72

What are the sample means? Are they significantly different from department to department at the level of significance 0.05?

9. To test the hypothesis that three assembly-line workers Jones, Brown, and Smith all perform a certain task, on the overall average, in the same amount of time, an efficiency analyst observes each worker nine times at random, timing the work in minutes, recording the following results:

Jones:	13	17	18	15	10	12	16	16	18
Brown:	9	10	12	8	8	13	11	10	9
Smith:	16	14	19	21	15	17	13	20	18

What are the sample means? Are they significantly different at the level 0.01?

10. It is claimed that the ancient Babylonian coins made in four different historical periods weigh, on the average, the same for all four periods. To test this hypothesis at the level 0.05, an archaeologist examines eight coins from each period, finding the following weights (in grams):

Period A:	2.8	3.1	3.0	2.7	3.3	2.6	3.0	2.7
Period B:	3.0	2.7	2.6	2.9	2.5	2.9	3.2	2.6
Period C:	3.4	2.9	3.0	3.6	3.2	3.0	2.8	3.7
Period D:	2.8	3.1	2.7	3.4	2.9	3.2	2.8	3.1

Compute the sample means. From an F-test, what can be concluded? (One of the d.f.'s will not be found in the table, but from the nearby values you should be able to tell what conclusion to draw; see also Section 5-D for interpolation.)

11. To test whether four experimental rocket engines all burn, on the average, a standard fuel package in the same amount of time, an engineer tests each engine seven times, obtaining the following times (in seconds):

Engine A:	11	13	12	11	12	13	12
Engine B:	10	12	11	13	11	10	10
Engine C:	12	13	11	13	13	12	10
Engine D:	9	10	11	10	11	9	10

Compute the sample means. Are they significantly different at the level $\alpha = 0.01$?

12. To test the hypothesis that three different kinds of maze can be run by laboratory mice in the same time, on the average, a laboratory technician samples each maze by running mice

through them and timing the mice in seconds, obtaining these results:

Maze A: 24 29 21 23 31 22 25

Maze B: 21 19 23 27 23 24 17

Maze C: 29 32 26 28 33 24 24

What are the sample mean times? Are they significantly different at the 0.05 level?

13. In Exercise 6-A.12, what if the F-test had been done at the level of significance 0.01? What does this result mean, compared with the result of Exercise 6-A.12?

14. To test the hypothesis that children of four different ethnic groups all start to walk, on the average, at the same age, a researcher randomly samples children in each group, recording the ages (in months) at which they learn to walk:

Group A: 12 13 11 14 12 13 15 13 14 16 10

Group B: 14 10 12 12 11 15 11 13 12 10 12

Group C: 11 12 10 12 13 15 11 12 13 13 10

Group D: 10 15 15 13 15 12 11 13 14 14 11

Compute the sample means. At the 0.05 level of significance, what can be concluded by running an F-test?

15. It is claimed that the employees of three different large research firms have, on the average, the same amount of graduate-degree training. To test this hypothesis at the level 0.05, a personnel analyst samples employees, recording how many years of graduate school they have:

Firm A: 1 3 2 3 3 4 1 5 3 2 2 2 3 6 4 4
3 1 2 5 4

Firm B: 2 5 4 3 2 4 4 6 4 5 4 3 3 2 5 4
5 5 3 1 3

Firm C: 1 5 2 4 3 4 5 5 3 4 2 2 2 6 3 4
3 4 1 4 3

What are the sample means? As analyzed by an F-test at the level 0.05, are they significantly different?

16. Someone has hypothesized that the used cars sold by the five used-car dealers in Fisher City all run, on the average, the same amount of time after purchase before needing repair, regardless of which dealer sells them. To test this hypothesis at the level $\alpha = 0.05$, a consumer-interest research group randomly samples cars purchased at the various dealers (A, B, C, D, E),

recording how many months they run without repairs:

A:	4	9	5	10	8	11	2
B:	5	1	3	9	7	7	3
C:	8	2	6	4	4	8	10
D:	3	9	7	3	4	8	1
E:	10	3	9	9	5	4	9

What are the sample means? At the 0.05 level of significance, what does an *F*-test show? What does the result mean?

17. For a rejection, what critical value would the computed *F*-ratio have to exceed, in an *F*-test at the level 0.05, for

(a) six samples, each of size 11?
(b) eight samples, each of size 6?
(c) four samples, each of size 16?
(d) ten samples, each of size 5?
(e) seven samples, each of size 4?

18. In Exercise 6-A.17, what critical value would have to be exceeded each time for a rejection at the level 0.01? Compare your answers with those of Exercise 6-A.17.

PONDERABLES 6-A

1. (*Factorial design and interaction.*) The analysis of variance procedure examined in this section is called "single-factor" or one-way ANOVA because it involves only one factor or consideration. For example, in the illustration at the beginning of the section, the factor was: *which route* the commuter drives to work. But suppose the analysis had taken into account not only that, but also which *day* the driving was done. One might set up a dual analysis as follows, with *two* factors (route and day):

	Monday	Tuesday	Wednesday	Thursday	Friday
Route *A*	—	—	—	—	—
Route *B*	—	—	—	—	—
Route *C*	—	—	—	—	—

with, say, six sample driving times in each of the 15 cells in the table. (The commuter drives Route *A* on six Mondays, Route *A* on six Tuesdays, and so on.) One could then ask (1) whether the routes are the same or significantly different, (2) whether the days of the week are the same for driving or significantly different, and (3) whether there is or is not any significant *joint* effect (called **interaction**) produced by the two factors taken together. Such analysis is called *two-way analysis of variance* **(two-way ANOVA).**

This sort of table, with one factor displayed against another factor, is called **factorial design.** We would call this particular one a three-by-five (3×5) factorial design, because the first factor

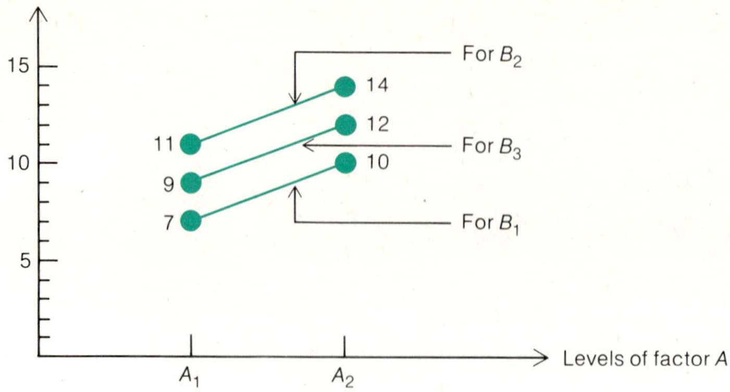

Figure 6.7

(route) has *three levels* (there are three routes) and the second factor (day) has *five levels* (there are five days of the week considered).

For a simpler working illustration, let us look at a 2×3 factorial design.

Suppose we are interested in testing how two greatly different room temperatures (A_1 and A_2) affect psychological examinees' timed responses in seconds to a certain test item *and* how certain experimental drugs (B_1, B_2, B_3) affect the responses. That is, we have a "factor A" (temperature) with two levels and a "factor B" (drug) with three levels. Suppose we test (say) four people in each combination (at temperature A_1 with drug B_1, at temperature A_1 with drug B_2, and so on) and get *sample means* in each category as follows:

	B_1	B_2	B_3
A_1	7	11	9
A_2	10	14	12

If we graph this information as shown in Figure 6.7, we see that there is a parallelism: the mean values of the levels of factor A behave alike, in their changes, no matter what B-level we are looking at: the change from temperature A_1 to temperature A_2 *always* is seen to involve an increase of 3 seconds, no matter which drug (B_1, B_2, or B_3) has been used. For that matter, we could have graphed the B-factor level changes as shown in Figure 6.8, where similarly this

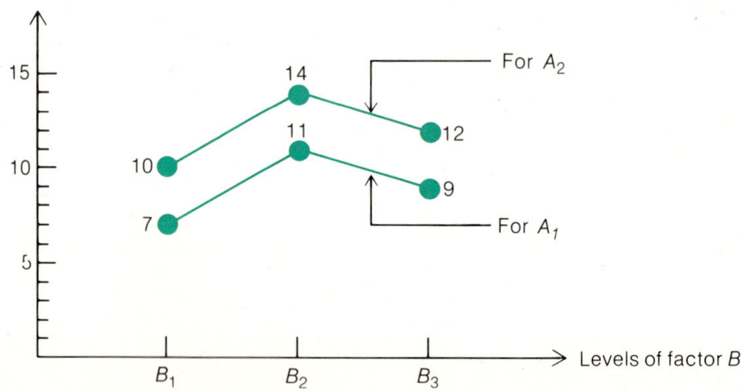

Figure 6.8

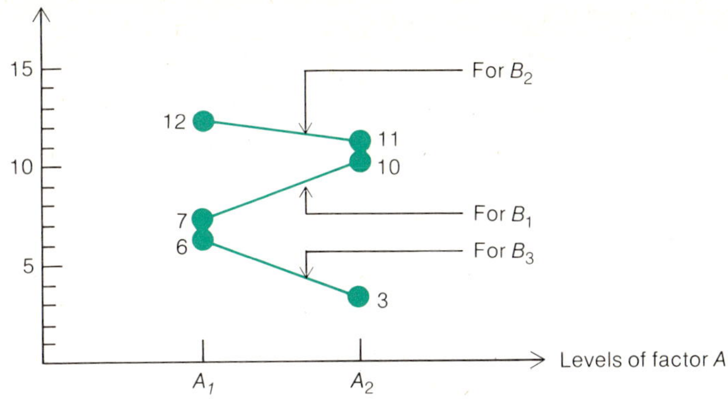

Figure 6.9

parallel "chevron"-type graph shows that the changes in the B-levels (that is, the changes in the mean times from one drug to another) behave similarly no matter which A-level we look at: When we change from drug B_1 to drug B_2 to drug B_3 the time always goes first up 4 seconds and then down 2 seconds, regardless of which room temperature (A_1 or A_2) is in effect. In this parallel graphical behavior, the factors A and B are said simply to behave "additively" toward each other, and we say that there is *no interaction* shown: Neither factor essentially seems to affect the manner in which the other factor exhibits changes from level to level.

But on the other hand, suppose the sample means had been:

	B_1	B_2	B_3
A_1	7	12	6
A_2	10	11	3

Then (for example) graphing the changes in the levels of factor A would produce a picture as shown in Figure 6.9, where the constituents of the graph are *not* parallel. It would appear that factor A's levels (sample means at different temperatures) do *vary differently*, depending on what drug was given. Or if we graph the other way around as in Figure 6.10, again we do *not* get

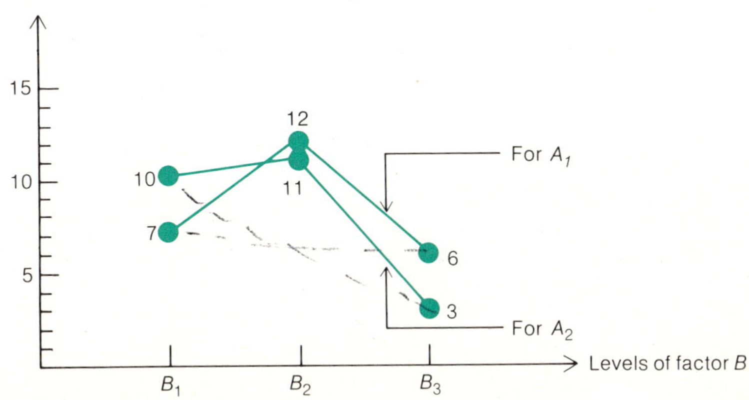

Figure 6.10

parallel behavior, and it would appear that factor B's levels (sample means with different drugs) do vary differently, depending on what the room temperature is.

Either way, to the extent that the graphs depart from parallel behavior, we say that there is an indication of *interaction* between the factors. That is, to the extent that to see how sample means vary from temperature to temperature we must know which drug was given, or to the extent that to see how sample means vary from drug to drug we must know which temperature is used—to the extent that these things matter, there is interaction between the factors.

The question always is: Do we have *significant* indication of interaction? Do the graphs depart *enough* from parallel behavior for us to say that the factors interact with each other significantly? By computing a certain kind of F-ratio as part of the procedure of two-way analysis of variance, we can tell (at a given level of significance) whether the indication of interaction is significant. The next Ponderables problem shows how this is done.

2. (*Two-way* ANOVA.) Suppose, as in problem 1, we have a two-factor experiment in which factor A is the room temperature (say A_1 cold, A_2 hot) and factor B is the drug administered (B_1, B_2, or B_3), so that, as in the previous problem, the factorial design is 2×3. In each of the six categories, suppose four examinees are timed (in seconds) by an experimental psychologist in their responses to a test item, with the sample data, together with the total and sample mean, for each cell in the table as in Figure 6.11.

Now we may ask three questions, computing a certain kind of F-ratio to answer each. (At the outset, let us assume that the level of significance is $\alpha = 0.05$.) The questions are:

(a) Are the results significantly different at different temperatures? (This question ignores factor B and calls for essentially a one-way ANOVA approach.)

(b) Are the results significantly different for different drugs? (This question ignores factor A and calls for essentially a one-way ANOVA approach, but a different computation than that for asking about temperature.)

(c) Do factors A and B significantly *interact*? That is, does the pattern of temperature-oriented variation depend on which drug is used, and does the pattern of drug-oriented variation depend upon what temperature is used? (This question calls for a two-way ANOVA computation.)

Factors (Temperature ↓, Drug →)	Drug B_1	Drug B_2	Drug B_3
Temperature A_1	8, 4, 6, 2; $T = 20$; $\bar{x} = 5.0$	9, 8, 11, 12; $T = 40$; $\bar{x} = 10.0$	3, 2, 4, 3; $T = 12$; $\bar{x} = 3.0$
Temperature A_2	8, 7, 10, 7; $T = 32$; $\bar{x} = 8.0$	9, 6, 12, 9; $T = 36$; $\bar{x} = 9.0$	3, 6, 2, 5; $T = 16$; $\bar{x} = 4.0$

Figure 6.11 Two-factor experiment with 2×3 factorial design.

	B_1	B_2	B_3	
A_1	x: 8, 4, 6, 2 x^2: 64, 16, 36, 4 $T = 20$ 120 $\bar{x} = \frac{20}{4} = 5.0$	x: 9, 8, 11, 12 x^2: 81, 64, 121, 144 $T = 40$ 410 $\bar{x} = \frac{40}{4} = 10.0$	x: 3, 2, 4, 3 x^2: 9, 4, 16, 9 $T = 12$ 38 $\bar{x} = \frac{12}{4} = 3.0$	$T_{A_1} = \Sigma\,(x_{A_1}) = 72$ $\Sigma\,(x^2_{A_1}) = 568$ $\bar{x}_{A_1} = \frac{72}{12} = 6.0$
A_2	x: 8, 7, 10, 7 x^2: 64, 49, 100, 49 $T = 32$ 262 $\bar{x} = \frac{32}{4} = 8.0$	x: 9, 6, 12, 9 x^2: 81, 36, 144, 81 $T = 36$ 342 $\bar{x} = \frac{36}{4} = 9.0$	x: 3, 6, 2, 5 x^2: 9, 36, 4, 25 $T = 16$ 74 $\bar{x} = \frac{16}{4} = 4.0$	$T_{A_2} = \Sigma\,(x_{A_2}) = 84$ $\Sigma\,(x^2_{A_1}) = 678$ $\bar{x}_{A_2} = \frac{84}{12} = 7.0$
	$T_{B_1} = \Sigma\,(x_{B_1}) = 52$ $\Sigma(x^2_{B_1}) = 382$ $\bar{x}_{B_1} = \frac{52}{8} = 6.5$	$T_{B_2} = \Sigma(x_{B_2}) = 76$ $\Sigma(x^2_{B_2}) = 752$ $\bar{x}_{B_2} = \frac{76}{8} = 9.5$	$T_{B_3} = \Sigma(x_{B_3}) = 28$ $\Sigma(x^2_{B_3}) = 112$ $\bar{x}_{B_3} = \frac{28}{8} = 3.5$	$\sum_{\text{all}} (x) = 156$ $\sum_{\text{all}} (x^2) = 1{,}246$ $\bar{x}_{\text{grand}} = \frac{156}{24} = 6.5$

Figure 6.12

To do all the necessary computation we need to record a number of things. First, as shown in Figure 6.12 (an expansion of Figure 6.11 to include the necessary items), we need, for each cell, the *squares* of the sample times and the sum of their squares. Then, looking horizontally, we need, at the A_1 level, the sum $T_{A_1} = \sum (x_{A_1})$, that is, the sum of the T's in those three cells: the sum of all twelve x's at that A-level. We also need $\sum (x^2_{A_1})$, the sum of the squares of all twelve x's at that A-level, which can be found by adding the three cell-sums of squares already computed. Also, we compute an A_1 *mean* ($\bar{x}_{A_1}$) by dividing T_{A_1} by the number of sample items (12) at that level. (It will be assumed essential that all cells contain the same number of sample items: in this case, 4. Sample sizes *can* vary, but it complicates the procedure for them to do so.) Then we repeat these computations for row A_2.

Next, looking vertically now, we do a similar kind of thing for each column. In column B_1 we add the cell-sums T, getting $T_{B_1} = \sum (x_{B_1})$, the sum of all the x's in that column (eight of them). We also compute $\sum (x^2_{B_1})$, the sum of the squares of all those x's in that column (obtained by adding the two cell-sums of squares already computed). Also we compute a mean $\bar{x}_{B_1}$ for column B_1 by dividing T_{B_1} by the number of sample items (8) in that column. And we repeat these computations for column B_2 and column B_3.

Finally, to complete the information in Figure 6.12, we compute $\sum_{\text{all}} (x)$, the sum of all 24 sample x's in the whole table; and $\sum_{\text{all}} (x^2)$, the sum of all the squares of the x's in the whole table (both computations, as you will see, can be done by adding row results or adding column results—in fact, do it both ways as a check); and the grand mean for the table

$$\bar{x}_{\text{grand}} = \frac{\sum_{\text{all}} (x)}{N_{\text{total}}}$$

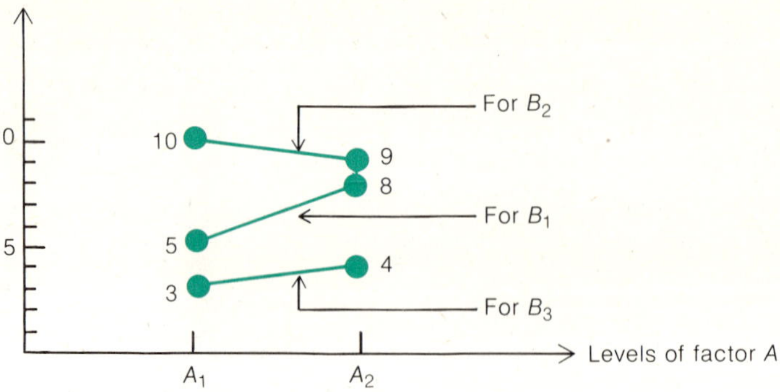

Figure 6.13

where N_{total} is the number of x's for the whole table (24). Study Figure 6.12 *very carefully* to be sure that you see where each computation comes from!

Now, as indicated before, we may compute three separate kinds of F-ratio from the table. We can compute F_A, the F-ratio that will test the null hypothesis $H_0(A)$ that the A-level sample means ($\bar{x}_{A_1} = 6.0$, $\bar{x}_{A_2} = 7.0$) are not significantly different. This is really a one-way ANOVA test in itself. We can also compute F_B, the F-ratio that will test the null hypothesis $H_0(B)$ that the B-level sample means ($\bar{x}_{B_1} = 6.5$, $\bar{x}_{B_2} = 9.5$, $\bar{x}_{B_3} = 3.5$) are not significantly different. Again, this is really in itself a one-way ANOVA. And we can compute the F-ratio F_{AB}, which will test the null hypothesis $H_0(AB)$ that states that the departure from parallel behavior in Figure 6.13 (the graph of the A-level mean changes at different B-levels) is not significant.

The computations bear a strong resemblance to those of this section in which we computed F-ratios for only one factor (one-way ANOVA). But because more than one factor will now be involved, we do change some of the notations. Where before we spoke of SS_{factor}, the variability (sum of squares) due to the single factor, now we will use the symbol $SS_{between}$ to denote variability arising between (among) the various samples. And instead of SS_{error}, we will use the symbol SS_{within} to suggest the variation arising "internally"—*within* the samples themselves. The symbols SS_A and SS_B will stand for the "factor"-type sums computed to derive F_A and F_B respectively, and the symbol SS_{AB} will be the sum computed to derive the "interaction" F-ratio F_{AB}. The letter n will stand for the sample size in each cell in the design: here $n = 4$. Also, n_A will be the number of sample items in each A-level: here $n_A = 12$. And n_B is the number of sample items in each B-level: here $n_B = 8$.

Study the following computations very closely, observing how all the details are taken from the 2×3 design table in Figure 6.12.

$$SS_{total} = \sum_{all} (x^2) - \frac{(\sum_{all} x)^2}{N_{total}}$$

$$= 1{,}246 - \frac{156^2}{24} = 1{,}246 - \frac{24{,}336}{24}$$

$$= 1{,}246 - 1{,}014 = 232.$$

$$SS_{between} = \frac{\sum(T^2)}{n} - \frac{(\sum_{all} x)^2}{N_{total}}$$

$$= \frac{20^2+40^2+12^2+32^2+36^2+16^2}{4} - 1{,}014$$

$$= \frac{4{,}720}{4} - 1{,}014 = 1{,}180 - 1{,}014 = 166.$$

$$SS_{within} = SS_{total} - SS_{between} = 232 - 166 = 66.$$

$$SS_A = \frac{\sum(T_A^2)}{n_A} - \frac{(\sum_{all} x)^2}{N_{total}}$$

$$= \frac{72^2+84^2}{12} - 1{,}014$$

$$= \frac{12{,}240}{12} - 1{,}014 = 1{,}020 - 1{,}014 = 6.$$

$$SS_B = \frac{\sum(T_B^2)}{n_B} - \frac{(\sum_{all} x)^2}{N_{total}}$$

$$= \frac{52^2+76^2+28^2}{8} - 1{,}014$$

$$= \frac{9{,}264}{8} - 1{,}014 = 1{,}158 - 1{,}014 = 144.$$

$$SS_{AB} = SS_{between} - SS_A - SS_B$$

$$= 166 - 6 - 144 = 16.$$

Now the various d.f. values and mean squares (MS) are computed as follows (where A stands for the number of A-levels, $A=2$, and B stands for the number of B-levels, $B=3$):

degrees of freedom

$$\text{d.f.}_A = A - 1 = 2 - 1 = 1$$

$$\text{d.f.}_B = B - 1 = 3 - 1 = 2$$

$$\text{d.f.}_{AB} = (A-1)(B-1) = 1 \cdot 2 = 2$$

$$\text{d.f.}_{within} = N_{total} - AB = 24 - 6 = 18$$

Mean Squares

$$MS_A = \frac{SS_A}{\text{d.f.}_A} = \frac{6}{1} = 6 \qquad MS_B = \frac{SS_B}{\text{d.f.}_B} = \frac{144}{2} = 72.$$

$$MS_{AB} = \frac{SS_{AB}}{\text{d.f.}_{AB}} = \frac{16}{2} = 8 \qquad MS_{within} = \frac{SS_{within}}{\text{d.f.}_{within}} = \frac{66}{18} \approx 3.667.$$

Finally, the F-ratios:

$$F_A = \frac{MS_A}{MS_{\text{within}}} = \frac{6}{3.667} \approx 1.64.$$

$$F_B = \frac{MS_B}{MS_{\text{within}}} = \frac{72}{3.667} \approx 19.63.$$

$$F_{AB} = \frac{MS_{AB}}{MS_{\text{within}}} = \frac{8}{3.667} \approx 2.18.$$

In each case, to check whether these computed F-ratios suggest significant differences at the level 0.05, we compare them with the critical values from the $F_{0.05}$ table, using in each instance the numerator and denominator d.f. values that we employed in computing the particular F-ratio. Thus we

Compare $F_A = 1.64$ with $F_{0.05}$ for d.f. 1, 8: $F_{0.05} = 4.41$.

Compare $F_B = 19.63$ with $F_{0.05}$ for d.f. 2,18: $F_{0.05} = 3.55$.

Compare $F_{AB} = 2.18$ with $F_{0.05}$ for d.f. 2,18: $F_{0.05} = 3.55$.

Because

1. $F_A = 1.64 < 4.41$,
2. $F_B = 19.63 > 3.55$, and
3. $F_{AB} = 2.18 < 3.55$,

we see that

a. We do *not* reject $H_0(A)$; the sample means for temperature conditions are not significantly different. As far as is known, people score essentially the same at either of the two temperatures.
b. We *do* reject $H_0(B)$; the sample means for various drugs *are* significantly different. That is, people do score differently for different drugs.
c. We do *not* reject $H_0(AB)$; as far as is known (that is, as far as has been shown), there is no significant *interaction*: no significant departure of Figure 6.13's graph from parallel behavior. As far as we know, the means *vary* in about the same manner from one temperature level to another regardless of the drug being administered; and the means vary in about the same manner from one drug condition to another regardless of which temperature level is used. That is, it has not been shown that temperature interacts with drug choice.

3. Using problem 2 as a working model, but taking into account the differences in the dimensions of the factorial design and the sample size, work the following two-way ANOVA problem, using a level of significance $\alpha = 0.05$. (Because of one of the d.f.'s you will not directly find the critical F-values in the table, but you can either tell from nearby values what conclusion to draw or can review Section 5-D if necessary for interpolation.)

	B_1	B_2	B_3	B_4
A_1	5, 8, 9, 6, 7	4, 6, 5, 3, 2	7, 8, 10, 9, 11	10, 6, 12, 7, 5
A_2	8, 8, 12, 7, 10	10, 9, 12, 11, 13	7, 3, 4, 8, 8	2, 4, 5, 2, 2
A_3	9, 7, 11, 12, 11	13, 11, 10, 12, 14	12, 8, 12, 11, 12	3, 6, 5, 10, 11

Figure 6.14

Again, as in the previous illustration, a psychologist wishes to time examinees in seconds on their responses to a test item under various circumstances. The first factor A is the amount of time that has elapsed since the examinees awoke from a night's sleep:

A_1: just awoke

A_2: awoke 4 hours before

A_3: awoke 8 hours before

The second factor B is which one of four experimental drugs (B_1, B_2, B_3, or B_4) has been administered. Five persons are tested in each category, with timed results as shown in Figure 6.14.

Draw a graph of the A-level variations. State the null hypotheses $H_0(A)$, $H_0(B)$, and $H_0(AB)$. Compute all the following as shown in Figure 6.15.

	B_1	B_2	B_3	B_4	
A_1	$T=?$ $\bar{x}=?$ $\Sigma(x^2)=?$	$T=?$ $\bar{x}=?$ $\Sigma(x^2)=?$	$T=?$ $\bar{x}=?$ $\Sigma(x^2)=?$	$T=?$ $\bar{x}=?$ $\Sigma(x^2)=?$	$T_{A_1}=?$ $\Sigma(x^2_{A_1})=?$ $\bar{x}_{A_1}=?$
A_2	$T=?$ $\bar{x}=?$ $\Sigma(x^2)=?$	$T=?$ $\bar{x}=?$ $\Sigma(x^2)=?$	$T=?$ $\bar{x}=?$ $\Sigma(x^2)=?$	$T=?$ $\bar{x}=?$ $\Sigma(x^2)=?$	$T_{A_2}=?$ $\Sigma(x^2_{A_2})=?$ $\bar{x}_{A_2}=?$
A_3	$T=?$ $\bar{x}=?$ $\Sigma(x^2)=?$	$T=?$ $\bar{x}=?$ $\Sigma(x^2)=?$	$T=?$ $\bar{x}=?$ $\Sigma(x^2)=?$	$T=?$ $\bar{x}=?$ $\Sigma(x^2)=?$	$T_{A_3}=?$ $\Sigma(x^2_{A_3})=?$ $\bar{x}_{A_3}=?$
	$T_{B_1}=?$ $\Sigma(x^2_{B_1})=?$ $\bar{x}_{B_1}=?$	$T_{B_2}=?$ $\Sigma(x^2_{B_2})=?$ $\bar{x}_{B_2}=?$	$T_{B_3}=?$ $\Sigma(x^2_{B_3})=?$ $\bar{x}_{B_3}=?$	$T_{B_4}=?$ $\Sigma(x^2_{B_4})=?$ $\bar{x}_{B_4}=?$	$\sum_{\text{all}}(x)=?$ $\sum_{\text{all}}(x^2)=?$ $N_{\text{total}}=?$ $\bar{x}_{\text{grand}}=?$

Figure 6.15

$SS_{\text{total}} = ?$ $SS_{\text{between}} = ?$ $SS_{\text{within}} = ?$

$SS_A = ?$ $SS_B = ?$ $SS_{AB} = ?$

$\text{d.f.}_A = ?$ $\text{d.f.}_B = ?$ $\text{d.f.}_{AB} = ?$ $\text{d.f.}_{\text{within}} = ?$ $\text{d.f.}_{\text{total}} = ?$

$MS_A = ?$ $MS_B = ?$ $MS_{AB} = ?$ $MS_{\text{within}} = ?$

$F_A = ?$ $F_B = ?$ $F_{AB} = ?$

Interpret each of these F-ratio results as rejection or nonrejection of the appropriate hypothesis at the level 0.05.

B CHI-SQUARE

We examined in the previous section whether the means of a number of different populations are equal or not. Such a question may be answered, within specified limits of confidence, by applying an F-test. In this section we will examine the question of whether a number of population *proportions* are equal or not, and we will look at a statistical procedure known as a **chi-square** test. (*Chi* is pronounced "ky" to rhyme with "my," and is the Greek letter χ, so that "chi-square" or "chi-squared" is often written χ^2.)

We illustrate the concept involved and the procedure to be followed in the following situation.

A carnival side-show features a spinning wheel with eight numbers, as shown in Figure 6.16. The patrons are invited to bet money on which number will turn up when the wheel is spun.

We immediately wonder, of course: Is the wheel *fair*? Intuitively, it is not hard to clarify what we mean by "fair"—the game is fair if all numbers on the wheel have equal probabilities of turning up. In fact, the probability of any one number's turning up should be exactly $\frac{1}{8}$. Each number should, ideally, turn up one-eighth of the time in the long run. On the other hand, the wheel may *not* randomly generate the numbers and may therefore *not* be fair: It may be

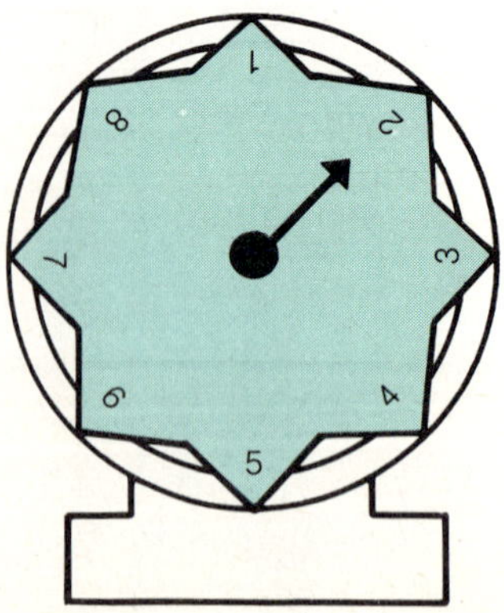

Figure 6.16

imbalanced or weighted so that some numbers turn up more easily than others, or the operator of the wheel may have secret controls by which the wheel can be made to stop at any number the operator desires.

How can we tell whether the wheel is fair or not? Let us state a null hypothesis: that the wheel *is* fair. And let us see how to test this hypothesis at a given level of significance.

We begin, naturally, by sampling: by observing the wheel in operation. Suppose that we watch the spinning of the wheel 800 times. Now, optimally, given the best theoretical behavior of the wheel if it is fair, we could expect it to turn up each number one-eighth of the time, that is:

Number on wheel:	1	2	3	4	5	6	7	8
Number of times observed:	100	100	100	100	100	100	100	100

If *this* happened precisely, we would certainly have *no* argument against the null hypothesis that the wheel is fair. But even if the wheel is fair, we realistically do not expect such perfection as exactly 100 occurrences of each number, because there is always, in sampling, some chance fluctuation. If the wheel turns up

1	2	3	4	5	6	7	8
99	101	101	100	99	100	99	101

we might well not be inclined to argue about its fairness. On the other hand, if the wheel were to turn up

1	2	3	4	5	6	7	8
480	280	8	7	7	7	6	5

then we would almost surely be correct in assuming that the wheel was "rigged" and was turning up on 1 or 2 a very suspicious amount of the time. But where do we draw the line? When is the wheel "close enough" to what we expect if it is fair, and when is it "too far off" for fairness to be believed? As usual in statistics, you can't tell by looking, and some specific statistical procedure is called for.

Let us suppose that our carnival spinner has turned up, in a sample of 800 spins, the following results:

1	2	3	4	5	6	7	8
91	110	89	108	112	93	107	90

We will test, at the level of significance $\alpha = 0.05$, the null hypothesis that the wheel is fair: if we watched it forever, instead of only 800 times in a sample, it would (says the null hypothesis) turn up 1 one-eighth of the time, 2 one-eighth of the time, and so on. That is, H_0 states that the *proportion* of 1's is the same as the proportion of 2's and so forth.

If the null hypothesis is true, then in each category (1, 2, 3, and so on) there is an *expected value* $\frac{1}{8}(800)=100$, and we should record both the **expected values (E)** and the actual **observed values (A)** as follows:

	1	2	3	4	5	6	7	8
Expected value E:	100	100	100	100	100	100	100	100
Observed value A:	91	110	89	108	112	93	107	90

Now the more *different* the A's are from the E's, the more we tend to suspect that H_0 may be false. But how different is *too* different? A certain amount of ragged-looking disparity may be tolerated and chalked up to mere chance fluctuation, but how much?

To proceed, we compute the statistic χ^2:

$$\chi^2=\sum\frac{(A-E)^2}{E}.$$

That is, for each category (1 through 8) we substract E from A, square the difference, and divide by E, then add all eight such results:

$$\begin{aligned}\chi^2&=\frac{(91-100)^2}{100}+\frac{(110-100)^2}{100}+\frac{(89-100)^2}{100}+\frac{(108-100)^2}{100}\\&\quad+\frac{(112-100)^2}{100}+\frac{(93-100)^2}{100}+\frac{(107-100)^2}{100}+\frac{(90-100)^2}{100}\\&=\frac{(-9)^2}{100}+\frac{10^2}{100}+\frac{(-11)^2}{100}+\frac{8^2}{100}\\&\quad+\frac{12^2}{100}+\frac{(-7)^2}{100}+\frac{7^2}{100}+\frac{(-10)^2}{100}\\&=\frac{81}{100}+\frac{100}{100}+\frac{121}{100}+\frac{64}{100}+\frac{144}{100}+\frac{49}{100}+\frac{49}{100}+\frac{100}{100}\\&=\frac{708}{100}=7.08.\end{aligned}$$

Now, the idea is that χ^2 measures *the extent to which the expected values E and the observed values A fail to agree.* (Agreement between the A's and the E's is called **goodness of fit.** χ^2 measures, then, in a manner of speaking, "badness of fit.") But is $\chi^2=7.08$ too large or not? To answer this question we must consult the chi-square table in the back of the book, and specifically use column $\chi^2_{.05}$, because the level of significance for this test, it was agreed from the beginning, is $\alpha = 0.05$. To determine the correct d.f. value (degrees of freedom) we simply subtract one from the number of categories being considered:

$8 - 1 = 7 = \text{d.f.}$

And in the table we find:

d.f. $\chi^2_{.05}$
⋮ ↓
$7 \rightarrow 14.067$

This *critical value* $\chi^2_{.05} = 14.067$ is the value which a computed χ^2 value must *exceed* in order to signal a rejection of the null hypothesis H_0. (That is, χ^2 can be up to 14.067 in size by chance fluctuation, without there being 95% confidence that the E's and the A's are different enough to reject H_0.) Because here we have only

$\chi^2_{\text{computed}} = 7.08 < 14.067,$

we do *not* reject H_0 at the level 0.05. We have not shown, by this sample of 800 spins, that the wheel is unfair: The difference between the expected and observed sets of values is not large enough to declare the wheel unfair with 95% confidence. As in previous situations, we are computing a statistic to see whether it falls within a rejection region, this time under a chi-square curve, as shown in Figure 6.17.

Thus we did not get the rejection that would have allowed us to make a case that the wheel was unfair. But consider the following examples.

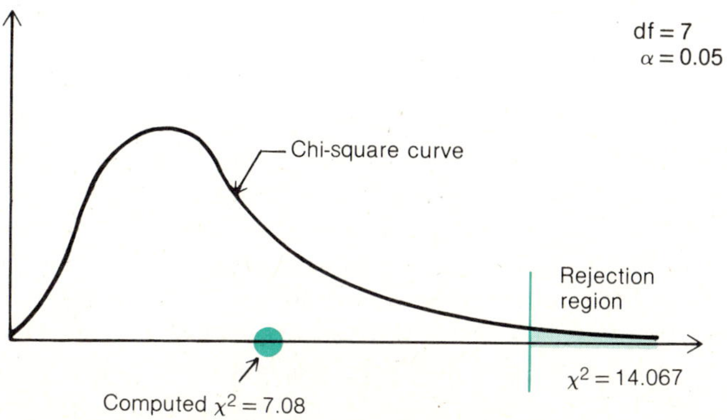

Figure 6.17

EXAMPLES

1. Suppose in the carnival game just described, the results of 800 spins of the wheel had been as follows:

	1	2	3	4	5	6	7	8
A:	87	86	88	90	92	117	119	121
E:	100	100	100	100	100	100	100	100

Now the statistic χ^2 from the sample data observed would be:

$$\chi^2 = \sum \frac{(A-E)^2}{E}$$

$$= \frac{(87-100)^2}{100} + \frac{(86-100)^2}{100} + \frac{(88-100)^2}{100} + \frac{(90-100)^2}{100}$$

$$+ \frac{(92-100)^2}{100} + \frac{(117-100)^2}{100} + \frac{(119-100)^2}{100} + \frac{(121-100)^2}{100}$$

$$= \frac{(-13)^2}{100} + \frac{(-14)^2}{100} + \frac{(-12)^2}{100} + \frac{(-10)^2}{100}$$

$$+ \frac{(-8)^2}{100} + \frac{17^2}{100} + \frac{19^2}{100} + \frac{21^2}{100}$$

$$= \frac{169}{100} + \frac{196}{100} + \frac{144}{100} + \frac{100}{100} + \frac{64}{100} + \frac{289}{100} + \frac{361}{100} + \frac{441}{100}$$

$$= \frac{1{,}764}{100} = 17.64.$$

Again comparing this result with the critical value from the table, for $\alpha = 0.05$ and d.f. $= 8 - 1 = 7$:

$$\chi^2_{\text{computed}} = 17.64 > \chi^2_{.05} = 14.067$$

and we *reject* the null hypothesis that the differences are tolerably small and the wheel is fair. We are 95% confident that the differences are *not* small enough to attribute to chance alone, and that the wheel is unfair.

Note (in the table, under column $\chi^2_{.01}$) that a rejection at the level of significance $\alpha = 0.01$ would have required exceeding 18.475, which our computation 17.64 does not do. Thus we would *not* have a rejection at the level 0.01. Although we can feel 95% confident that the game is unfair, the sample evidence is not quite strong enough to warrant feeling 99% confident in such a conclusion.

2. In the carnival game described in the text, the *proportions* observed for the various numbers were:

Number:	1	2	3	4
Proportion:	$\frac{91}{800}$	$\frac{110}{800}$	$\frac{89}{800}$	$\frac{108}{800}$
	≈ 0.114	≈ 0.138	≈ 0.111	$= 0.135$
	$= 11.4\%$	$= 13.8\%$	$= 11.1\%$	$= 13.5\%$

Number:	5	6	7	8
Proportion:	$\frac{112}{800}$	$\frac{93}{800}$	$\frac{107}{800}$	$\frac{90}{800}$
	$= 0.14$	≈ 0.116	≈ 0.134	≈ 0.112
	$= 14.0\%$	$= 11.6\%$	$= 13.4\%$	$= 11.2\%$

It turned out in this case that these various proportions and percentages did not vary significantly from the uniformly expected $\frac{100}{800} = 0.125 = 12.5\%$, so that we did not reject the null hypothesis. But for the data in Example 1, the reader may verify that the percentages involved are:

1	2	3	4	5	6	7	8
10.9%	10.8%	11.0%	11.2%	11.5%	14.6%	14.9%	15.1%

and the chi-square analysis has shown that at the level 0.05 these percentages *are* significantly different from the uniformly expected 12.5%, so that we *did* reject H_0. These proportions or percentages, at this particular level of significance and at this sample size, cannot be said to differ from the expected this much by chance alone.

Notice that we are really working, then, with *proportions*, and not just with how many too few or too many occurrences of each number we observe. In the data of Example 1 we had:

1	2	3	4	5	6	7	8
87:	86:	88:	90:	92:	117:	119:	121:
13	14	12	10	8	17	19	21
too	too	too	too	too	too	too	too
few	few	few	few	few	many	many	many

We got a rejection here when we computed χ^2. But these same amounts "too few" or "too many" might well *not* be significant at a larger sample size. For example, if the wheel were observed 1,600 times (as in Figure 6.18), and if the same numbers "too few" or "too many" were observed, then a chi-square

	1	2	3	4	5	6	7	8
A:	187: 13 too few	186: 14 too few	188: 12 too few	190: 10 too few	192: 8 too few	217: 17 too many	219: 19 too many	221: 21 too many
E:	200	200	200	200	200	200	200	200

Figure 6.18

computation would give

$$\chi^2 = \frac{(187-200)^2}{200} + \frac{(186-200)^2}{200} + \frac{(188-200)^2}{200} + \frac{(190-200)^2}{200}$$

$$+\frac{(192-200)^2}{200} + \frac{(217-200)^2}{200} + \frac{(219-200)^2}{200} + \frac{(221-200)^2}{200}$$

$$= \frac{169}{200} + \frac{196}{200} + \frac{144}{200} + \frac{100}{200} + \frac{64}{200} + \frac{289}{200} + \frac{361}{200} + \frac{441}{200}$$

$$= \frac{1{,}764}{200} = 8.82 < 14.067$$

with *no* rejection of the null hypothesis at the level 0.05. The idea here is that with so many observations (1,600 instead of 800) it is easier for there to be a given amount of variation, in any observation, from the expected, even if the wheel is fair. With this larger sample, the *proportions* are different from the ones obtained earlier:

Number:	1	2	3	4
Proportion:	$\frac{187}{1600}$	$\frac{186}{1600}$	$\frac{188}{1600}$	$\frac{190}{1600}$
	≈ 0.117	≈ 0.116	≈ 0.118	≈ 0.119
	$= 11.7\%$	$= 11.6\%$	$= 11.8\%$	$= 11.9\%$

Number:	5	6	7	8
Proportion:	$\frac{192}{1600}$	$\frac{217}{1600}$	$\frac{219}{1600}$	$\frac{221}{1600}$
	$= 0.12$	≈ 0.136	≈ 0.137	≈ 0.138
	$= 12.0\%$	$= 13.6\%$	$= 13.7\%$	$= 13.8\%$

and the chi-square test shows that at the level 0.05 these proportions do *not* differ significantly from the uniformly expected $\frac{200}{1{,}600} = 0.125 = 12.5\%$.

Chi-square analysis also applies to a type of format called a **contingency table.**

Suppose we go to three different cities (say Boston, Chicago, and Los Angeles) and randomly sample people in each city, asking them which of two candidates, *C* or *D*, they would vote for in an election. In each city we record the number of responses in each category:

	Boston	Chicago	Los Angeles		
Prefer *C*:	130	180	290	→	600
Prefer *D*:	70	120	210	→	400
	total 200	total 300	total 500		Grand total 1,000

This display of data is our contingency table: a two-by-three table (two rows, three columns). We wish to know (as political analysts, say) whether the *C* or *D* choice tends to be made the same or differently, proportionally, in the different cities. The null hypothesis H_0 will state that the *C* or *D* choice is, overall, made the same way in all three cities; that is, in terms of proportions of people who would prefer *C* or would prefer *D*, the relative popularity of each candidate is uniform from city to city. In effect, the people sampled have been classified by two different modes of classification: political and geographical—and the null hypothesis states that the two modes of classification are independent of each other.

To test this hypothesis at (say) the level 0.05, we do the following computations.

First, to go along with the *observed* values (*A*) already shown in our contingency table, we must compute an *expected* value (*E*) for each cell in the table. To do so, we *act as if the null hypothesis were known to be true.* That is, we assume that the same proportion of people in Boston, Chicago, and Los Angeles prefer choice *C*, and similarly for choice *D*. Looking at the horizontal totals at the right of the table, we see that 600 people out of a total of 1,000 sampled would prefer choice *C*: Thus $\frac{600}{1{,}000}=\frac{3}{5}$ of the total group of people sampled prefer choice *C*. Then *if* the null hypothesis is true, it should make no difference whether we are talking about Boston, Chicago, or Los Angeles voters, and we theoretically would "expect" $\frac{3}{5}$ *of each group* to prefer choice *C*.

Thus, with 200 persons sampled in Boston, the expected number of choice-*C* people in Boston is $\frac{3}{5}(200)=120$. (We enter this and other expected values in the contingency table in parentheses—see Figure 6.19.) The idea is that if the null hypothesis is true and candidate choice has nothing to do with location, then the 200 people sampled in Boston might just as well be 200 people taken

$\frac{3}{5}(200)$ $\frac{3}{5}(300)$ $\frac{3}{5}(500)$

	Boston	Chicago	Los Angeles	Row totals
Choice *C*	130 (120)	180 (180)	290 (300)	600
Choice *D*	70 (80)	120 (120)	210 (200)	400
Column totals	200	300	500	Grand total: 1,000

$\frac{2}{5}(200)$ $\frac{2}{5}(300)$ $\frac{2}{5}(500)$

Figure 6.19

$\frac{600}{1{,}000}=\frac{3}{5}$ prefer *C*. $\frac{400}{1{,}000}=\frac{2}{5}$ prefer *D*.

at random from the total of 1,000 so that we ideally expect $\frac{3}{5}$ of them to prefer choice *C*.

Similarly, with 300 persons sampled in Chicago, the expected value for choice *C* is $\frac{3}{5}(300)=180$. And with 500 persons sampled in Los Angeles, the expected value for choice *C* is $\frac{3}{5}(500)=300$. As a check, we note that these three *E*'s do add up to 600 as they should.

Similarly, we may compute the expected values in the *D* row by noticing that $\frac{400}{1{,}000}=\frac{2}{5}$ of the total crowd prefers choice *D*, so that we should theoretically be able to say that $\frac{2}{5}$ of each city's sample would prefer *D*:

$$\text{Boston: } \frac{2}{5}(200) = 80$$

$$\text{Chicago: } \frac{2}{5}(300) = 120$$

$$\text{Los Angeles: } \frac{2}{5}(500) = \underline{200}$$

400 total for choice *D*

or we may simply arrive at these results (with *E*'s, you might say) by subtracting the row *C* expected values from the column (city) totals. Thus, either way, the expected values are as shown in Figure 6.19.

Now, we notice that the actual *observed* values are—and generally will be in any event—a bit different from the expected values. The question is whether they are different *enough* to warrant rejecting the null hypothesis; after all, even if the null hypothesis is true we would expect a certain amount of chance variation. To settle the question we compute the statistic χ^2, by the same formula as before, where this time the formula means to compute $\frac{(A-E)^2}{E}$ for *each cell* (six of them) in the contingency table:

$$\chi^2 = \sum \frac{(A-E)^2}{E}$$

The computation goes as follows:

$$\chi^2 = \frac{(130-120)^2}{120} + \frac{(180-180)^2}{180} + \frac{(290-300)^2}{300}$$

$$+ \frac{(70-80)^2}{80} + \frac{(120-120)^2}{120} + \frac{(210-200)^2}{200}$$

$$= \frac{100}{120} + 0 + \frac{100}{300}$$

$$+ \frac{100}{80} + 0 + \frac{100}{200}$$

$$= \frac{5}{6} + \frac{1}{3} + \frac{5}{4} + \frac{1}{2} = \frac{10}{12} + \frac{4}{12} + \frac{15}{12} + \frac{6}{12}$$

$$= \frac{35}{12} \approx 2.917.$$

(Very often, finding a common denominator for all the fractions involved may be very unwieldy, and one would instead change the fractions individually to decimal form before adding.)

Now, again, you should realize that χ^2 measures the extent to which the observed values (A) and the expected values (E) *fail to agree*: the extent to which we are tempted to reject the null hypothesis. The question now is: is the computed $\chi^2 = 2.917$ large enough to justify a rejection at (say) the level of significance $\alpha = 0.05$? To find out, we of course have to check $\chi^2 = 2.917$ against a critical value from the chi-square table, and in this situation, with a contingency table analysis, we use

$$\text{d.f.} = (\text{number of rows} - 1)(\text{number of columns} - 1)$$

which in this case is

$$\text{d.f.} = (2-1)(3-1) = 1 \cdot 2 = 2.$$

And in the chi-square table under column heading $\chi^2_{.05}$ and across from d.f. = 2 we find: $\chi^2_{.05} = 5.991$. Thus the computed χ^2 must exceed 5.991 for a rejection of H_0 at this level. But because

$$\chi^2_{\text{computed}} = 2.917 < 5.991$$

we do *not* reject H_0 at the level 0.05. The difference between the set of A's and the set of E's is not significant at the level 0.05, and we do not have enough sample evidence to reject the claim that person C's and person D's relative proportions of popularity are uniform from city to city. If the relative popularities of the two candidates *do* vary from city to city, then that fact will have to be demonstrated by some evidence other than the sample data shown here—in these sample data which we have analyzed, the difference measured by $\chi^2 = 2.917$ is small enough to attribute to chance, and the three "different-looking" proportion breakdowns (for the three cities)

	Boston	Chicago	Los Angeles
C:	$\frac{130}{200} = 0.65$	$\frac{180}{300} = 0.60$	$\frac{290}{500} = 0.58$
D:	$\frac{70}{200} = 0.35$	$\frac{120}{300} = 0.40$	$\frac{210}{500} = 0.42$

are not significantly different. There is no clear evidence that C and D stack up differently against each other in different cities. The fact that in the *sample* data they compare 65% to 35% in Boston, 60% to 40% in Chicago, and 58% to 42% in Los Angeles, is not, under the circumstances, strong enough evidence to argue that they really do compare differently from city to city with the entire voting populations.

EXAMPLES

3. Suppose, in the situation just described (sampling three cities to determine the uniformity of relative popularity of two political hopefuls C and D) the responses had been as follows:

	Boston	Chicago	Los Angeles		
C:	108	171	321	→	total 600
D:	92	129	179	→	total 400
Totals	200	300	500		Grand total 1,000

We will again test, at the level 0.05, the null hypothesis H_0 that the proportions of people who would prefer each candidate are uniform from city to city. Again, as in the discussion in the text, because the row and column totals are still the same, the expected values (E) in parentheses are:

	Boston	Chicago	Los Angeles
C:	108(120)	171(180)	321(300)
D:	92(80)	129(120)	179(200)

but the *observed* values A are different. Now the χ^2 computation goes as follows:

$$\chi^2 = \sum \frac{(A-E)^2}{E}$$

$$= \frac{(108-120)^2}{120} + \frac{(171-180)^2}{180} + \frac{(321-300)^2}{300}$$

$$+ \frac{(92-80)^2}{80} + \frac{(129-120)^2}{120} + \frac{(179-200)^2}{200}$$

$$= \frac{144}{120} + \frac{81}{180} + \frac{441}{300} + \frac{144}{80} + \frac{81}{120} + \frac{441}{200}$$

$$= 1.200 + 0.450 + 1.470 + 1.800 + 0.675 + 2.205$$

$$= 7.800$$

And again, as in the previous discussion, the critical $\chi^2_{.05}$ table value is 5.991, so that for

$$\chi^2_{\text{computed}} = 7.800 > 5.991$$

we *reject* the null hypothesis H_0 at the level 0.05. We are 95% confident that the political hopefuls C and D do *not* have uniform popularity proportions from city to city: the "different-looking" proportion breakdowns

	Boston	Chicago	Los Angeles
C:	$\frac{108}{200} = 0.540$	$\frac{171}{300} = 0.570$	$\frac{321}{500} = 0.642$
D:	$\frac{92}{200} = 0.460$	$\frac{129}{300} = 0.430$	$\frac{179}{500} = 0.358$

are significantly different. The way C and D stack up against each other (54.0% to 46.0%, or 57.0% to 43.0%, or 64.2% to 35.8%) depends significantly on which city we look at. The two modes of classification, political preference and city of residence, are not independent of each other.

Note carefully that this result does *not* have to do only with the fact that $\frac{3}{5}$ of the people sampled prefer candidate C and $\frac{2}{5}$ prefer D: That was true also in the discussion preceding this example, but we failed to reject H_0 that time and did reject H_0 this time, even at the same sample size. The difference is that the first time around it did not matter what city we talked about, but this time, as it turns out, it does matter—we are not just talking about the relative popularity of C and D, but rather about the extent to which that relative popularity pattern changes from place to place. In this present example, *C*'s and *D*'s campaign managers would now probably concentrate their efforts in specific places.

Note, also, in the chi-square table, that a rejection at the 0.01 level would require the computed χ^2 to exceed $\chi^2_{.01} = 9.210$, which it does not. Thus we are 95% confident in rejecting what H_0 says about the population of voters but not quite 99% confident.

4. A market researcher wishes to determine whether there is any connection between the amount of college education people in a certain region have and the breakfast cereal they would prefer if they had to choose among three brands (*X, Y, Z*). The researcher identifies four levels of education and records random sample responses as follows, in a 3×4 contingency table:

	No college	1–2 years	3–4 years	More than 4 years		
X:	22	55	32	16	→	total 125
Y:	49	96	47	33	→	total 225
Z:	29	49	41	31	→	total 150
	100	200	120	80		grand total 500

The null hypothesis H_0 states that there is no relation between cereal preference and education level: the proportion breakdowns among *X, Y,* and *Z* are (over the population) uniform from education level to education level. This hypothesis will be chi-square tested, and let us say that the level of significance will be 0.01. The expected values *E* are computed as follows.

Because $\frac{125}{500} = \frac{1}{4}$ of the responses were for cereal *X*, then the null hypothesis would suggest, theoretically, that $\frac{1}{4}$ of each education-level group should be expected to prefer cereal *X*, because H_0 asserts that cereal preference is unrelated to education level: When we look, for instance, at 100 noncollege-educated people we might as well be taking these 100 people at random, without regard for their education level. Thus to get the *E* values in the first row we multiply each column total by $\frac{1}{4}$:

$$\frac{1}{4}(100) = 25, \quad \frac{1}{4}(200) = 50, \quad \frac{1}{4}(120) = 30, \quad \frac{1}{4}(80) = 20.$$

Similarly, because $\frac{225}{500} = \frac{9}{20}$ of the responses were for cereal *Y*, we obtain the

E values in the second row by multiplying the column totals by $\frac{9}{20}$:

$$\frac{9}{20}(100)=45,\quad \frac{9}{20}(200)=90,\quad \frac{9}{20}(120)=54,\quad \frac{9}{20}(80)=36.$$

Likewise, because $\frac{150}{500}=\frac{3}{10}$ of the responses were for cereal Z, we obtain the expected values E in the third row by multiplying the column totals by $\frac{3}{10}$:

$$\frac{3}{10}(100)=30,\quad \frac{3}{10}(200)=60,\quad \frac{3}{10}(120)=36,\quad \frac{3}{10}(80)=24.$$

(Or possibly we could have added the previous expected values in each column and subtracted from the column total to obtain the Z-row expected values; but now we have a check on our computations, because all independently computed E's in each column should add up to the column total.)

Now the observed values A and expected values E (in parentheses) look like this:

	No college	1–2 years	3–4 years	More than 4 years
X:	22(25)	55(50)	32(30)	16(20)
Y:	49(45)	96(90)	47(54)	33(36)
Z:	29(30)	49(60)	41(36)	31(24)

and we may proceed with the chi-square computation:

$$\begin{aligned}\chi^2 &= \sum\frac{(A-E)^2}{E}\\ &= \frac{(22-25)^2}{25}+\frac{(55-50)^2}{50}+\frac{(32-30)^2}{30}+\frac{(16-20)^2}{20}\\ &\quad+\frac{(49-45)^2}{45}+\frac{(96-90)^2}{90}+\frac{(47-54)^2}{54}+\frac{(33-36)^2}{36}\\ &\quad+\frac{(29-30)^2}{30}+\frac{(49-60)^2}{60}+\frac{(41-36)^2}{36}+\frac{(31-24)^2}{24}\\ &= \frac{(-3)^2}{25}+\frac{5^2}{50}+\frac{2^2}{30}+\frac{(-4)^2}{20}\\ &\quad+\frac{4^2}{45}+\frac{6^2}{90}+\frac{(-7)^2}{54}+\frac{(-3)^2}{36}\\ &\quad+\frac{(-1)^2}{30}+\frac{(-11)^2}{60}+\frac{5^2}{36}+\frac{7^2}{24}\end{aligned}$$

$$=\frac{9}{25}+\frac{25}{50}+\frac{4}{30}+\frac{16}{20}$$

$$+\frac{16}{45}+\frac{36}{90}+\frac{49}{54}+\frac{9}{36}$$

$$+\frac{1}{30}+\frac{121}{60}+\frac{25}{36}+\frac{49}{24}$$

$$=\frac{9}{25}+\frac{1}{2}+\frac{2}{15}+\frac{4}{5}$$

$$+\frac{16}{45}+\frac{2}{5}+\frac{49}{54}+\frac{1}{4}$$

$$+\frac{1}{30}+\frac{121}{60}+\frac{25}{36}+\frac{49}{24}$$

The common denominator of these fractions is hardly worth pursuing (it happens to be 1,080); we change the fractions to decimals:

$$\chi^2 = 0.360+0.500+0.133+0.800$$

$$+0.356+0.400+0.907+0.250$$

$$+0.033+2.017+0.694+2.042=8.492.$$

To see whether this computed χ^2 value is large enough to warrant rejecting the null hypothesis at the level 0.01 we use

$$\text{d.f.} = (\text{rows}-1)(\text{columns}-1)=(3-1)(4-1)$$

$$=2\cdot 3=6$$

and find in the chi-square table the critical value

d.f. $\chi^2_{.01}$

⋮ ↓

$6 \rightarrow 16.812$

Because we see that

$$\chi^2_{\text{computed}} = 8.492 < 16.812$$

we do *not* reject the null hypothesis that cereal choice is independent of education level. By this sample, no relation between cereal choice and education level has been shown. The observed values are different from the expected values only by an amount small enough to attribute to chance, and too small to make a rejection. (Note that in the chi-square table even a weaker rejection at the level 0.05 would have required the computed χ^2 to exceed 12.592, but it does not. We cannot reject H_0 even at the level 0.05.)

It should be mentioned that for the chi-square test to be reliable, each cell in the contingency table should contain at least 5 for its observed value A. If this is not the case, then probably the experiment can be restructured (by redoing the categories of classification, possibly merging some categories) so that each cell does contain at least $A = 5$.

NEW TERMS IN THIS SECTION

chi-square (χ^2)
expected values (E)
observed values (A)
goodness of fit
contingency table
***Yates' Correction**

EXERCISES 6-B

1. A gambler wishes to test a die to see whether or not to reject the null hypothesis that the die is fair. A level of significance $\alpha = 0.05$ is decided on. The gambler observes 300 tosses of the die, recording the number of times each face turns up as follows:

	1	2	3	4	5	6
A:	48	46	52	47	52	55

What are the expected values E? Compute χ^2. What is the critical value at the level 0.05? What is the conclusion, then, concerning the null hypothesis H_0?

2. In Exercise 6-B.1, suppose the observed values A had been:

(a)

	1	2	3	4	5	6
A:	47	46	53	46	52	56

(b)

	1	2	3	4	5	6
A:	46	45	53	45	54	57

(c)

	1	2	3	4	5	6
A:	45	44	54	45	55	57

(d)

	1	2	3	4	5	6
A:	44	43	55	44	56	58

(e)

	1	2	3	4	5	6
A:	43	42	56	42	58	59

(f)

	1	2	3	4	5	6
A:	41	42	57	41	59	60

(g)

	1	2	3	4	5	6
A:	41	41	58	40	59	61

(h)

	1	2	3	4	5	6
A:	40	41	59	40	58	62

3. An English instructor claims that there is no bias in the manner in which he calls on students in class. One student, however, suspects that what *row* one sits in has a lot to do with how often one gets called on. To test (at the level of significance 0.05) the instructor's hypothesis that all rows get equal treatment in the long run, the student observes 140 "callings," at random, noting the following observed values *A* (numbers of calls):

Row:	1	2	3	4	5	6	7
A:	27	28	24	23	15	10	13

What is the expected value in each case? Compute χ^2. What is the critical value at level 0.05, and what is the conclusion? What if the level of significance had been chosen, from the beginning, to be 0.01?

4. At a carnival a game operator draws single cards out of a deck (always with replacement before drawing again), and the patrons are asked to bet money on what *suit* of card (spades, clubs, hearts, or diamonds) will be drawn each time. If the drawing is fair (and not determined by some sleight-of-hand), each suit should turn up one-fourth of the time in the long run. To test (at the level 0.05) the null hypothesis that the game is fair, an investigator observes 500 drawings at random, recording the following numbers of observations:

	Spades	Clubs	Hearts	Diamonds
A:	112	135	137	116

What are the expected values *E*? Compute χ^2 and compare it with the appropriate critical value; what is the result?

5. In the same situation as in Exercise 6-B.4, suppose the investigator had made 1,000 observations, as follows:

	Spades	Clubs	Hearts	Diamonds
A:	224	270	274	232

Now compute χ^2. What is the result at the level 0.05? Considering that the *proportions* this time are the same as before $\left(\frac{224}{1{,}000} = \frac{112}{500}, \text{ and so on}\right)$, what has made the difference?

6. The manager of a supermarket notices that out of 400 randomly observed purchases of soap, of which there are five brands (*A, B, C, D, E*) available, the distribution of purchases is as

follows:

	A	B	C	D	E
Purchases:	68	99	95	67	71

At the level of significance 0.05, test the null hypothesis (by computing χ^2) that the proportions of purchases overall are essentially the same for all the brands.

7. A market analyst wishes to know whether the relative popularities of three brands of soft drink (*R, S, T*) vary significantly from one populous region of the country to another. Sampling in four regions of interest, the analyst finds response data as follows:

	Northeast	South	Midwest	West
Prefer *R*:	28( )	18( )	30( )	24( )
Prefer *S*:	32( )	66( )	71( )	31( )
Prefer *T*:	20( )	36( )	23( )	21( )

What are the expected values that should go into the parentheses? At the level 0.01, test the hypothesis that soft drink preference is not regional, by computing χ^2 and comparing it with the appropriate critical value.

8. A program analyst wishes to know whether there is a relation between the age brackets of people in a given area and which televised local news (channel 3, 5, or 7) they prefer to watch. The researcher samples people in five age brackets and records the following numbers of channel preferences:

	Less than 18 years	18 to 32 years	33 to 47 years	48 to 62 years	More than 62 years
Channel 3:	85( )	74( )	79( )	99( )	63( )
Channel 5:	60( )	55( )	74( )	79( )	32( )
Channel 7:	75( )	41( )	57( )	62( )	65( )

What are the expected values which belong in the parentheses? At the level of significance $\alpha = 0.05$, test the null hypothesis, by a chi-square test, where the null hypothesis states that channel preference is independent of age bracket—that is, each channel enjoys uniform proportions across the age brackets.

9. In Exercise 6-B.8, what if the null hypothesis (chosen from the beginning) had been $\alpha = 0.01$?

10. A psychologist wonders whether there is any connection (however unlikely it may sound) between hair color and intelligence as measured by a standard IQ test. She categorizes 600

subjects tested as follows:

Hair:	Black	Brown	Red	Blond
IQ<100	34( )	33( )	20( )	33( )
100≤IQ<110	60( )	64( )	53( )	63( )
110≤IQ<120	43( )	56( )	33( )	48( )
IQ>120	23( )	17( )	14( )	6( )

Compute the expected values, and compute χ^2 to test, at level 0.05, the hypothesis that IQ is unrelated to hair color.

11. A researcher, who divides college students into three categories according to their major subject-matter interests (arts, sciences, or business), wonders if classifications by majors are connected to the students' astrological signs. (For simplicity's sake here we shall consider only four Zodiac signs.) The sample data are as follows:

	Leo	Scorpio	Capricorn	Aries
Arts:	71( )	92( )	103( )	34( )
Science:	52( )	66( )	68( )	14( )
Business:	127( )	142( )	179( )	52( )

Compute the expected values and do a chi-square test at the level of significance 0.05, testing the null hypothesis that subject-matter inclination is independent of astrological sign.

12. Suppose we wish to determine, on the basis of 500 sample observations of unicorns, whether their fur color seems to be related to their eye color, with sample data as follows:

	Brown eyes	Green eyes	Blue eyes
Black fur:	141( )	50( )	59( )
Brown fur:	64( )	15( )	21( )
White fur:	55( )	45( )	50( )

Compute the expected values. At the level 0.01, what can be said, from a chi-square test, about the null hypothesis that fur color and eye color are independent?

13. A sociologist wishes to find out whether the type of television program people most enjoy is related to their living environment (urban, suburban, or country), and finds sample responses as follows:

	Comedy	Western	Adventure	Mystery	Documentary
Urban:	208( )	166( )	187( )	157( )	82( )
Suburban:	268( )	204( )	224( )	186( )	118( )
Country:	54( )	40( )	49( )	37( )	20( )

What are the expected values? Compute χ^2. What can be said at the level 0.05 concerning the null hypothesis that program choice has nothing to do with living environment?

14. A medical researcher wishes to know whether there is any connection between the mode of transportation commuters in a certain area take to work and the number of sick days which they take over a year's time (light, moderate, or heavy). The sample data are:

	Car	Bus	Train
Light:	117( )	41( )	82( )
Moderate:	211( )	81( )	108( )
Heavy:	102( )	38( )	20( )

What are the expected values? Compute χ^2. At the 0.01 level of significance, what can be said about the null hypothesis H_0 that mode of transportation is independent of sick-day classification?

15. A psychologist wants to see whether there is any relation between a supermarket shopper's first purchase in the store and the dominant color of the store's interior signs. Different-colored signs are used at different times, and the sample data observed are as follows:

	Fruits & vegetables	Meats	Dairy products	Paper goods	Other
Red:	91( )	62( )	76( )	97( )	74( )
Green:	19( )	21( )	20( )	23( )	17( )
Blue:	62( )	48( )	50( )	74( )	66( )
Black:	48( )	39( )	34( )	36( )	43( )

Compute the expected values. At the level $\alpha = 0.05$, what can be said about the null hypothesis H_0 that sign color is independent of first purchase?

16. An analyst desires to know whether there is any relation between an adult person's education level and preference among television networks (network X, Y, or Z). Suppose people are sampled at random and the following data are gathered:

	Less than high school	High school grad	Some college	College grad	Graduate study
X:	20( )	21( )	38( )	39( )	32( )
Y:	101( )	104( )	120( )	72( )	53( )
Z:	41( )	61( )	88( )	57( )	53( )

What are the expected values? Compute χ^2. Using a level of significance $\alpha = 0.01$, what can be said, from the sample data, about the null hypothesis H_0? (What does the null hypothesis state?)

17. A statistician samples people and classifies them two ways: by mode of habitation (renters, houseowners, and so on) and by which auto insurance company they use (A, B, C, or

D), with sample data as follows:

	A	*B*	*C*	*D*
Rent apartment:	120( )	161( )	155( )	164( )
Rent house:	99( )	147( )	120( )	134( )
Own condominium:	47( )	50( )	47( )	56( )
Own house:	49( )	62( )	53( )	36( )

What does the null hypothesis H_0 say? Test it at the 0.05 level by computing the expected values and χ^2. What is the result?

18. A manufacturer wishes to know if the manner in which purchasers originally found out about the product is related to an age classification. Purchasers are sampled with the following results:

	Newspaper ad	Magazine ad	Radio	TV	Word of mouth
Age<20:	76( )	69( )	132( )	156( )	67( )
Age 20–40:	131( )	111( )	203( )	235( )	120( )
Age>40:	153( )	140( )	145( )	169( )	93( )

What does the null hypothesis say? What are the expected values? At the level of significance $\alpha = 0.01$, test H_0 by computing χ^2. What is the result?

PONDERABLES 6-B

1. In a chi-square analysis, if each of the observed values turned out to be exactly equal to the corresponding expected value (a most unlikely event, to be sure), then what would be the value of χ^2?

2. (*Yates' Correction*) If d.f. = 1 in a chi-square analysis (for example, with a 2×2 contingency table), it turns out that χ^2, computed by the usual formula

$$\chi^2 = \sum \frac{(A-E)^2}{E},$$

may tend to exaggerate the difference between the observed values *A* and the expected values *E*. As a hedge against this exaggerative tendency, many statisticians employ **Yates' Correction:** 0.5 is subtracted from each difference before squaring. That is, the formula becomes

$$\chi^2 = \sum \frac{(|A-E|-0.5)^2}{E}$$

The vertical bars (absolute value bars) essentially mean that the difference $A-E$ is regarded as nonnegative, and the 0.5 is subtracted from that nonnegative value, the result then being squared. For example, from the 2×2 contingency table

22(28) 48(42)
18(12) 12(18)

the usual χ^2 formula would give

$$\chi^2 = \sum \frac{(A-E)^2}{E}$$

$$= \frac{(-6)^2}{28} + \frac{6^2}{42} + \frac{6^2}{12} + \frac{(-6)^2}{18}$$

$$= \frac{36}{28} + \frac{36}{42} + \frac{36}{12} + \frac{36}{18}$$

$$\approx 1.2857 + 0.8571 + 3.0000 + 2.0000$$

$$= 7.1428 \approx 7.143,$$

but the adjusted formula, using Yates' Correction, would give instead

$$\chi^2 = \sum \frac{(|A-E|-0.5)^2}{E}$$

$$= \frac{(6-0.5)^2}{28} + \frac{(6-0.5)^2}{42}$$

$$+ \frac{(6-0.5)^2}{12} + \frac{(6-0.5)^2}{18}$$

$$= \frac{5.5^2}{28} + \frac{5.5^2}{42} + \frac{5.5^2}{12} + \frac{5.5^2}{18}$$

$$= \frac{30.25}{28} + \frac{30.25}{42} + \frac{30.25}{12} + \frac{30.25}{18}$$

$$\approx 1.0804 + 0.7202 + 2.5208 + 1.6806$$

$$= 6.002,$$

a value noticeably smaller than the first result, 7.143. Because d.f. $= 1$, the value 6.002 may be regarded here as being more reliable than 7.143. By comparison with the critical χ^2 value at the level 0.01, what would the computation have suggested, in terms of rejection or nonrejection of H_0,

(a) without Yates' Correction? (b) with Yates' Correction?

3. A psychologist asks a number of subjects which color they like better, red or green, and wonders whether there is a relation between color preference and the subject's sex. The sample

data are as follows:

	Red	Green
Male:	81( )	69( )
Female:	111( )	139( )

Compute χ^2 using Yates' Correction. At the level 0.05, what can be said about the null hypothesis that color preference is independent of sex classification?

4. A coin is flipped a number of times with the following results:

H	H	T	T	H	T	T	H	T	T
T	T	H	T	H	T	H	H	T	H
T	H	H	T	H	T	T	T	H	T
T	H	T	T	T	H	T	H	H	T
H	T	H	T	H	H	H	T	H	T
T	T	H	T	H	T	T	H	T	T
T	T	H	H	H	T	H	T	T	H
H	T	H	T	T	H	T	T	T	T
T	T	T	H	T	H	T	H	H	T
T	T	H	H	T	T	H	T	T	T
T	T	T	H	T	T	T	H	T	H
T	H	H	T	H	T	H	T	T	H

Fill in the data:

	Heads	Tails
Observed:	_____	_____
Expected:	( )	( )

Note that this problem is similar in structure to the "spin the wheel" and "toss the die" problems examined earlier in this section and in the exercises. Thus d.f. $= 2 - 1 = 1$. Suppose we wish to test, with χ^2 and at the level 0.05, the null hypothesis H_0 that the coin is fair. Using Yates' Correction, what is the result?

5. In problem 4, if Yates' Correction had been neglected, what would χ^2 have come out to be, and what would have been the result in terms of conclusions drawn about H_0 at the level of significance 0.05?

SUGGESTED READINGS

Statistics: With a View Toward Applications, by Leo Breiman (Houghton Mifflin Company, 1973). Chapter 10 provides a very general treatment of analysis of variance.

Everything You Always Wanted To Know About Elementary Statistics (*but were afraid to ask*), by Jerald G. Schutte (Prentice-Hall, Inc., 1977). Chapter 13 explores the ways in which the statistics t, F, and χ^2 are related.

Basic Statistics: Tales of Distributions, by Chris Spatz and James O. Johnston (Brooks/Cole Publishing Company, 1976). Chapters 10 and 11 provide, respectively, excellent discussion of one-way and two-way ANOVA.

7 Linear Correlation and Regression

"Knowledge once gained casts a light beyond its own immediate boundaries."—Tyndall

A COEFFICIENT OF CORRELATION

We will now look at a concept called *correlation* and a means of measuring it. Specifically, we will examine the question of *linear* correlation, and we will compute a statistic called the *coefficient of correlation*, denoted r.

The type of sample data that will concern us here is called **paired data,** and the nature of it will be clarified by the following illustrations and examples.

Suppose we are in a laboratory testing an experimental drug on lab mice. The drug is supposed to be a sort of depressant, slowing down and inhibiting the mice's reactions and perceptions. We give the drug to each of a group of 12 mice (all equally healthy and as similar to each other in all respects as possible), administering different amounts to different mice, and running each mouse then through a maze to see how long it takes it to get through. The idea is to see if the amount of the drug administered seems to be related to the time which is then required to run the maze—our expectation, in general, being that *more* of the drug should be associated with *longer* times getting through the maze.

Suppose, then, for each mouse we have recorded *two* things x and y: the amount x (in milligrams) of the drug administered, and the amount of time y (in seconds) required to run through the maze. And suppose our sample data are as follows:

		x (milligrams)	y (seconds)
Mouse	1:	10	10
	2:	10	15
	3:	20	10
	4:	20	25
	5:	30	25
	6:	30	30
	7:	40	25

8:	50	30
9:	60	35
10:	70	30
11:	80	35
12:	80	40

(Although we have referred to "Mouse 1," "Mouse 2," and so on, the idea of order is unimportant here, as long as we keep the right y associated with each x.) We could plot these pairs of numbers on a graph, obtaining a point (we may call them "data points") for each mouse's results, as shown in Figure 7.1. We may represent the amount of the drug on the horizontal axis (x-axis) and the number of seconds on the vertical axis (y-axis), and then we plot each point (pair of values x, y) by moving from the point of intersection of the axes, x units to the right and y units up. For example, to plot the point (70, 30) we move 70 units over to the right and 30 units up to plot the point.

Now, looking at the graph of these 12 points, we see that they appear, at least very roughly and vaguely, to suggest a *line:* They do not actually lie on a line, obviously, but there would be a line running through their midst that would not miss the points, by and large, by very much. If the points *had* arranged themselves perfectly on a line, as in Figure 7.2, we would have said that there was a perfect **positive correlation** shown between the x-values and the y-values: a perfect linear relationship by which the increasing x-values would "predict" increasing y-values in a systematic fashion. (In this discussion, "correlation" will mean **linear correlation**—the tendency of data points to form a line. Other types of correlation are known to statisticians, but are beyond the scope of this text.)

Or if the data points had arranged themselves on a line leaning the other way, as in Figure 7.3, we would have said that there was a perfect **negative correlation** shown between the x-values and the y-values: a perfect linear

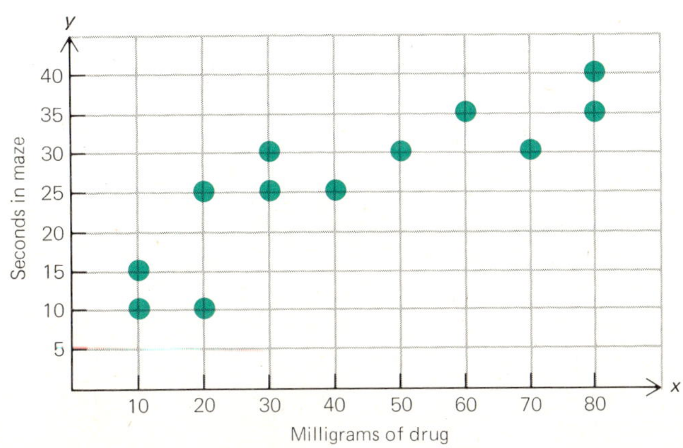

Figure 7.1

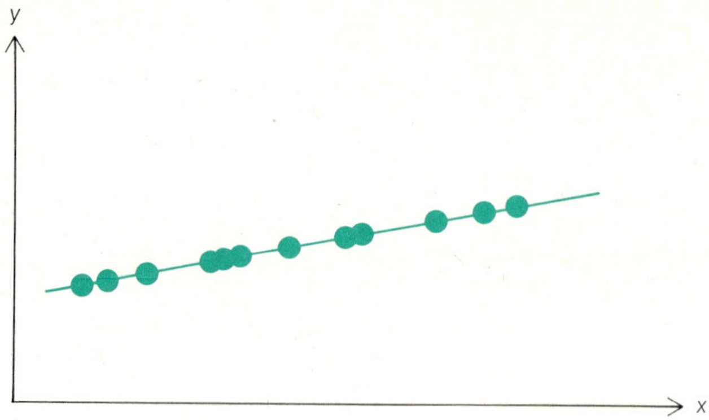

Figure 7.2 Positive correlation

relationship by which the increasing x-values would suggest *decreasing* y-values in a systematic fashion. (For that to happen here, increasing the drug dosage would have to correspond to decreasing the time required to run the maze.)

In our actual example (Figure 7.1 again), there only appears to be, however, a *possible* linear relationship suggested between x and y, and we ask a question in the spirit of many previous statistical questions: do the points *nearly enough* tend to lie on a line for us to conclude that in general there is indeed a linear relationship—a linear *correlation*—between x and y? If there is a good correlation, the line running "through" the points will signal a *positive* correlation, as you may see by comparing with Figure 7.2—but we cannot "tell by looking" whether the points really do suggest a linear arrangement. Perhaps they do not—perhaps they too much resemble the points in Figure 7.4, which do not show *any* good correlation at all.

To settle this question, we compute a specific statistic called the **coefficient of correlation,** denoted r. This computation, with a given set of data points

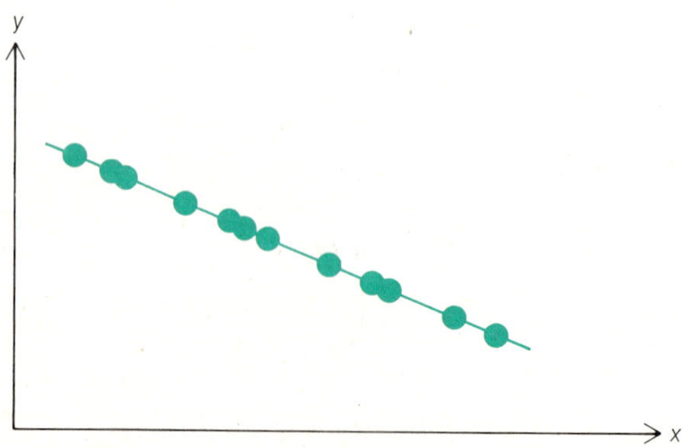

Figure 7.3 Negative correlation

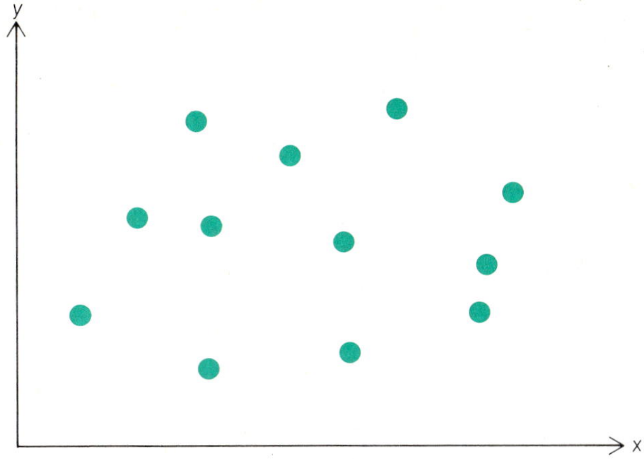

Figure 7.4 No correlation suggested

(x, y), requires first that we compute seven specific quantities, as follows:

$\sum x$: the sum of all the x-coordinates in the data
$\sum y$: the sum of all the y-coordinates in the data
$(\sum x)^2$: the square of the sum $\sum x$
$(\sum y)^2$: the square of the sum $\sum y$
$\sum (x^2)$: the sum of the squares of the individual x's
$\sum (y^2)$: the sum of the squares of the individual y's
$\sum (xy)$: the sum of the products $x \cdot y$ (multiplied individually for each point)

Note carefully the difference between $(\sum x)^2$ (in which the x's are added first and then that sum is squared) and $\sum (x^2)$, in which each x is first squared and then the squares resulting are added. Also note that $\sum (xy)$ does *not* mean $(\sum x) \cdot (\sum y)$; rather, in $\sum (xy)$ the product xy is computed for each point first and then the results are added.

When these sums are ready, we compute the coefficient of correlation r by the rather formidable-looking but (as you will see) not really difficult formula

$$r = \frac{n \sum (xy) - (\sum x)(\sum y)}{\sqrt{n \sum (x^2) - (\sum x)^2}\sqrt{n \sum (y^2) - (\sum y)^2}}$$

where n is the number of data points.

First, let us look at this computation for the given "mouse in a maze" data, and then we will discuss how the result r is interpreted.

For the given data

x:	10	10	20	20	30	30	40	50	60	70	80	80
y:	10	15	10	25	25	30	25	30	35	30	35	40

you may verify that the required sums are as follows:

$$\sum x = 10+10+20+20+30+30+40+50$$
$$+60+70+80+80=500$$

$$\sum y = 10+15+10+25+25+30+25+30$$
$$+35+30+35+40=310$$

$$\left(\sum x\right)^2 = 500^2 = 250{,}000$$

$$\left(\sum y\right)^2 = 310^2 = 96{,}100$$

$$\sum (x^2) = 10^2+10^2+20^2+20^2+30^2+30^2$$
$$+40^2+50^2+60^2+70^2+80^2+80^2$$
$$=100+100+400+400+900+900$$
$$+1{,}600+2{,}500+3{,}600+4{,}900+6{,}400+6{,}400$$
$$=28{,}200$$

$$\sum (y^2) = 10^2+10^2+15^2+25^2+25^2+30^2$$
$$+25^2+30^2+35^2+30^2+35^2+40^2$$
$$=100+100+225+625+625+900+625$$
$$+900+1{,}225+900+1{,}225+1{,}600$$
$$=9{,}050$$

$$\sum (xy) = (10\cdot 10)+(20\cdot 10)+(10\cdot 15)+(20\cdot 25)$$
$$+(30\cdot 25)+(30\cdot 30)+(40\cdot 25)+(50\cdot 30)$$
$$+(60\cdot 35)+(70\cdot 30)+(80\cdot 35)+(80\cdot 40)$$
$$=100+200+150+500+750+900$$
$$+1{,}000+1{,}500+2{,}100+2{,}100+2{,}800$$
$$+3{,}200=15{,}300.$$

Now we plug all these results into the formula for r, noting that $n = 12$ data points:

$$r = \frac{n\sum(xy) - (\sum x)(\sum y)}{\sqrt{n\sum(x^2) - (\sum x)^2}\sqrt{n\sum(y^2) - (\sum y)^2}}$$

$$= \frac{12(15{,}300) - (500)(310)}{\sqrt{12(28{,}200) - 250{,}000}\sqrt{12(9{,}000) - 96{,}100}}$$

$$= \frac{183{,}600 - 155{,}000}{\sqrt{338{,}400 - 250{,}000}\sqrt{108{,}600 - 96{,}000}}$$

$$= \frac{28{,}600}{\sqrt{88{,}400}\sqrt{12{,}500}}$$

which, from the square root table, is approximately

$$\frac{28{,}600}{(297.3)(111.8)} = \frac{28{,}600}{33{,}238.14} \approx 0.860.$$

Now, how do we interpret this result? Does it or does it not mean that the relation between drug amount (x) and maze time (y) is a good positive correlation?

It can be shown mathematically that r *must* lie on a scale between -1 and 1, and the whole idea is this: if r turns out (as is seldom the case) to be exactly 1, then the correlation is a *perfect positive correlation*, as was illustrated in Figure 7.2. And if r turns out (as is seldom the case) to be exactly -1, then the correlation is a *perfect negative correlation*, as illustrated in Figure 7.3.

More often, however, we will have to be content with less than perfect correlations. If r is *close* to 1 (and we must find out what that means), then we have a good positive correlation indicated, though not perfect. If r is close to -1, then we have a good, though not perfect, negative correlation. See Figure 7.5.

But what do we mean by "close to" 1, or "close to" -1?

First, as usual in statistics, we must be working at a specified level of significance, say $\alpha = 0.05$ or $\alpha = 0.01$, and it is helpful to frame our problem in terms of hypothesis testing. Here, the null hypothesis H_0 will be that there is *no* good correlation between x and y: any appearance of correlation is coincidental. Thus if we *reject* the null hypothesis H_0, we are saying yes, on

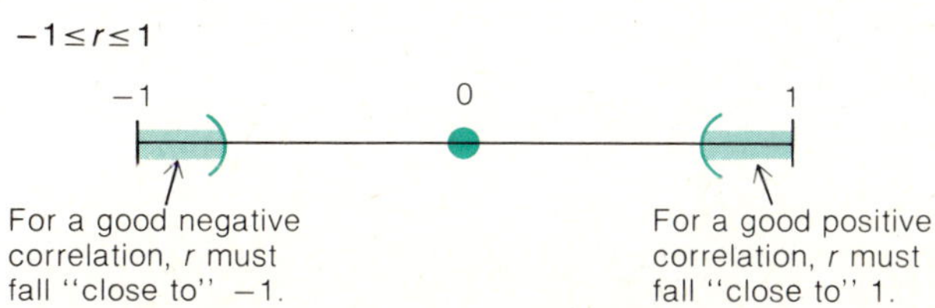

Figure 7.5

the contrary, there *is* a good correlation, at the level of significance specified. (If $\alpha = 0.05$, for example, then in that case we would be asserting the existence of correlation with 95% confidence.)

Suppose in our mouse-in-a-maze problem the level of significance is 0.01 (so that we would reject H_0 and claim correlation only if we could do so with 99% confidence), and suppose we wish to tell whether our result $r = 0.860$ shows a good correlation at that level or not. The null hypothesis is that there is really no correlation shown between the drug amounts and the maze times. We consult the table of *critical values of r* in the back of the book, using

$$\begin{aligned} \text{d.f.} &= (\text{sample size}) - 2 \\ &= n - 2, \end{aligned}$$

where by sample size we mean the number of data points, the number of pairs (x, y) of observations involved. In each case, because we have 12 mice, and a pair (x, y) for each, we have $n = 12$, so that d.f., the number of degrees of freedom to refer to in the table, is $12 - 2 = 10$. Looking, then, in the table, across from d.f. = 10 and under the heading $r_{.01}$ (because the level of significance is 0.01), we find:

d.f.		$r_{.01}$
⋮		⋮
10	⟶	0.708

that is, the critical value is $r_{.01} = 0.708$. This means that the "closeness" to 1 (or closeness to -1) needed for a good correlation must be defined by the value 0.708 as shown in Figure 7.6. The computed r-value, at this sample size of 12 and level of significance 0.01, must lie between 0.708 and 1 (or, symmetrically, on the other side, between -0.708 and -1) for the sample data to indicate a good correlation. The sample data points on sight do *suggest* (see Figure 7.1 again) a possible correlation, but that apparent indication is significant at the level 0.01 only if the computed r lies close enough to -1 or 1, and the

For $n = 12$, df $= 12 - 2 = 10$; at the level $\alpha = 0.01$, critical table value $r_{.01} = 0.708$.

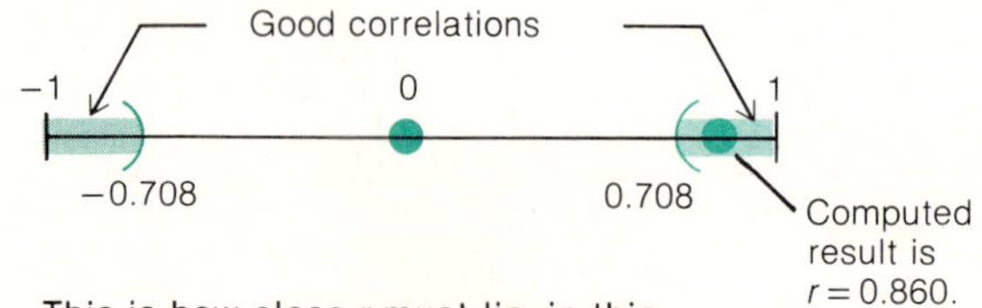

This is how close r must lie, in this situation, for the suggested correlation to be significant at the level 0.01.

Figure 7.6

table-value 0.708 defines in this case what we mean by "close." Note that because

$$r_{\text{computed}} = 0.860 > 0.708,$$

our computed result does lie close enough to 1, and we conclude that the correlation *is* good at the level 0.01. That is, we reject the null hypothesis H_0 that there was really no correlation. (Note, then, that the dark regions in Figure 7.6 are rejection regions; getting a computed r-value inside those regions means that H_0 can be rejected so that there *is* a significant indication of correlation. We accept a 1% risk of being wrong in this rejection.)

Thus we may conclude now that our sampling indicates a positive correlation between the drug amounts x and the maze times y. It is critically important, however, to understand that this result alone does not establish a *cause-and-effect* connection between x and y. *Correlation* alone does not necessarily mean *causation*. It is true that we *suspect* that the increased drug amounts have caused the maze times to increase, but we have no absolute proof. We have seen that the maze times do increase proportionally *while* the drug amounts increase, but we have not proven that the maze times increase this way *because* the drug amounts increase. Such proof is, strictly speaking, beyond the province of statistics; statistics can only suggest a *possible* cause-and-effect relation.

Philosophically, cause-and-effect is a difficult problem anyway. We never actually *see* anything cause an effect; we only infer such things. (When you move the pool cue across the table and strike a ball, you *assume* that the action "caused" the ball to move; you cannot prove absolutely that the ball would not have moved anyway—you can only call it extremely unlikely that it would have!)

Correlation can exist *without* cause-and-effect connections. For example, there is probably some sort of correlation between the population of a city and the annual number of crimes committed there, but it is another matter altogether to claim (in a cause-and-effect way) that increased urban population size, as such, *causes* crime to increase. There may indeed be a causal connection there, but the mere correlation alone does not prove it.

Similarly, there is probably some sort of correlation between the shoe size and the hat size of young children, but it would be ludicrous to claim that "the head's getting bigger causes the feet to get bigger," or vice-versa—rather, one would know that *both* phenomena are jointly part of the simple fact that the child's whole body is growing.

Returning to the mouse-in-a-maze idea, we look now at an example in which the sample data are different.

EXAMPLE

1. Suppose, again, each of 12 mice is given a certain amount x (in milligrams) of a drug and is tested to see how long (y, in seconds) it takes to get through a

maze. Suppose the following sample data are obtained:

x:	10	10	20	30	30	40	40	50	60	60	70	80
y:	15	30	10	15	25	30	25	20	35	20	40	25

We wish to test (say at the level of significance $\alpha = 0.05$ this time) the hypothesis that there is no correlation between x and y. We tabulate the necessary quantities:

x^2:	100	100	400	900	900	1600	1600	2500	3600	3600	4900	6400
y^2:	225	900	100	225	625	900	625	400	1225	400	1600	625
xy:	150	300	200	450	750	1200	1000	1000	2100	1200	2800	2000

and by adding the rows in these tables we find (as you may verify):

$$\sum x = 500, \quad \sum y = 290, \quad \sum (x^2) = 26{,}600, \quad \sum (y^2) = 7{,}850,$$

$$\sum (xy) = 13{,}150; \quad \text{and} \quad \left(\sum x\right)^2 = 500^2 = 250{,}000$$

$$\left(\sum y\right)^2 = 290^2 = 84{,}100.$$

We have $n = 12$, because there are 12 pairs of numbers. Now we compute the coefficient of correlation r:

$$r = \frac{n \sum (xy) - (\sum x)(\sum y)}{\sqrt{n \sum (x^2) - (\sum x)^2}\sqrt{n \sum (y^2) - (\sum y)^2}}$$

$$= \frac{12(13{,}150) - 500(290)}{\sqrt{12(26{,}600) - 250{,}000}\sqrt{12(7{,}850) - 84{,}100}}$$

$$= \frac{157{,}800 - 145{,}000}{\sqrt{319{,}200 - 250{,}000}\sqrt{94{,}200 - 84{,}100}}$$

$$= \frac{12{,}800}{\sqrt{69{,}200}\sqrt{10{,}100}}$$

$$\approx \frac{12{,}800}{(263.1)(100.5)} = \frac{12{,}800}{26{,}441.55} \approx 0.484.$$

This computed r-value must be compared with a critical value from the table, and noting that d.f. $= n - 2 = 12 - 2 = 10$, we find the necessary critical value as follows:

d.f.	$r_{.05}$
⋮	⋮
10 ⟶	0.576

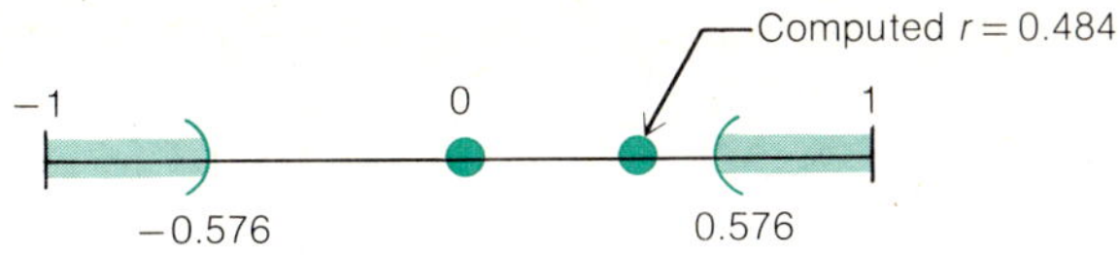

Figure 7.7 No correlation indicated at level 0.05, sample size 12.

Because $0.484 < 0.576$ (see Figure 7.7), our computed r-value has not fallen within a rejection region, and we do not reject the hypothesis H_0 that there is no correlation. That is, "close enough to 1" now means "between 0.576 and 1," and our result $r = 0.484$ is not close enough to 1 to argue for a good correlation. There may ultimately be some connection between x and y, but this sample information does not show it, at the level 0.05; we cannot claim a correlation and feel 95% confident. (See Figure 7.8: The data points do not strongly suggest a line graph; but because we cannot tell by looking, it was necessary to compute r.)

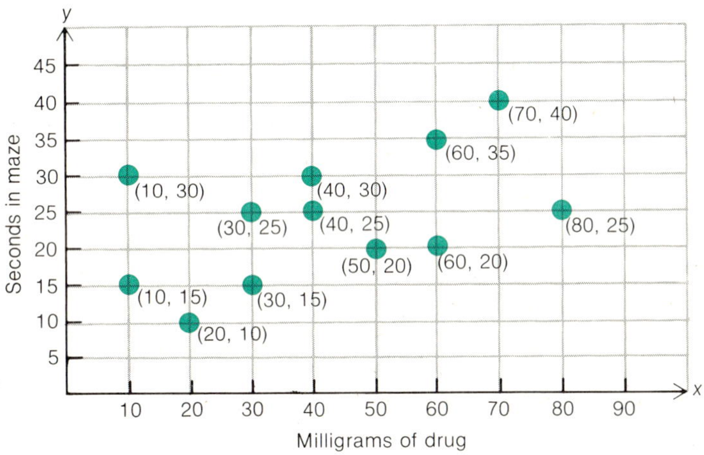

Figure 7.8 No correlation at the level 0.05.

The following examples should further illustrate the procedure. It should be pointed out here that *if* there is a good correlation indicated between x and y, then we may profitably think of the variables as generating points which approximately lie on some specific line, and it will be the business of the next section to show how that line is determined.

EXAMPLES

2. In a certain European city, an investigator wishes to determine if there is a correlation between the number of storks inhabiting a section of the city and the number of births recorded there during a given week. The investigator observes seven sections of the city and finds the following sample data ($n=7$);

Storks:	1	2	2	3	3	4	5
Babies:	1	3	4	4	5	5	7

At the level of significance $\alpha=0.01$, the investigator will test the null hypothesis H_0 that there is no correlation between storks (x) and babies (y). The necessary sums are as follows:

$$\sum x = 1+2+2+3+3+4+5=20$$

$$\sum y = 1+3+4+4+5+5+7=29$$

$$\begin{aligned}\sum (x^2) &= 1^2+2^2+2^2+3^2+3^2+4^2+5^2 \\ &= 1+4+4+9+9+16+25=68\end{aligned}$$

$$\begin{aligned}\sum (y^2) &= 1^2+3^2+4^2+4^2+5^2+5^2+7^2 \\ &= 1+9+16+16+25+25+49=141\end{aligned}$$

$$\begin{aligned}\sum (xy) &= 1\cdot 1+2\cdot 3+2\cdot 4+3\cdot 4+3\cdot 5+4\cdot 5+5\cdot 7 \\ &= 1+6+8+12+15+20+35=97\end{aligned}$$

with also

$$\left(\sum x\right)^2 = 20^2 = 400, \quad \left(\sum y\right)^2 = 29^2 = 841.$$

The coefficient of correlation, computed from these results, is:

$$\begin{aligned} r &= \frac{n\sum (xy)-(\sum x)(\sum y)}{\sqrt{n\sum (x^2)-(\sum x)^2}\sqrt{n\sum (y^2)-(\sum y)^2}} \\ &= \frac{7(97)-20(29)}{\sqrt{7(68)-400}\sqrt{7(141)-841}} \\ &= \frac{679-580}{\sqrt{476-400}\sqrt{987-841}} = \frac{99}{\sqrt{76}\sqrt{146}} \\ &\approx \frac{99}{(8.718)(12.08)} = \frac{99}{105.31344} \approx 0.94. \end{aligned}$$

To see whether to reject the null hypothesis H_0 at the level 0.01 (and thus conclude with 99% confidence that there is an indicated correlation between x

and y) the investigator checks this computed r-value against the critical table-value $r_{.01}$, which is found (since d.f. $= n-2=7-2=5$) by looking in the table across from d.f. $=5$ and under $r_{.01}$, to find $r_{.01}=0.874$. Because $r=0.94>0.874$, the null hypothesis is *rejected* at the level 0.01, and the indication of correlation is significant; the sample data suggest an actual correlation. (But remember, correlation does not necessarily imply causation! What would a cause-and-effect relation mean here?) You should sketch the graph and note the appearance of the seven data points; the suggested correlation is a *positive* correlation.

3. A researcher wishes to know whether there is any correlation, in a certain town, between the number (x) of credit cards a family maintains and the number (y) of times during the past 6 months the family has failed to make the minimal monthly payment to a credit card company. Ten families ($n=10$) are sampled at random, and the sample data observed are as follows:

x:	1	2	3	3	4	4	5	6	7	7
y:	2	3	1	2	3	5	2	4	2	5

The squares and products needed are:

x^2:	1	4	9	9	16	16	25	36	49	49
y^2:	4	9	1	4	9	25	4	16	4	25
xy:	2	6	3	6	12	20	10	24	14	35

The level of significance will be $\alpha=0.05$. The necessary sums (as you are encouraged to verify) are as follows:

$$\sum x=42,\quad \sum y=29,\quad \sum (x^2)=214,\quad \sum (y^2)=101,$$

$$\sum (xy)=132,\quad \left(\sum x\right)^2=1{,}764,\quad \left(\sum y\right)^2=841$$

so that the coefficient of correlation is

$$r=\frac{n\sum(xy)-(\sum x)(\sum y)}{\sqrt{n\sum(x^2)-(\sum x)^2}\sqrt{n\sum(y^2)-(\sum y)^2}}$$

$$=\frac{10(132)-42(29)}{\sqrt{10(214)-1{,}764}\sqrt{10(101)-841}}$$

$$=\frac{1{,}320-1{,}218}{\sqrt{2{,}140-1{,}764}\sqrt{1{,}010-841}}$$

$$=\frac{102}{\sqrt{376}\sqrt{169}}\approx\frac{102}{(19.39)(13)}$$

$$=\frac{102}{252.07}\approx 0.405.$$

This computed r must be compared with the critical table-value $r_{.05} = 0.632$ (from d.f. $= n - 2 = 10 - 2 = 8$), and because $r = 0.405 < 0.632$, the null hypothesis H_0 (which states that there is no correlation) is *not* rejected at the level 0.05. Then the researcher cannot conclude with 95% confidence that x and y are correlated—that is, that the number of credit cards and the number of failures to make payments are correlated.

4. A statistician wishes to determine whether, at a certain factory, there is a correlation between the number of miles a worker commutes to work and the number of days the worker has volunteered to work overtime during the past two months. The level of significance will be $\alpha = 0.01$. A sample of size $n = 11$ is taken, with x the number of miles each worker commutes and y the number of volunteer overtime days; the sample data are as follows (see Figure 7.9):

x:	3	3	4	5	5	6	7	8	9	10	10
y:	10	9	7	8	6	8	6	9	5	3	4

We tabulate the needed squares and products:

x^2:	9	9	16	25	25	36	49	64	81	100	100
y^2:	100	81	49	64	36	64	36	81	25	9	16
xy:	30	27	28	40	30	48	42	72	45	30	40

The sums needed are as follows:

$$\sum x = 70, \quad \sum y = 75, \quad \sum (x^2) = 514, \quad \sum (y^2) = 561,$$

$$\sum (xy) = 432, \quad \left(\sum x\right)^2 = 70^2 = 4{,}900, \quad \left(\sum y\right)^2 = 75^2 = 5{,}625$$

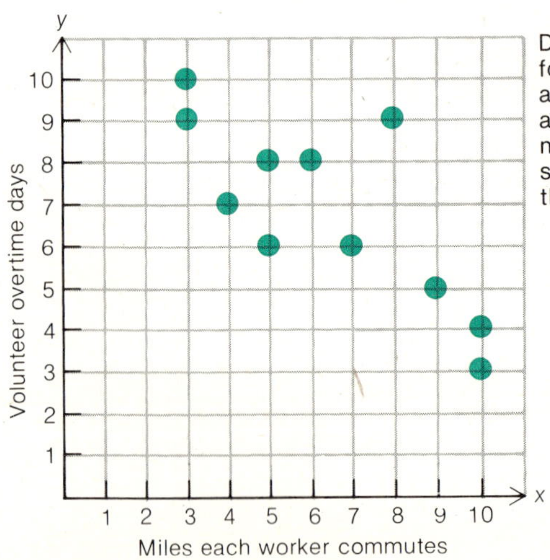

Data points for Example 4; as it turns out, an indication of negative correlation significant at the level 0.01.

Figure 7.9

so that the coefficient of correlation is:

$$r=\frac{11(432)-70(75)}{\sqrt{11(514)-4{,}900}\sqrt{11(561)-5{,}625}}$$

$$=\frac{4{,}752-5{,}250}{\sqrt{5{,}654-4{,}900}\sqrt{6{,}171-5{,}625}}$$

$$=\frac{-498}{\sqrt{754}\sqrt{546}}\approx\frac{-498}{(27.46)(23.37)}$$

$$=-\frac{498}{641.7402}\approx-0.776.$$

Checking this result against the table value (critical value), which, from d.f. $= n-2=9$, we find to be $r_{.01}=-0.735$, we note that $r=-0.776$ is between -0.735 and -1 (see Figure 7.10). Therefore, at the level of significance 0.01, we *do* claim a *negative* correlation between x and y. The more miles a worker commutes, the less often the worker volunteers for overtime. There may or may not be any *causal* relation between the two, even so.

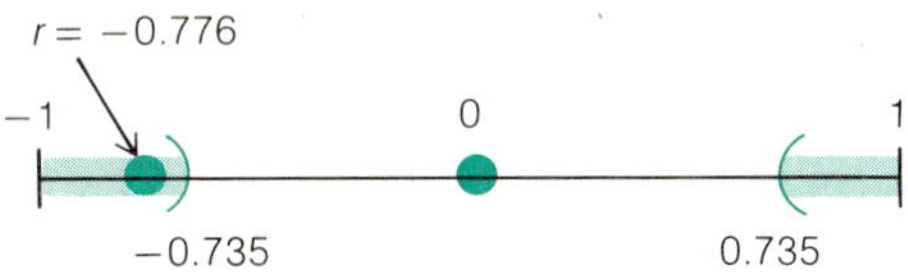

Figure 7.10

NEW TERMS IN THIS SECTION

paired data
(linear) correlation
positive correlation
negative correlation
coefficient of correlation

EXERCISES 7-A

(Keep your results and your work on these exercises; some of them will be used in Exercises 7-B.)

1. A researcher wishes to determine whether there is a correlation (at the level of significance $\alpha=0.05$), in a given locale, between the average number of hours (x) a schoolchild watches television nightly and the number of percentage points (y) below a certain reference point the student falls on a standardized reading exam. Eleven students are sampled at random, with results as follows:

x:	2	2	2	3	3	3	4	4	4	5	5
y:	8	7	9	9	8	10	9	11	10	9	12

Sketch the graph of these data points. Compute the coefficient of correlation and compare it with the critical value $r_{.05}$. What is the conclusion? What would have been the conclusion at the level of significance 0.01?

2. A health researcher at a certain factory complex wishes to know whether there is a correlation between the number of miles (x) a commuting employee drives to work every day and the annual number of sick days (y) the employee has taken during the previous year. A level of significance $\alpha = 0.05$ will be used. The researcher randomly samples 13 employees, with the following sample results:

x:	5	6	6	7	7	7	8	8	8	9	9	10	10
y:	3	4	2	3	6	5	2	5	7	4	1	6	3

Sketch the graph, compute r, and compare r with the critical value $r_{.05}$. What is the result?

3. An English professor at Miskatonic University wishes to know whether there is any correlation between the number (x) of novels assigned in her literature course and the number (y) of students making an A in the course. She has taught the course a great many times, sometimes assigning more novels and sometimes less, and she randomly chooses 10 offerings of the course from her record files, with these results:

x:	3	3	3	4	4	4	5	5	6	6
y:	4	5	6	4	1	3	2	3	1	2

The level of significance will be $\alpha = 0.01$. Graph the data points. What kind of possible correlation do the points suggest? Compute r and compare it with $r_{.01}$. What is the conclusion?

4. A publisher is curious to know whether there is any correlation between the number (x) of chapters in a best-selling novel and the number (y) of weeks the novel remains on the best-seller list. The level of significance will be $\alpha = 0.05$. Twelve novels are chosen at random, with these results:

x:	4	4	5	5	5	6	6	6	7	7	8	9
y:	8	6	9	12	14	7	10	12	10	6	9	15

Graph these data points. Compute r and interpret the results.

5. A medical researcher wonders if there is a correlation, for patients with a certain medical condition, between the number (x) of degrees above normal a patient's temperature is and the number (y) of millimeters the blood pressure exceeds a certain established norm. The level of significance used is $\alpha = 0.01$. Fourteen such patients are randomly sampled, with the following results:

x:	0.0	0.5	1.0	1.0	1.0	1.5	1.5	2.0	2.0	2.5	2.5	3.0	3.5	4.0
y:	1.0	2.0	1.5	1.0	2.0	1.0	2.0	2.0	3.0	3.0	3.5	3.0	4.0	3.5

Draw the graph of these data points. What kind of possible correlation do they suggest? Compute r and interpret the result.

6. A researcher in a parapsychology laboratory wants to know whether there is a correlation between the number (x) of days randomly chosen test subjects have been involved in an ESP (extrasensory perception) experiment and the number (y) of correct responses they achieve on a certain sensitivity test. Fifteen test subjects are chosen at random, with these results:

x:	3	4	4	4	5	5	5	6	6	6	7	7	7	8	8
y:	7	9	6	5	5	7	8	6	7	5	6	3	8	5	4

Graph the data points. Compute r and interpret the result, using a level of significance $\alpha = 0.05$.

7. A psychologist wishes to determine whether there is a correlation between the number (x) of days before the next full moon and the number (y) of violent crimes reported in a certain large city. Observations are made for 10 different days, with results as follows:

x:	0	1	2	3	4	5	6	7	8	9
y:	8	6	7	6	8	5	4	4	2	2

Draw the graph of these data points, compute the coefficient of correlation r, and interpret the result at the level 0.01.

8. A physiologist interested in the question of correlation between a certain kind of physical agility and mental quickness samples 13 people at random, measuring for each one the number (x) of logic and mathematics problems the person can solve on a standard exam and the number (y) of specified physical tasks each can perform, all within a given time control. The results are:

x:	4	4	5	5	6	6	7	8	8	8	9	9	10
y:	5	8	6	10	4	9	5	10	7	9	6	7	11

Sketch the graph, compute r, and tell what the conclusion is at the level 0.05.

9. To investigate the question of whether there is correlation (at a certain age) between long-term smokers' heights and the average daily number of cigarettes smoked, a researcher randomly chooses 16 such smokers, ascertaining for each the average daily number (x) of cigarettes and the height in centimeters (y). (Intuitively, the researcher is probably wondering about a possibly significant negative correlation.) The sample data pairs (x, y) are:

(12, 175), (15, 170), (16, 180), (20, 162),
(22, 177), (28, 167), (30, 182), (33, 160),
(35, 155), (38, 173), (40, 165), (42, 171),
(45, 156), (47, 150), (50, 163), (57, 164)

Sketch the graph of these data points, compute r, and tell (at the level of significance 0.05) whether a correlation is indicated. What would be the result at the level 0.01?

10. An employer is interested in determining (at the level of significance 0.05) if the employees' ages (x) has any correlation with the number of times (y) they have been late to work

during a 1-year period. Sampling 25 employees at random, he makes the following pairs of observations:

(23, 3), (25, 1), (29, 5), (30, 4), (31, 6),
(32, 2), (34, 3), (36, 1), (37, 5), (39, 0),
(40, 2), (40, 4), (41, 4), (41, 3), (41, 6),
(42, 1), (42, 5), (42, 4), (44, 5), (45, 7),
(47, 4), (47, 2), (50, 1), (53, 0), (59, 2)

Graph these points, compute r, and interpret the result.

11. A researcher wants to see if there is a correlation (at the level 0.01) between x, the daily air pollution level measured on a scale from 0 to 20 in a certain city and y, the daily number of violent crimes reported there. Fifteen daily observations are made, with results as follows:

x:	2	4	5	5	7	9	10	10	10	11	11	13	13	14	16
y:	5	6	5	7	8	7	9	8	7	9	7	11	10	11	10

Graph the points, compute r, and tell what the result is.

PONDERABLES 7-A

1. In Example 2 in Section 7-A, what would a cause-and-effect relation between the correlated variables mean? Is it possible that x and y are correlated because some third, unconsidered factor causes *both*, or has a causal relation with both?

2. What would happen, in the computation of the coefficient of correlation r by formula, if all the y-values were the same? For example, if the data points were points such as

(1, 3), (2, 3), (4, 3), (5, 3), (7, 3), (10, 3)?

3. Similarly, what would happen if the x-values were all the same?

4. In problems 2 and 3, what conclusions can be drawn then?

5. Compute r for the data points

(1, 3), (2, 5), (3, 7), (4, 9), (5, 11),
(6, 13), (7, 15), (8, 17), (9, 19), (10, 21)

Look again at the data points; is there a pattern or mathematical relation between x and y?

B LINEAR REGRESSION

In Section 7-A we examined samples of data points (x, y) whose graphs may or may not approximate or suggest significantly a *linear* arrangement—that is,

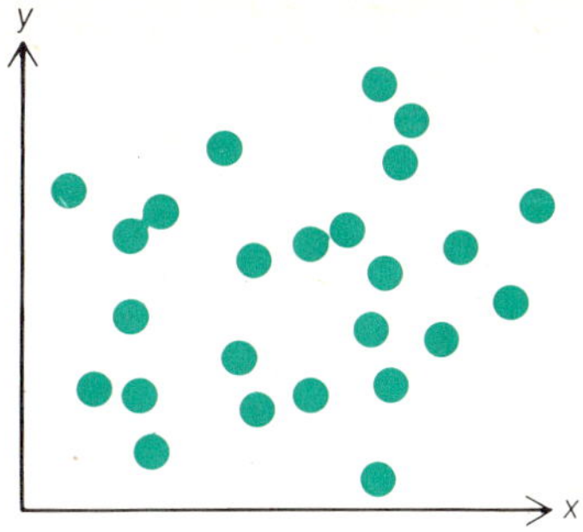

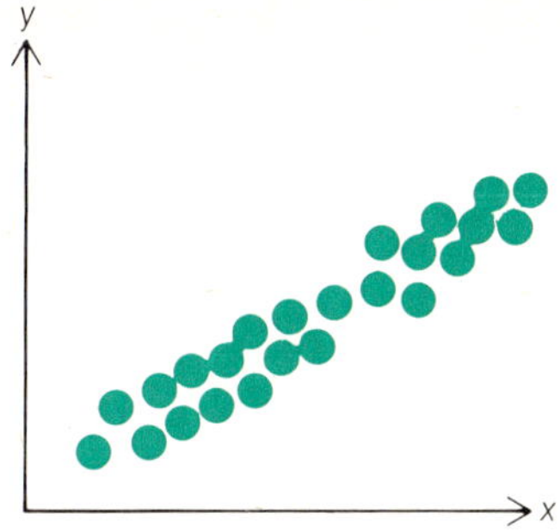

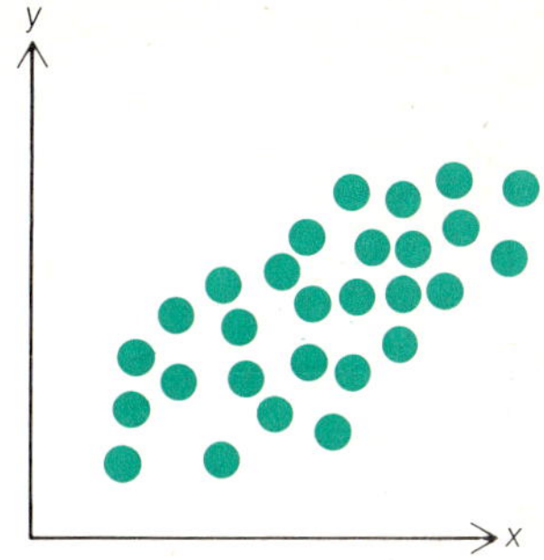

Figure 7.11

whose graphs may more or less tend to look as if the data points are grouped roughly onto a line. For example, although the points in the first part of Figure 7.11 do not suggest such a linear configuration, the points in the second part do strongly suggest one, while the points in the third part are very much open to question either way. It is always necessary to compute r to settle the question with any confidence, because looks can deceive.

Now, at least in the case of the second part of Figure 7.11, where the suggestion of a linear correlation between x and y is very strong, we may well want to know: If there is a *line* which fairly well "fits" the sample points, what line is it? That is, what is its *equation?*

To consider this question properly, one must be aware that in general, such lines have equations of the form

$$y = bx + a,$$

where b and a are constants. For example, the equation

$$y = 2x + 1$$

(with $b = 2$ and $a = 1$) represents a line. To graph this line, we are required only to "plot" two points, and we do so by arbitrarily choosing some two x-values, plugging them in, and determining in each case the corresponding y-value. For instance, we could let x be 1 and then let x be 4:

If $x = 1$, then $y = 2x + 1 = 2(1) + 1 = 2 + 1 = 3$.
If $x = 4$, then $y = 2x + 1 = 2(4) + 1 = 8 + 1 = 9$.
(See Figure 7.12.)

Now, given a set of paired data (x, y), *if* we have computed the coefficient of correlation

$$r = \frac{n \sum (xy) - (\sum x)(\sum y)}{\sqrt{n \sum (x^2) - (\sum x)^2}\,\sqrt{n \sum (y^2) - (\sum y)^2}}$$

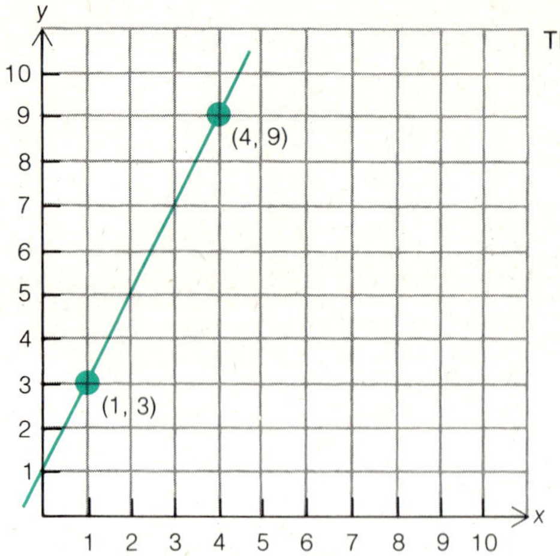

Figure 7.12

and have found that the sample points do suggest significantly a linear correlation, then we will often wish to know what line $y = bx + a$ "best fits" the data points. There is always a "line of best fit" as shown in Figure 7.13, and there are formulas for finding the line's equation—specifically, formulas for producing the constants b and a that are appropriate.

It should be mentioned that when statisticians say **line of best fit,** they mean the line such that if we look at the *vertical distances* by which the line misses the given points, the *sum of the squares* of those distances is *minimal.* There will always be some line such that if all those squared "misses" were added up, the sum would be smaller than it would have been for any other line. (See Figure 7.14.)

The line satisfying this condition, that is, the line of *best fit in the sense of least squares,* is called the **regression line** for the data. Its equation $y = bx + a$ is found (that is to say, the constant amounts b and a are found) by the

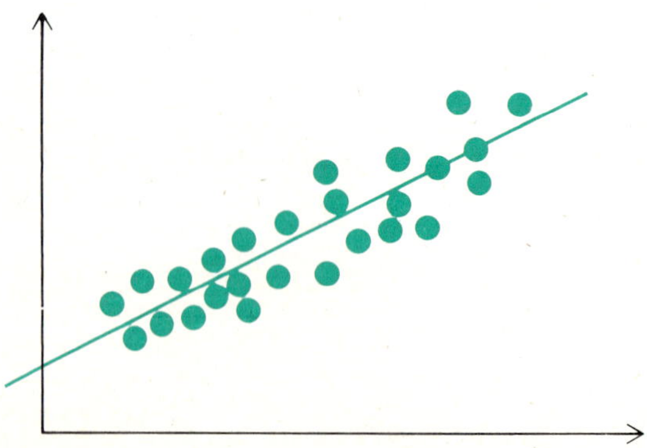

Figure 7.13

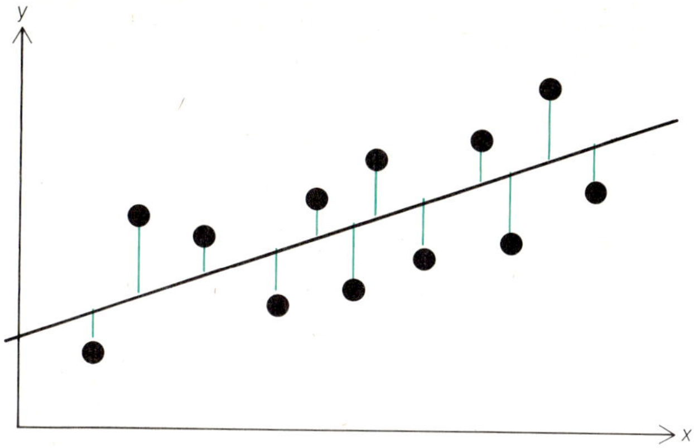

The regression line: the sum of the squares of the vertical distances shown must be smaller than for any other line.

Figure 7.14

following formulas:

$$b = \frac{n\sum(xy) - (\sum x)(\sum y)}{n\sum(x^2) - (\sum x)^2}$$

$$a = \frac{\sum y \sum(x^2) - \sum x \sum(xy)}{n\sum(x^2) - (\sum x)^2}$$

Remarks: The regression line $y = bx + a$ can be computed (that is, b and a can be found) whether or not the data points have been shown to indicate a good correlation—but if there is no significant indication of correlation, the regression line is of little or no interest. Imagine a regression line "fitted" to the random-looking points of the "no correlation" part of Figure 7.11: the line exists mathematically, but who cares? We generally bother to compute $y = bx + a$ only when r has been computed and has shown a good correlation. In that case, in fact, having computed r saves us a lot of the computational trouble involved in finding b and a, because a close look at the formulas for r, b, and a shows that the parts of the fractions for b and a (except for the numerator of a) have already been computed in the process of finding r. See Figure 7.15.

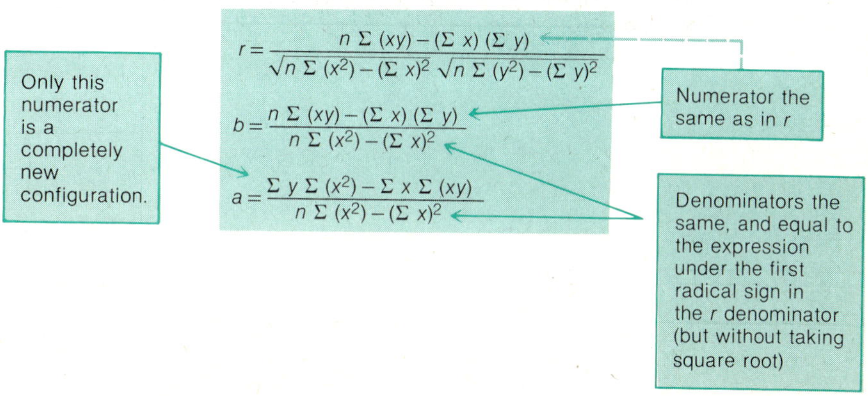

Figure 7.15

EXAMPLES

1. An experimental drug is administered to 10 subjects to see if there is a correlation (at the level 0.01) between the number (x) of milligrams of the drug given and the number (y) of seconds before subjects react to a certain stimulus. The observed data are as follows:

x:	10	10	15	15	20	20	25	30	35	40
y:	4	5	5	7	6	7	8	8	9	9

from which you may verify that the necessary sums are

$$\sum x = 220, \quad \sum y = 68, \quad \left(\sum x\right)^2 = 220^2 = 48{,}400,$$

$$\left(\sum y\right)^2 = 68^2 = 4{,}624, \quad \sum (x^2) = 5{,}800,$$

$$\sum (y^2) = 490, \quad \sum (xy) = 1{,}645$$

so that the computation of the coefficient of correlation r goes as follows:

$$r = \frac{n \sum (xy) - (\sum x)(\sum y)}{\sqrt{n \sum (x^2) - (\sum x)^2}\sqrt{n \sum (y^2) - (\sum y)^2}}$$

$$= \frac{10(1{,}645) - (220)(68)}{\sqrt{10(5{,}800) - 48{,}400}\sqrt{10(490) - 4{,}624}}$$

$$= \frac{16{,}450 - 14{,}960}{\sqrt{58{,}000 - 48{,}400}\sqrt{4{,}900 - 4{,}624}}$$

$$= \frac{1{,}490}{\sqrt{9{,}600}\sqrt{276}} \approx \frac{1{,}490}{(97.98)(16.61)} \approx 0.916.$$

The critical value from the table (with d.f. $= n - 2 = 8$) is $r_{.01} = 0.765$, and because 0.916 is between 0.765 and 1, the indication of correlation is significant at the level 0.01. Therefore it does make sense to derive the equation of the regression line $y = bx + a$, computing the constants b and a from their formulas, making use (where possible, as shown in Figure 7.15) of the computations already made in finding r:

$$b = \frac{n \sum (xy) - (\sum x)(\sum y)}{n \sum (x^2) - (\sum x)^2} = \frac{1{,}490}{9{,}600} \approx 0.16$$

$$a = \frac{\sum y \sum (x^2) - \sum x \sum (xy)}{n \sum (x^2) - (\sum x)^2}$$

$$= \frac{68(5{,}800) - 220(1{,}645)}{9{,}600} = \frac{394{,}400 - 361{,}900}{9{,}600}$$

$$= \frac{32{,}500}{9{,}600} \approx 3.39.$$

Thus the desired regression line is

$$y = 0.16x + 3.39.$$

We may plot this line by choosing some two x-values arbitrarily and computing the corresponding y-values, thus plotting two points and sketching the line through them. For example, taking $x = 0$ and $x = 30$:

$$\text{If } x = 0, \quad y = 0.16x + 3.39$$
$$= 0.16(0) + 3.39 = 3.39.$$
$$\text{If } x = 30, \quad y = 0.16x + 3.39$$
$$= 0.16(30) + 3.39$$
$$= 8.19.$$

The points $(0, 3.39)$ and $(30, 8.19)$ are plotted and the regression line is drawn through them as in Figure 7.16; the two points plotted from the regression line equation are circled to distinguish them from the 10 actually observed data points, which are also shown in Figure 7.16.

Now, this regression line represents the approximate linear relationship between the variables x and y as suggested by the sample data. That is, given an x-value (an amount of the drug being administered), the *expected* y-value (reaction time to the stimulus) is $0.16x + 3.39$ seconds. This y-value would be, given a particular x, the most intelligent *prediction* we could make about what y might correspondingly be: The most intelligent prediction we could make about what reaction time y might be observed for a drug dosage x, on the basis of sample observations. It is critically important to realize, however, that *such predictions are safe to make only for x-values within* **the range of**

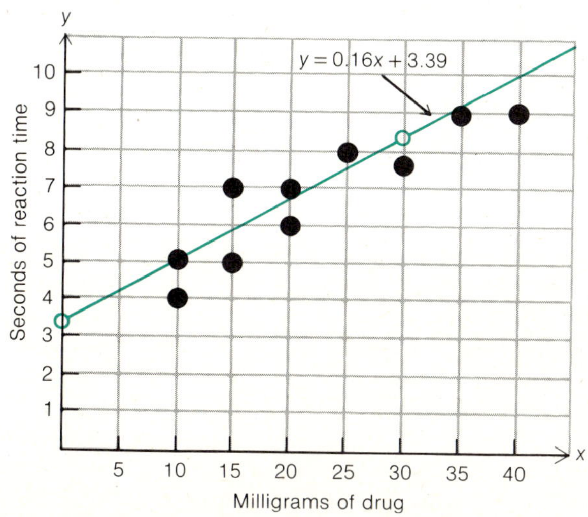

Figure 7.16

sample observations. That is, in this example, we can make *cautious* predictions, predictions about y for given x-values between 10 and 40, because the x's (drug dosages) were *observed* only between 10 and 40 milligrams. Thus, for instance, it would be reasonable to make a prediction about what reaction time one would expect at (say) a drug dosage of $x = 23$ milligrams, and our best prediction would be (putting 23 for x in the line equation):

$$\begin{aligned} y &= 0.16(23) + 3.39 \\ &= 7.07 \text{ seconds.} \end{aligned}$$

Even though a drug dosage of $x = 23$ milligrams was never actually tried in the experiment, our best educated guess at the reaction time that it would be likely to give is slightly more than 7 seconds. We are actually finding, at $x = 23$, the height (vertical height) from the horizontal x-axis at $x = 23$ up to the regression line in the graph of Figure 7.16.

But it would be folly to try to make predictions for x-values (drug dosages) very far *outside* the range of observation, because we have no way of knowing if the relation between x and y still tends to be linear out there, "off the graph" as it were. Besides, there are usually limits of sensibility placed on the variables used. For example, if we try to predict what reaction time "should" follow from a drug dosage of $x = 1{,}000$ milligrams (and predict $0.16(1{,}000) + 3.39 = 163.39$ seconds), we might be overlooking the fact that 1,000 milligrams could be a fatal overdose of the drug. If the subject is given enough of the drug to cause death, needless to say the "predicted" reaction time of 163.39 seconds is of little practical interest or meaning—especially to the subject.

2. In Example 4 in Section 7-A, we looked at a situation in which a statistician wished to determine if there seemed to be a correlation between the number of miles a commuter travels to work (at a certain factory) and the number of times the worker volunteered to work overtime during the past two months. A level of significance $\alpha = 0.01$ was used, and the sample data for 11 workers were:

x:	3	3	4	5	5	6	7	8	9	10	10
y:	10	9	7	8	6	8	6	9	5	3	4

The computation of r went as follows:

$$\begin{aligned} r &= \frac{11(432) - 70(75)}{\sqrt{11(514) - 4{,}900}\,\sqrt{11(561) - 5{,}625}} \\ &= \frac{-498}{\sqrt{754}\,\sqrt{546}} \approx -0.776, \end{aligned}$$

and because this value was seen to be between -1 and the critical value

$r_{.01} = -0.735$, the conclusion was that the sample data suggest a good negative correlation at the level 0.01. Thus the regression line would be reasonable to compute, and again in finding b and a we make use of parts of the formula for r already computed:

$$b = \frac{-498}{754} \approx -0.66$$

$$a = \frac{\sum y \sum (x^2) - \sum x \sum (xy)}{754} = \frac{75(514) - 70(432)}{754}$$

$$= \frac{8{,}310}{754} \approx 11.02$$

so that the equation of the regression line is

$$y = -0.66x + 11.02.$$

By plotting some such two points as these:

$$x = 2: \quad y = -0.66x + 11.02 = -0.66(2) + 11.02$$
$$= -1.32 + 11.02 = 9.70$$

$$x = 10: \quad y = -0.66x + 11.02 = -0.66(10) + 11.02$$
$$= -6.6 + 11.02 = 4.42$$

(shown circled in Figure 7.17) we may sketch the regression line, which graphically shows the *negative* correlation: As x increases, y decreases.

Again, cautious prediction could be made from this regression line. For example, we might say that if a worker lived $x = 7.5$ miles from work, our best

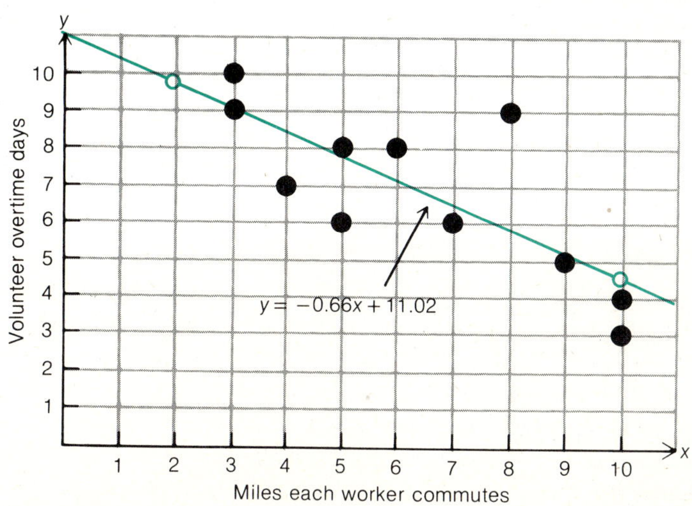

Figure 7.17

guess at the number of times the worker might volunteer to work overtime (which must be a whole number) would be

$$y = -0.66(7.5) + 11.02$$
$$= -4.95 + 11.02 = 6.07 \approx 6 \text{ times.}$$

As always, predictions for x-values *outside* the range of observation would be suspect. It might well be unfair to a worker who lived $x = 17$ miles from work (17 being outside the range of observed x-values) to assume that since the corresponding y would be $-0.66(17) + 11.02 = -11.22 + 11.02 \approx 0$, the worker could *never* be expected during such a time period to volunteer for overtime. (You may even notice that for x's larger than 17, the y-values are negative, and that makes no sense at all in this setting.)

NEW TERMS IN THIS SECTION

regression line
"best fit" in the sense of least squares
range of sample observation

EXERCISES 7-B

1. Using the regression line $y = 0.16x + 3.39$ derived in Example 1 of Section 7-B, find the y-value (the number of seconds until subjects react to the stimulus) for the drug dosage x (in milligrams) equal to

(a) 13 (b) 16 (c) 22 (d) 24 (e) 27 (f) 29 (g) 31
(h) 32 (i) 37 (j) 38

2. In Exercise 1 of Section 7-A, a researcher wished to know whether there was a level 0.05 correlation between the average number (x) of hours of television that school children watched nightly and the number (y) of percentage points they scored below a certain norm on a standardized reading exam. The sample data were:

x:	2	2	2	3	3	3	4	4	4	5	5
y:	8	7	9	9	8	10	9	11	10	9	12

and the computation of r went essentially as follows:

$$r = \frac{120}{\sqrt{138}\sqrt{222}} \approx 0.685 > 0.602$$

so that at the level 0.05 a good positive correlation was indicated. Compute the regression line equation by finding the constants b and a, plot the data points and two arbitrary points for the regression line, and sketch the line.

3. In Exercise 7-A.5, a medical researcher's sample observations indicated a significant correlation at the level 0.01 between the number (x) of degrees above normal certain patients' temperature was and the number (y) of millimeters their blood pressure exceeded a certain established norm. The sample data were:

x:	0.0	0.5	1.0	1.0	1.0	1.5	1.5	2.0	2.0	2.5	2.5	3.0	3.5	4.0
y:	1.0	2.0	1.5	1.0	2.0	1.0	2.0	2.0	3.0	3.0	3.5	3.0	4.0	3.5

and r was calculated as

$$r = \frac{180.5}{\sqrt{241}\sqrt{186.25}} \approx 0.853 > 0.661.$$

Compute the equation of the regression line and sketch the graph of the line and the observed data points.

4. Using the equation of the regression line obtained in the previous exercise, find the best "expectable" y-value to two decimal places (number of millimeters of blood pressure above the norm) for $x =$
(a) 0.7 (b) 1.2 (c) 1.8 (d) 2.3 (e) 2.4 (f) 2.6 (g) 2.9
(h) 3.3 (i) 3.6 (j) 3.8

5. In Exercise 7-A.9, a researcher found indication of a good 0.05-level negative correlation between average numbers of cigarettes smoked daily and the heights in centimeters for people sampled at a certain age who had been long-term smokers. The sample data points were:

(12, 175), (15, 170), (16, 180), (20, 162), (22, 177), (28, 167), (30, 182), (33, 160),
(35, 155), (38, 173), (40, 165), (42, 171), (45, 156), (47, 150), (50, 163), (57, 164).

The computation of r was

$$r = \frac{-16{,}236}{\sqrt{43{,}868}\sqrt{20{,}412}} \approx -0.543$$

which fell between -1 and the critical value -0.497. Compute the equation of the regression line (what should it look like, for a negative correlation?) and sketch the graph.

6. Using the regression line obtained in the previous exercise, what height (to the nearest centimeter) would we best "predict" for a long-term smoker, in the population sampled, whose average daily number of cigarettes was $x =$
(a) 14? (b) 18? (c) 25? (d) 29? (e) 34? (f) 37? (g) 43?
(h) 46? (i) 53? (j) 55?

7. Still with regard to the situation of Exercises 7-B.5 and 7-B.6, the range of observation runs from 12 to 57 cigarettes per day; it would be foolish to use the regression line to theorize about a height that one would "expect" for $x = 484$ cigarettes per day. But if such a prediction were computed, what height (y) would it be?

8. In Exercise 7-A.11, someone found a good correlation, level 0.01, between the daily air pollution level x in a certain city (measured on a scale from 0 to 20) and the daily number (y) of violent crimes reported there. The data observed were:

x:	2	4	5	5	7	9	10	10	10	11	11	13	13	14	16
y:	5	6	5	7	8	7	9	8	7	9	7	11	10	11	10

and the computation of r was

$$r=\frac{1{,}440}{\sqrt{3{,}380}\sqrt{810}}\approx 0.870>0.641.$$

Find the equation of the regression line and draw the graph.

9. From the regression line found in Exercise 7-B.8, how many violent crimes would we best predict (rounded off, of course, to the nearest whole number) for a day when the air pollution reading was
(a) 3? (b) 6? (c) 8? (d) 12? (e) 15?

10. A chemist wishes to see if there is a linear correlation at the level 0.01 between the number of grams (x) of an experimental gasoline additive used in cars and the number (y) of miles per gallon their mileages are improved over the previous averages. The sample data from 14 cars are as follows:

x:	5	5	7	9	9	11	13	13	14	16	18	20	20	25
y:	1.5	1.0	1.2	2.0	1.8	2.5	2.3	2.7	2.6	3.2	3.0	3.4	3.0	3.8

(a) Compute r, the coefficient of correlation, and show that a significant indication of correlation exists at the level 0.01.
(b) Find the equation of the regression line and graph the data points and the line.

11. Using the regression line obtained in the previous exercise, what improvement in gas mileage (in miles per gallon to two decimal places) would be predicted using an amount of gasoline additive equal to
(a) 6 grams? (b) 6.5 grams? (c) 7.5 grams? (d) 8 grams? (e) 9.5 grams?
(f) 10 grams? (g) 12 grams? (h) 15 grams? (i) 17 grams? (j) 19 grams?
(k) 21 grams? (l) 22 grams? (m) 23 grams? (n) 24 grams?

PONDERABLES 7-B

1. In Example 1 in Section 7-B, where the regression line turned out to be $y=0.16x+3.39$ (for x the number of milligrams of a drug given and y the number of seconds taken to respond to a given stimulus), it has been pointed out that at least for x-values within the range of observation (between 10 and 40), one may cautiously predict the y-values that most reasonably would be expected for given x-values. Consider the problem of prediction the other way around. Given a y-value, can we make some cautious predictions about the expected x-value corresponding to

the given y? For what sorts of y-values could such predictions reasonably be made, considering the observed data which gave rise to the regression line? For example, can we safely make a prediction about $y=20$? Find the best determination of an x-value (drug dosage) given that the time y for response to the stimulus is to be each of the following values; express x to the nearest tenth of a milligram.

(a) 4.5 seconds (b) 4.7 seconds (c) 5.3 seconds (d) 5.6 seconds (e) 6.5 seconds
(f) 6.8 seconds (g) 7.2 seconds (h) 7.5 seconds (i) 8.4 seconds (j) 8.6 seconds

2. It is noticed in a laboratory that the time y in seconds required for a certain number of milligrams (x) of a chemical salt to go into solution in a given amount of a liquid is always three more than twice x. What y-values (times) would be obtained from these x-values:

2, 3, 4, 5, 6, 7, 8?

For the resulting data points what do you suppose r should be? To verify your supposition, compute r, then compute b and a and write the equation of the regression line. Compare what it says, algebraically, to the given fact about the way x and y are related.

3. Find an example of four data points for which $r=0$. Compute the regression line anyway (even though for $r=0$ it is of little interest), sketch the graph, and verify in a few cases by looking at your line and a few other lines that even in this situation your line "best fits" the data points in the sense of least squares.

SUGGESTED READINGS

Finite Mathematics From Sets to Game Theory, by Adelbert F. Hackert (D. C. Heath & Company, 1974). Chapter 8 provides interesting worked examples of near-perfect positive and negative correlation.

Statistics in the Real World: A Book of Examples, by Richard J. Larsen and Donna Fox Stroup (Macmillan Publishing Co., Inc., 1976). Chapter 5 gives well-illustrated discussions and very interesting examples of linear correlation and regression.

Statistics and Probability in Modern Life, by Joseph Newmark (Holt, Rinehart & Winston, 1977). Chapter 9 provides some useful applied examples of linear correlation and regression.

8 Nonparametric Statistical Tests

There, though last, not least.—Spenser

A THE SIGN TEST

We come now to a category in the science of statistics known as **nonparametric statistics,** sometimes called **distribution-free statistics.** This subject is a vast area of study that contains a very large number of different statistical tests, and no one user of statistics is likely to use all of them routinely. In this chapter we will look at several of the most important and most often used nonparametric tests. With this background, it should be relatively easy to learn other tests independently.

First, a few remarks are in order about the term *nonparametric statistics.*

Some of the regular statistical tests previously given require the user of the tests to make certain *assumptions* about the populations from which the sample data are taken. For instance, if we test a null hypothesis

$$H_0\colon \mu_1 = \mu_2$$

by sampling two populations with *small* samples and computing a t-value for the difference $\bar{x}_1 - \bar{x}_2$ of the sample means (see Section 5-B, Ponderables problem 4), we have to assume that:

1. The populations are at least approximately normally distributed;
2. The populations have at least approximately the same variance; and
3. The samples are independent of each other.

If we cannot safely make these assumptions, the t-test may be inappropriate or of doubtful validity, and instead a *nonparametric* test (for example, the Mann–Whitney U-test) may be applied instead. It is because such "safer" tests do *not*, for example, assume equality of population variances (variance σ^2 being a population *parameter*) that such tests are called "nonparametric." Also, it is because such tests do not assume such things as population normal distribution that these tests are also sometimes called "distribution-free"—we may apply the tests free from the assumption that populations involved follow a particular distribution.

Nonparametric statistical tests are often referred to as being **robust** or **sturdy** because they are not so fragile as to have their validity destroyed or called into doubt by the absence of such assumptions. (Strictly speaking, the *chi-square* technique studied in Chapter 6 is a nonparametric technique, but it was placed earlier in the text for convenience of coverage.)

Also, these tests are quite useful in their own right, having many applications, as this chapter will show.

The first nonparametric test we will examine here is a simple procedure called the **sign test,** and it will be illustrated by an example as we go.

Suppose we have a group of 15 people involved in a short course designed to improve their reading comprehension. Each person, before the course, is given a thoroughgoing exam which is graded on a scale from 0 to 500. After the course, each person is tested again by a similar but new exam, also graded on a scale from 0 to 500. Let us suppose their scores are as follows; if the score has gone up we will write a *plus* (+), to show that the difference "after minus before" is positive; and if the score has gone down we will write a minus (−); writing nothing if the score has remained unchanged:

Person:	1	2	3	4	5	6	7	8
Before:	385	296	332	240	377	275	412	326
After:	389	280	357	240	401	260	461	315
Sign:	+	−	+		+	−	+	−

Person:	9	10	11	12	13	14	15
Before:	349	220	261	348	242	391	406
After:	388	220	293	355	220	433	428
Sign:	+		+	+	−	+	+

(Actually, the subtraction can be done in either order, and either the number of plus signs or the number of minus signs can be concentrated on.)

It might be tempting to think that we might compare the two situations "before" and "after" by, say, looking at the means

$$\bar{x}_{\text{before}} = \frac{4{,}860}{15} = 324.0$$

$$\bar{x}_{\text{after}} = \frac{5{,}040}{15} = 336.0$$

and doing some standard comparative test (say a *t*-test for small samples) to see if the two sample means are significantly different. But there is a problem with this: The samples are *not independent*. That should be clear from the fact that it was the *same* person who made 385 "before" and 389 "after," and the *same* person who made 296 and then 280, and so on. The samples, then, can hardly be called independent, and such an ordinary testing procedure as a *t*-test may well be inappropriate.

Instead, we conduct a *sign* test with the plus and minus signs as given. (Actually, the sign test is often employed even when a t-test might have been possible, simply because the sign test is computationally much easier.) First, we note that in two cases, Person 4 and Person 10, the score did not change, so that we record neither a plus nor a minus either time. In the sign test we ignore such a nonchange, ignoring Persons 4 and 10 altogether and reducing the sample size accordingly:

$$n = 13.$$

Now, let us say that the null hypothesis H_0 states: There is no overall *improvement* in the "after" results as compared with the "before" results. We do not expect the reading comprehension course to make people's reading comprehension *worse*; we look only at the possibilities of improvement or lack of improvement, hence in this situation we will be dealing with a one-sided test. We can reject H_0 only by arguing that there *is* significant indication of improvement.

And there is such improvement if the number of plus signs is *significantly greater than half* the number of examinees being counted. Because we are counting $n = 13$ people, we will reject the null hypothesis of no improvement only if the *nine* plus signs obtained are significantly better than $\frac{1}{2}n = 6.5$. (We will use a level of significance $\alpha = 0.05$.)

Because each person can turn up either a plus or minus sign, the number (x) of plus signs possible follows a binomial distribution with $\pi = \frac{1}{2}$, but as long as $n > 10$ we may instead use a *normal* distribution whose mean is

$$\mu = \frac{1}{2}n$$

and whose standard deviation is

$$\sigma = \sqrt{n\pi(1-\pi)} = \sqrt{n\left(\frac{1}{2}\right)\left(1-\frac{1}{2}\right)}$$

$$= \sqrt{n\left(\frac{1}{4}\right)} = \frac{1}{2}\sqrt{n}.$$

Thus to apply the sign test, we change the *appropriate* x-value (which the logic of the problem will determine) into a z-score by computing

$$z = \frac{x - \frac{1}{2}n}{\frac{1}{2}\sqrt{n}}$$

which (if we multiply through by 2) may equivalently be written

$$z = \frac{2x - n}{\sqrt{n}}$$

and we check this result in the usual manner against a critical z-value, in this case against $z = 1.64$ because the test is one-sided at the level 0.05.

The null hypothesis here is that there has been *no* improvement in the reading scores after the reading comprehension course; that is, the sample result $\bar{x}_{\text{after}} = 336.0$ is not significantly better, at the level 0.05, than $\bar{x}_{\text{before}} = 324.0$. We will reject this hypothesis only if the probability of getting as many as nine plus signs (that is, getting 9 or more) is less than 0.05. Look at Figure 8.1, and remember that because x is a *discrete* variable (the number of plus signs must be an integer), the continuity correction is called for.

In Figure 8.1 we are checking to see if the shaded area is less than 0.05, and this is equivalent to changing the value $x = 8.5$ into a z-score and seeing whether it exceeds the critical value $z = 1.64$. We compute the z-score:

$$z = \frac{2x - n}{\sqrt{n}} = \frac{2(8.5) - 13}{\sqrt{13}} = \frac{17 - 13}{\sqrt{13}}$$

$$= \frac{4}{\sqrt{13}} \approx \frac{4}{3.605} \approx 1.11 < 1.64$$

so that we do *not* reject the null hypothesis. There is a bigger-than-0.05 chance at random that the number of plus signs would be nine or more, and we cannot state with 95% confidence that the "after" reading results are generally better than the "before" results.

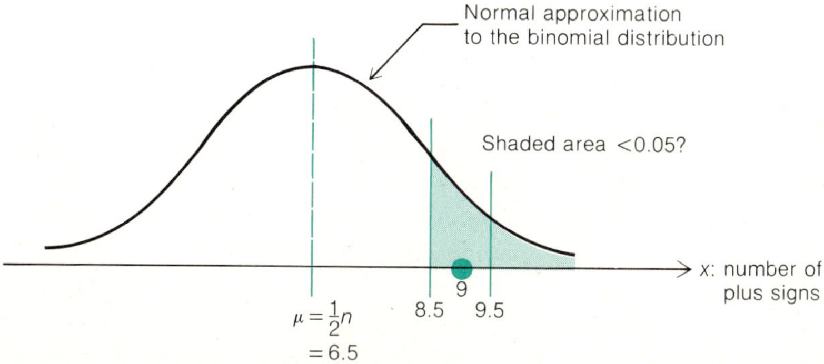

Figure 8.1

EXAMPLES

1. Suppose, in the illustration just discussed, the results of the reading comprehension tests were:

Person:	1	2	3	4	5	6	7	8
Before:	385	296	332	240	377	275	412	326
After:	389	280	357	240	401	290	461	315
Sign:	+	−	+		+	+	+	−

Person:	9	10	11	12	13	14	15
Before:	349	220	261	348	242	391	406
After:	388	220	293	355	220	433	428
Sign:	+		+	+	−	+	+

This time (with $n=13$ again, ignoring the two nonchanges) we have 10 plus signs, and to reject the null hypothesis that there is no significant indication of improvement we must show that the probability of getting as many as 10 (10 or more) plus signs is less than 0.05. That is, we reject the null hypothesis of no improvement only if the z-value for $x=9.5$ exceeds 1.64. We compute

$$z=\frac{2x-n}{\sqrt{n}}=\frac{2(9.5)-13}{\sqrt{13}}$$

$$=\frac{19-13}{\sqrt{13}}=\frac{6}{\sqrt{13}}\approx\frac{6}{3.605}\approx 1.66>1.64$$

and at the level 0.05 we reject the null hypothesis, concluding with 95% confidence that there *is* general improvement in the reading results.

2. Each of 16 persons is tested twice on elementary mathematics: once at normal room temperature and once at a somewhat colder temperature, the tests themselves being very similar in form and difficulty. The results are as follows:

Person:	1	2	3	4	5	6	7	8
Normal:	72	81	66	88	75	63	90	52
Colder:	67	83	66	80	71	58	85	50

Person:	9	10	11	12	13	14	15	16
Normal:	77	86	94	67	78	57	55	80
Colder:	80	80	87	62	71	65	47	80

We will test, at the level 0.05, the hypothesis that on the average there is no difference (either way) between the "normal temperature" results and the "colder temperature" results. (That is, there is no difference between students' knowing their math and knowing it cold!) We note first that there are two unchanged scores, Person 3 and Person 16, so that $n=16-2=14$. If we

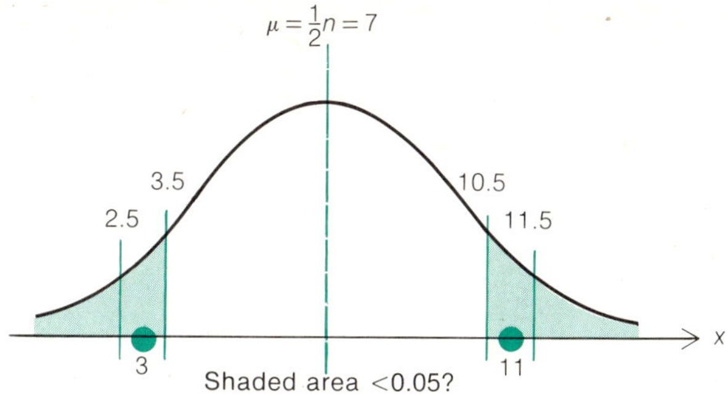

Figure 8.2

subtract the "cold" scores from the "normal" scores, we see that the signs of the differences are:

1	2	3	4	5	6	7	8	9	10	11	12	13	14	15	16
+	−		+	+	+	+	+	−	+	+	+	+	−	+	

Thus there are 11 plus signs. If there is indeed no difference between the two situations, we would theoretically expect *half* the signs to be plus, and the question becomes: Are the eleven plus signs significantly different from $\frac{1}{2}(14)=7$? That is, can we reject the null hypothesis by finding that the probability at random of getting a number of plus signs *that far away* from 7 is less than 0.05? (The test is two-sided, because we have no particular expectations about whether the lower temperature "should" produce better or worse results.) See Figure 8.2: We must consider the probability that the number of plus signs would be as many as 11 (four more than seven) *or* as few as 3 (four less than seven, symmetrically).

The total shaded area in Figure 8.2 would be less than 0.05 (right and left combined) if the z-score for $x=10.5$ were beyond 1.96. We compute

$$z=\frac{2x-n}{\sqrt{n}}=\frac{2(10.5)-14}{\sqrt{14}}=\frac{21-14}{\sqrt{14}}$$

$$=\frac{7}{\sqrt{14}}\approx\frac{7}{3.742}\approx 1.87<1.96$$

so that we do not quite have a case for rejecting the null hypothesis at the level 0.05. The probability at random that the number of plus signs would differ from seven by as much as four on either side is somewhat greater than 0.05.

NEW TERMS IN THIS SECTION

nonparametric (distribution-free) statistics
robust
sturdy
sign test

EXERCISES 8-A

1. Each of 20 student athletes is observed doing a broad jump, once having taken (for a period of time) an experimental vitamin supplement V_1 and once (after a different period of time) having taken a vitamin supplement V_2. Their jumping results, in inches, are as follows:

Student:	1	2	3	4	5	6	7	8	9	10
After V_1:	70	73	68	78	75	80	71	77	81	83
After V_2:	73	72	66	80	77	77	71	75	74	80

Student:	11	12	13	14	15	16	17	18	19	20
After V_1:	76	74	82	71	69	75	85	70	69	74
After V_2:	73	72	78	73	67	75	80	74	66	77

The null hypothesis is that the overall results are not different for the two vitamin supplements. (Is the test one-sided or two-sided? Why?) Test this hypothesis at the 0.05 level with the sign test. Is the number of plus signs (for V_1 results minus V_2 results) significantly different from half?

2. In an experiment similar to that of Exercise 8-A.1, each of 20 student athletes is observed at the broad jump, once having had no vitamin supplement and once having had an extended use of supplement V_1. Their jumping results, in inches, are as follows:

Student:	1	2	3	4	5	6	7	8	9	10
After V_1:	70	73	68	78	75	80	71	77	81	83
None:	68	70	65	79	74	80	72	75	79	80

Student:	11	12	13	14	15	16	17	18	19	20
After V_1:	76	74	82	71	69	75	85	70	69	74
None:	76	75	79	69	66	77	82	68	66	72

The null hypothesis will be that the results with V_1 are no better overall than the results without a vitamin supplement. (Is the test one-sided or two-sided? Why?) Test H_0 at the level 0.05 by the sign test.

3. Seventeen high school students compare the results that they have achieved on an exam from a certain English teacher with the results that they have achieved on an exam from a certain history teacher, as follows:

Student:	1	2	3	4	5	6	7	8	9
English:	78	85	70	91	82	77	63	77	95
History:	83	90	74	88	87	70	69	83	90

Student:	10	11	12	13	14	15	16	17
English:	80	68	74	84	93	81	92	79
History:	85	74	82	91	90	88	86	87

The null hypothesis, to be tested at the level 0.05, is that the history teacher grades no more harshly than the English teacher. Using the sign test, what can you conclude about the null hypothesis? (Concentrate on the minus signs—that is, English results minus history results.)

4. In Exercise 8-A.3, suppose Student 16 had scored 95 in history. Then what is the result of such a test of H_0?

5. An experiment is conducted on 25 cigarette smokers who all smoke Brand *A*. Each one is observed, silently, to see how many cigarettes of Brand *A* he or she smokes in a day. Then, unknown to the smokers, their Brand *A* cigarettes are replaced by Brand *B* cigarettes disguised as Brand *A*, and each smoker is observed again for a day. The results are as follows:

Smoker:	1	2	3	4	5	6	7	8	9
Brand *A*:	22	31	28	35	40	43	26	27	38
Brand *B*:	25	33	26	30	43	40	25	28	32

Smoker:	10	11	12	13	14	15	16	17	18
Brand *A*:	35	25	27	39	44	26	23	33	35
Brand *B*:	31	20	22	35	48	20	20	30	32

Smoker:	19	20	21	22	23	24	25
Brand *A*:	43	28	36	32	30	41	32
Brand *B*:	45	26	32	35	27	35	29

The null hypothesis is that the overall amount of smoking is no different for the two situations. Test this hypothesis at the level 0.05 by the sign test.

6. Eighteen students are tested on a certain topic in mathematics, both before the institution of a tutoring program for them and afterward. Their results are:

Student:	1	2	3	4	5	6	7	8	9
Before:	75	70	62	68	74	80	73	82	85
After:	83	75	60	73	70	75	80	90	81

Student:	10	11	12	13	14	15	16	17	18
Before:	72	80	65	60	71	83	78	84	78
After:	86	92	71	54	88	92	71	76	83

Subtract the "after" scores from the "before" scores and look at the plus signs. The null hypothesis is that the scores overall are no better after tutoring than before. Test this hypothesis with a sign test at the 0.05 level.

7. In the situation in Exercise 8-A.6, suppose the test scores had been:

Student:	1	2	3	4	5	6	7	8	9
Before:	75	70	62	68	74	80	73	82	85
After:	83	75	60	73	78	91	80	90	81

Student:	10	11	12	13	14	15	16	17	18
Before:	72	80	65	60	71	83	78	84	78
After:	86	92	71	54	88	92	71	76	83

What would have been the result of a sign test at the level 0.05? Again, take the null hypothesis to be that the scores overall are no better after tutoring than before.

8. In an experimental psychology laboratory, two different mazes are constructed, called Maze *A* and Maze *B*. Then 30 laboratory mice are made to run through Maze *A*, then at a separate time made to run through Maze *B*, all runs being timed in seconds. The results are as follows:

Mouse:	1	2	3	4	5	6	7	8	9	10
Maze *A*:	28	25	31	34	23	36	27	30	28	38
Maze *B*:	24	30	34	30	27	31	29	24	26	31

Mouse:	11	12	13	14	15	16	17	18	19	20
Maze *A*:	35	35	21	29	34	30	36	22	24	30
Maze *B*:	30	27	23	20	31	22	34	20	28	35

Mouse:	21	22	23	24	25	26	27	28	29	30
Maze *A*:	37	27	20	26	28	29	23	33	27	34
Maze *B*:	32	31	25	24	26	25	29	30	30	31

The null hypothesis is that overall the times are no different for the two mazes. By a sign test at the level of significance 0.05, test this hypothesis.

9. In the same situation as in Exercise 8-A.8, what if Mouse 2 had taken 20 seconds in Maze B and Mouse 29 had taken 24 seconds in Maze B? What does the result mean at the level 0.05? At the level 0.01?

10. In a sign test situation with $n = 50$, what would be the result of a two-sided hypothesis test at the level 0.05 if the number of plus signs was

(a) 27? (b) 29? (c) 31? (d) 33?

PONDERABLES 8-A

1. For a set of matched-pair data, it is possible that if a *t*-test were done on the data it might lead to a rejection even though the sign test did not. Why might this be so?

2. (*Concerning the median.*) The sign test may be used to test an hypothesis that says the median of a population is a certain value. For example, suppose H_0 asserts that the median of a population equals 45.0, and suppose a random sample of size 25 is taken:

53	38	41	32	35	56	40	45	43
37	34	56	48	51	55	37	36	40
41	43	44	52	57	56	57		

A sign test can be done if we consider the differences between these numbers and the claimed median:

Number:	53	38	41	32	35	56	40	45	43
Claimed median:	45	45	45	45	45	45	45	45	45
Difference:	+8	−7	−4	−13	−10	+11	−5	0	−2

Number:	37	34	56	48	51	55	37	36	40
Claimed median:	45	45	45	45	45	45	45	45	45
Difference:	−8	−11	+11	+3	+6	+10	−8	−9	−5

Number:	41	43	44	52	57	56	57
Claimed median:	45	45	45	45	45	45	45
Difference:	−4	−2	−1	+7	+12	+11	+12

We note that there is one instance (the number 45 in the sample) in which the difference is zero, counting for neither a plus nor a minus sign; discounting this instance we have $n = 24$. *If* the null hypothesis is true and the population median is 45.0, we would theoretically expect an equal number of plus and minus signs: 12 of each. There are actually 10 plus signs and 14 minus signs. The question is whether this result is *significantly* different (at, say, the level 0.05) from the expected 12 and 12. Looking at, say, the minus signs, for $x = 13.5$ the z-score is

$$\frac{2(13.5)-24}{\sqrt{24}} = \frac{27-24}{\sqrt{24}}$$

$$\approx \frac{3}{4.899} \approx 0.61 < 1.96$$

so that we do not reject the null hypothesis that the median is 45.0. (This does not "prove," of course, that the median *is* 45.0; we merely cannot with 95% confidence reject the claim that it is.) It has turned out that two plus signs too few and two minus signs too many are not, at the level 0.05, significantly different from the expected. See the following problem.

3. Using the sample data of the previous problem, test at the level 0.05 the hypothesis that the population median is

(a) 42 (b) 40 (c) 38

B THE MANN–WHITNEY *U*-TEST

In the previous section we examined the sign test, which is often applied when a t-test is of doubtful use because (in a matched-pair situation) the two samples are not independent. Now we look at another nonparametric test, called the **Mann–Whitney *U*-test,** which is often applied instead of a t-test when it is not certain whether other t-test conditions are really met: that the two populations sampled have about the same variance and are normally distributed. (Also, a Mann–Whitney U-test may be applied simply because it is computationally simpler to do than a *t*-test.) We will assume that the hypothesis to be tested has the form

$$\mu_1 = \mu_2.$$

Suppose, for example, we wish to test (at the level of significance $\alpha = 0.05$) the hypothesis that two different strains *A* and *B* of laboratory mice have the same (population) mean reaction times, $\mu_1 = \mu_2$, to a given stimulus, and suppose 12 strain *A* mice and 10 strain *B* mice are observed with results (reaction times in milliseconds) as follows:

A: 43 54 38 62 58 48 45 58 40 52 58 55

B: 48 58 63 41 44 55 42 55 57 68

Now, a *t*-test might be done on these sample data to see whether the two sample means $\left(\text{for strain } A, \bar{x}_A = \frac{613}{12} \approx 51.1 \text{ and for strain } B, \bar{x}_B = \frac{531}{10} = 53.1\right)$ are significantly different. But if we instead do a Mann–Whitney *U*-test, the procedure is as follows.

First, the two samples are "pooled" or lumped together, making in this case 22 sample items together. And then these items are *ranked*, arranged in order, from smallest to largest:

38 40 41 42 43 44 45 48 48 52 54

55 55 55 57 58 58 58 58 62 63 68

We will assign what can be called **preliminary ranks** from 1 to 22 to these numbers: rank 1 for 38, rank 2 for 40, and so on. *But*: Then we change these ranks to what may be called **adjusted ranks** by looking at any instance where the sample data are *repeated*, and reassigning to those items the *mean* of the preliminary ranks originally assigned. (No change in ranking is made for sample items that are not repeated.) For instance, both the eighth and the ninth ranked items are 48, and because it hardly seems sensible to act as if one of these 48s is larger than the other, we average the ranks 8 and 9 and assign the rank 8.5 to *each* 48. Figure 8.3 shows the 22 sample items with their preliminary and adjusted ranks. Attention is drawn to those places where "tied" scores (like 48 and 48) have been reranked in the manner described.

Then, having determined ranks (adjusted ranks) for all the data, we sort the data back out into their original two samples, with their ranks attached. In other words, we simply return to the original sample listings, and record by each number the rank that has been assigned to it. (See Figure 8.4.)

We then *add* the ranks associated with each sample, producing two **rank sums** denoted R_1 and R_2, as Figure 8.4 shows.

Now we are ready to compute **the statistic called *U*.** Actually, there are two such statistics, which we will denote ***U*** and ***U′***, and in general they do not

Data	Preliminary ranks	Adjusted ranks	
38	1	1	
40	2	2	
41	3	3	
42	4	4	
43	5	5	
44	6	6	
45	7	7	
48	8	8.5	$\frac{8+9}{2}=8.5$
48	9	8.5	
52	10	10	
54	11	11	
55	12	13	$\frac{12+13+14}{3}=13$
55	13	13	
55	14	13	
57	15	15	
58	16	17.5	$\frac{16+17+18+19}{4}=17.5$
58	17	17.5	
58	18	17.5	
58	19	17.5	
62	20	20	
63	21	21	
68	22	22	

Figure 8.3

have the same value but (in a manner to be explained) will both lead to the same conclusions. The formulas are:

$$U = n_1 n_2 + \frac{n_1(n_1+1)}{2} - R_1$$

$$U' = n_1 n_2 + \frac{n_2(n_2+1)}{2} - R_2$$

where n_1 and n_2 are the sizes of the samples, and R_1 and R_2 are the rank sums as just described.

Strain *A* reaction times	Ranks	Strain *B* reaction times	Ranks
43	5	48	8.5
54	11	58	17.5
38	1	63	21
62	20	41	3
58	17.5	44	6
48	8.5	55	13
45	7	42	4
58	17.5	55	13
40	2	57	15
52	10	68	22
58	17.5	Rank sum $R_2 = 123.0$	
55	13		
Rank sum $R_1 = 130.0$			

Figure 8.4

If we choose to compute the formula for U, we have (with $n_1 = 12$ and $n_2 = 10$):

$$U = 12(10) + \frac{12(12+1)}{2} - 130$$

$$= 120 + 78 - 130 = 68.$$

Now, provided both $n_1 \geq 8$ and $n_2 \geq 8$, this statistic U is approximately *normally distributed*, with mean

$$\mu_U = \frac{n_1 n_2}{2}$$

and standard deviation

$$\sigma_U = \sqrt{\frac{n_1 n_2 (n_1 + n_2 + 1)}{12}}.$$

In this case,

$$\mu_U = \frac{12(10)}{2} = 60$$

$$\sigma_U = \sqrt{\frac{12(10)(12+10+1)}{12}}$$

$$= \sqrt{230} \approx 15.17.$$

Then to test the given hypothesis $\mu_1 = \mu_2$ at the level 0.05, we change U into a z-score

$$z = \frac{U - \mu_U}{\sigma_U}$$

and treat it as in an ordinary two-sided test, rejecting the null hypothesis only if the z-score obtained is out beyond 1.96 or -1.96. In this example we have

$$z = \frac{68 - 60}{15.17} = \frac{8}{15.17} \approx 0.53 < 1.96$$

so that we do *not* reject the null hypothesis $\mu_1 = \mu_2$.

Note that if we had computed U' instead of U, and changed *that* to a z-score, the result would have been

$$U' = 12(10) + \frac{10(10+1)}{2} - 123$$

$$= 120 + 55 - 123 = 52,$$

$$z = \frac{52-60}{15.17} = \frac{-8}{15.17} \approx -0.53$$

which is not beyond -1.96 and leads to no rejection of H_0.

In general, U and U', though they are different in value, have z-scores which are the same except for sign—the same in absolute value. Thus either one can be used in the Mann–Whitney test.

It is not necessary to compute the sample means $\bar{x}_1$ and $\bar{x}_2$ to do the Mann–Whitney test; but it is generally desirable to know what they are, because after all this test's purpose is to determine whether the means are significantly different.

EXAMPLES

1. Suppose, in the situation just discussed, the lab mice's reaction times (in milliseconds) had been:

A:	46	50	39	49	41	53	44	50	61	49	48	50
B:	61	56	64	59	67	67	63	59	61	68		

It may be noted that $\bar{x}_A = \frac{580}{12} \approx 48.3$ and $\bar{x}_B = \frac{625}{10} = 62.5$. Again the null hypothesis, which we will test at the level 0.05, is that $\mu_1 = \mu_2$: The true overall means for strains A and B are equal, and the apparent difference showing up between the sample means is not significant. We proceed to rank the combined data from the two samples:

	39	41	44	46	48	49	49	50
Preliminary rank:	1	2	3	4	5	6	7	8
Adjusted rank:	1	2	3	4	5	6.5	6.5	9

	50	50	53	56	59	59	61	61
Preliminary rank:	9	10	11	12	13	14	15	16
Adjusted rank:	9	9	11	12	13.5	13.5	16	16

	61	63	64	67	67	68
Preliminary rank:	17	18	19	20	21	22
Adjusted rank:	16	18	19	20.5	20.5	22

and then copy these ranks back beside the numbers in their original samples:

A:	46	50	39	49	41	53	44	50	61	49	48	50
Rank:	4	9	1	6.5	2	11	3	9	16	6.5	5	9

B:	61	56	64	59	67	67	63	59	61	68
Rank:	16	12	19	13.5	20.5	20.5	18	13.5	16	22

so that by addition of the ranks for each sample the rank sums are

$R_1 = 82, \quad R_2 = 171.$

If we choose R_1 to work with, we may compute

$$U = n_1 n_2 + \frac{n_1(n_1+1)}{2} - R_1$$

$$= 12(10) + \frac{12(12+1)}{2} - 82$$

$$= 120 + 78 - 82 = 116.$$

And from $\mu_U = 60$ and $\sigma_U = \sqrt{230} \approx 15.17$ as in the previous discussion, we have

$$z = \frac{U - \mu_U}{\sigma_U} = \frac{116-60}{15.17} = \frac{56}{15.17}$$

$$\approx 3.69 > 1.96$$

so that at the level 0.05 we reject the null hypothesis that $\mu_1 = \mu_2$. The two strains of mice do not have the same overall mean reaction times. (Actually we have here a strong enough result even for a rejection at the level 0.01, because $3.69 > 2.58$.) You may verify that if U' had been computed, it would have turned out to be 4, with then $z = -3.69$ and the same conclusion as before.

2. Employees at two different large factories are sampled at random to see how many years they have been at their present jobs, with results as follows:

Factory 1:	11	7	8	9	11	12	9	8	7	10	7
Factory 2:	6	6	2	4	8	5	7	4	3		

$\left(\text{with sample means } \bar{x}_1 = \frac{99}{11} = 9.0,\ \bar{x}_2 = \frac{45}{9} = 6.0\right)$. To determine whether the sample means are different enough to reject the hypothesis $\mu_1 = \mu_2$ at the level of significance $\alpha = 0.01$, we will do a Mann–Whitney test. The 20 sample values are pooled, listed in order, and ranked as previously described:

	2	3	4	4	5	6	6	7	7	7
Preliminary rank:	1	2	3	4	5	6	7	8	9	10
Adjusted rank:	1	2	3.5	3.5	5	6.5	6.5	9.5	9.5	9.5

	7	8	8	8	9	9	10	11	11	12
Preliminary rank:	11	12	13	14	15	16	17	18	19	20
Adjusted rank:	9.5	13	13	13	15.5	15.5	17	18.5	18.5	20

We apply these ranks to the original samples and determine the rank sums:

Factory 1:	11	7	8	9	11	12	9	8	7	10	7
Rank:	18.5	9.5	13	15.5	18.5	20	15.5	13	9.5	17	9.5

Factory 2:	6	6	2	4	8	5	7	4	3
Rank:	6.5	6.5	1	3.5	13	5	9.5	3.5	2

$R_1 = 159.5, \quad R_2 = 50.5.$

Choosing (say) R_1 to work with, and computing U:

$$U = 11(9) + \frac{11(11+1)}{2} - 159.5$$

$$= 99 + 66 - 159.5 = 5.5$$

and with

$$\mu_U = \frac{n_1 n_2}{2} = \frac{(11)(9)}{2} = \frac{99}{2} = 49.5$$

and
$$\sigma_U = \sqrt{\frac{n_1 n_2 (n_1 + n_2 + 1)}{12}}$$

$$= \sqrt{\frac{11(9)(11+9+1)}{12}}$$

$$= \sqrt{\frac{11(3)(11+9+1)}{4}} = \frac{\sqrt{11(3)(21)}}{\sqrt{4}}$$

$$= \frac{\sqrt{693}}{2} \approx \frac{26.32}{2} = 13.16$$

we compute

$$z = \frac{U - \mu_U}{\sigma_U} = \frac{5.5 - 49.5}{13.16}$$

$$= -\frac{44}{13.16} \approx -3.34$$

and because -3.34 is beyond -2.58 under the normal curve, we reject the null hypothesis that the two overall mean longevity figures (μ_1 and μ_2) are the same. You may verify that if R_2 had been used instead of R_1, for computing U' instead of U, the result would have been

$$U' = 93.5$$

$$z = \frac{93.5 - 49.5}{13.16} = \frac{44}{13.16} \approx 3.34 > 2.58.$$

NEW TERMS IN THIS SECTION

- **Mann–Whitney *U*-test**
- **preliminary ranks**
- **adjusted ranks**
- **rank sums**
- **the *U* statistics: *U* and *U'***

EXERCISES 8-B

1. The prices charged for a certain item in various stores are randomly sampled in Gauss City and in Alpha Junction, with results (prices in cents) as follows:

G.C.: 49 53 47 55 51 45 47 49 51 53 44

A.J.: 46 43 56 48 42 57 50 42 49 43 44 40

What are the sample means? The rank sums? Apply the Mann–Whitney *U*-test to tell whether to reject (at the level 0.05) the null hypothesis that the population means (overall mean prices for the item in question) are the same in the two places.

2. In Exercise 8-B.1, suppose the prices sampled had been:

G.C.: 49 54 57 50 59 48 54 45 60 56 59

A.J.: 44 43 50 45 39 40 42 42 41 44 40 38

What are the sample means now? The rank sums? Again, apply the Mann–Whitney test at the level 0.05 to test the hypothesis that the true population mean prices are the same ($\mu_1 = \mu_2$). What if the level of significance had been $\alpha = 0.01$?

3. Two taxicab companies (*A* and *B*) are sampled randomly to see how many minutes it takes a cab to arrive each time when ordered by telephone. The results, in minutes, are as follows:

A: 13 21 18 15 25 17 14 20 15 14 18 23 10 13

B: 22 17 14 15 21 19 24 13 10 11 16 16 12 23

What are the sample mean times? Use the Mann–Whitney procedure to test, at the level 0.05, the hypothesis that the overall mean response times for the two companies are the same.

4. In Exercise 8-B.3, would the Mann–Whitney test result be any different if the fifth sample time for company *A* had been (instead of 25 minutes) 50 minutes? 150 minutes? 300 minutes? What does this suggest?

5. Two groups of students are sampled at random: One group has been taught a course by the lecture method and one group has been taught the same course by an experimental audiovisual method. Their test scores are recorded as follows:

Lecture: 81 70 88 90 84 93 71 88 68 73 75 95 80 90 79 71 81

A-V: 83 70 66 75 71 69 64 74 66 74 62 75 70 80 56 60 75

Compute the sample means. What are the rank sums? Use the Mann–Whitney technique to test the hypothesis that there is no overall difference in the mean scores for the two teaching techniques. Use the level of significance $\alpha = 0.01$.

6. A long-distance commuter, who has two possible routes to drive to work, randomly samples each route, recording the number of times each trip it is necessary to stop for traffic or to wait for traffic lights. The results are as follows:

Route *A*: 7 5 4 8 6 5 7 8 5 4 4 6 5

Route *B*: 4 4 3 3 5 2 6 4 6 3 4 3 2 5 7 3

Use the Mann–Whitney technique to test the null hypothesis that the mean numbers of stops (μ_A and μ_B) are no different overall for the two routes. Use the level of significance $\alpha = 0.01$.

7. Two major newspapers, the *Normal Times* and the *Sigma*, are sampled randomly to see how long their lead editorials are, to the nearest 25 words. The results are:

Normal Times:	350	400	325	375	425	400	450	300	500	425
	450	375	400	525	325	425	350	325	500	375
Sigma:	325	300	500	425	400	475	375	350	350	525
	550	275	300	400	525	550	500	250	525	500

Test, with a Mann–Whitney U-test at the level 0.05, the hypothesis that the true overall mean editorial lengths are the same for the two newspapers.

8. In Exercise 8-B.7, suppose (as unlikely as it might be in reality) all 20 *Normal Times* editorials sampled had been 400 words long, while all 20 *Sigma* editorials sampled had been 425 words long. If the Mann–Whitney test were applied here, what would be the rank sums? U and U'? The z-score(s)? The conclusion?

9. The employees of two different firms are sampled at random to see how many minutes it takes them to perform a certain task at work. Their times are as follows:

Firm 1: 14 12 18 13 15 15 16 11 17 11 18 15 13 16 19 10

Firm 2: 19 17 16 20 14 18 15 20 13 16 17 19 21 12 19

Do a Mann–Whitney U-test at the level 0.05 to determine whether to reject the hypothesis that the overall mean performance times are the same for the two firms. What if the level of significance had been 0.01?

10. A given chemical reaction is timed a dozen times (in tenths of a second: For example, a recording of "17" means 1.7 seconds) in ordinary light and in ultraviolet light, with times as follows:

Ordinary:	17	18	18	16	17	18	16	19	18	18	19	16
Ultraviolet:	16	15	17	16	17	18	17	17	18	15	15	17

Concerning the null hypothesis that the overall mean reaction times are equal regardless of the light used, what is the result of a Mann–Whitney U-test at the level $\alpha = 0.05$?

PONDERABLES 8-B

1. Given that for any positive integer M,

$$1+2+3+\ldots+M = \frac{M(M+1)}{2}$$

prove that the z-scores for U and U' are always the same in absolute value but have opposite signs: $z_U + z_{U'} = 0$.

2. What procedures may the preceding problem suggest as ways to check your computations in the process of doing a Mann–Whitney U-test?

C THE KRUSKAL–WALLIS *H*-TEST

We examined, in Section 8-B, a nonparametric way of testing an hypothesis of the form $\mu_1 = \mu_2$, involving two samples. Now we look at an extension of this idea: a nonparametric way of testing an hypothesis of the form $\mu_1 = \mu_2 = \mu_3 = \ldots$, involving more than two samples. The nonparametric test that will be employed here is called the **Kruskal–Wallis *H*-test.**

First, let us note that an hypothesis of the form $\mu_1 = \mu_2 = \mu_3 = \ldots$ would often be tested by an F-test, as discussed in Chapter 6 under the topic "analysis of variance" (ANOVA). However, the F-test, strictly speaking, involves some underlying assumptions: that the parent populations from which the samples are taken are at least approximately normally distributed, and that they have at least approximately the same variance σ^2. If these assumptions cannot safely be made, then a nonparametric test *not* depending on such assumptions may be more appropriate, so that the F-test may be replaced by the Kruskal–Wallis H-test. For that reason, the latter test is sometimes referred to as **nonparametric analysis of variance.** Also, the H-test is computationally fairly convenient and straightforward, though sometimes a little tedious. And it can readily be done even when the samples are of different sizes. We turn now to a look at how this test is done; the procedure will be shown by way of an example.

Suppose the sixth-grade school children in three different locales are randomly sampled to determine how many hours during a certain week they spend watching television. The results of the sampling are as follows:

Locale 1:	20	19	24	31	17	26	37	15	28	13	
Locale 2:	18	16	21	20	23	22	17	19	21		
Locale 3:	38	26	17	19	16	29	13	27	15	30	28

If we computed the sample means, we would find $\bar{x}_1 = 22.7$, $\bar{x}_2 = 19.7$, and $\bar{x}_3 = 23.5$. The null hypothesis to be tested (say at the level 0.05) is that the true population means, which these sample means estimate, are all equal, so that the children in the three locales watched, on the average, the same amount of television from one place to another during the week in question: $\mu_1 = \mu_2 = \mu_3$.

Now, besides the fact that the samples are of different sizes, the dispersions of the populations may not be uniform, because the sample from locale 2 appears considerably less dispersed than the others. This fact alone would not

Data	Preliminary ranks	Adjusted ranks
13	1	1.5
13	2	1.5
15	3	3.5
15	4	3.5
16	5	5.5
16	6	5.5
17	7	8
17	8	8
17	9	8
18	10	10
19	11	12
19	12	12
19	13	12
20	14	14.5
20	15	14.5
21	16	16.5
21	17	16.5
22	18	18
23	19	19
24	20	20
26	21	21.5
26	22	21.5
27	23	23
28	24	24.5
28	25	24.5
29	26	26
30	27	27
31	28	28
34	29	29
38	30	30

Figure 8.5

Locale #1		Locale #2		Locale #3	
Data	Ranks	Data	Ranks	Data	Ranks
20	14.5	18	10	38	30
19	12	16	5.5	26	21.5
24	20	21	16.5	17	8
31	28	20	14.5	19	12
17	8	23	19	16	5.5
26	21.5	22	18	29	26
34	29	17	8	13	1.5
15	3.5	19	12	27	23
28	24.5	21	16.5	15	3.5
13	1.5		$R_2=120.0$	30	27
	$R_1=162.5$			28	24.5
					$R_3=182.5$

Figure 8.6.

necessarily have invalidated an F-test, because that test is "robust" enough to withstand considerable relaxing of the underlying conditions. But here we will proceed with a Kruskal–Wallis test.

First, just as in the Mann–Whitney U-test of the previous section, the data from the various samples are "pooled" together and ranked, with the ranks of repeated items being adjusted just as in the Mann–Whitney test. See Figure 8.5. Then the data are "sorted back out" into their original samples, with their assigned ranks, and the rank sum is computed for each sample; see Figure 8.6.

Now, the **test statistic** to be computed for this test is denoted H, and is given by the formula

$$H=\frac{12}{n(n+1)}\left(\frac{R_1^2}{n_1}+\frac{R_2^2}{n_2}+\ldots+\frac{R_k^2}{n_k}\right)-3(n+1)$$

where n_1, n_2, and so on (up to n_k) are the sizes of the individual samples, the subscript k merely standing for the *number* of samples. The symbol n stands for the *total* number of sample items: $n=n_1+n_2+\ldots+n_k$.

In our current example, we have *three* samples ($k=3$) of sizes $n_1=10$, $n_2=9$, and $n_3=11$, respectively, so that $n=10+9+11=30$. Plugging these values, together with the rank sums obtained, into the H-formula, we have:

$$H=\frac{12}{30(30+1)}\left(\frac{162.5^2}{10}+\frac{120.0^2}{9}+\frac{182.5^2}{11}\right)-3(30+1)$$

$$=\frac{12}{30(31)}\left(\frac{26{,}406.25}{10}+\frac{14{,}400.00}{9}+\frac{33{,}306.25}{11}\right)-3(31)$$

$$\approx\frac{12}{930}(2{,}640.625+1{,}600+3{,}027.841)-93$$

$$=\frac{12}{930}(7{,}268.466)-93\approx93.787-93=0.787.$$

Let us see what conclusion this value $H=0.787$ leads to.

The H-statistic, it turns out, tends to be "chi-square distributed" (provided each sample size is at least 5), so that we use the chi-square table of critical values to decide whether a computed H-value leads to a rejection of the null hypothesis or not. The d.f. value (degrees of freedom) is

$$\text{d.f.} = k - 1,$$

that is, one less than the number of samples involved.

In our present example, we have d.f. $= k - 1 = 3 - 1 = 2$, so that we look in the chi-square table, under the level of significance 0.05, across from d.f. $= 2$, finding

$$\chi^2_{.05} = 5.991.$$

In the Kruskal–Wallis test, the null hypothesis is rejected only if the computed value of H exceeds the critical chi-square value from the table. Because, in this case, $H = 0.787 < 5.991$, the null hypothesis $\mu_1 = \mu_2 = \mu_3$ is *not* rejected at the level of significance 0.05. It has not been shown that the mean television watching times are other than uniform from locale to locale.

EXAMPLE

1. The drugstore prices charged for a certain item are randomly sampled in four different cities, with results (prices in cents) recorded as follows:

Los Angeles	Dallas	Chicago	Boston
79	68	72	84
83	65	74	82
85	65	70	88
76	70	73	89
78	67	74	87
80	58	56	92
77	55	68	88
79		77	85
		52	

The sample means can be seen to be $\bar{x}_1 = 79.6$, $\bar{x}_2 = 64.0$, $\bar{x}_3 = 68.4$, and $\bar{x}_4 = 86.9$. The null hypothesis in question is that the true, overall mean prices are uniform from city to city:

$$\mu_1 = \mu_2 = \mu_3 = \mu_4.$$

We will test this hypothesis by a Kruskal–Wallis H-test at the level 0.01. The data are combined from the various samples and ranked as follows:

52	55	56	58	65	65	67	68
1	2	3	4	5	6	7	8
1	2	3	4	5.5	5.5	7	8.5

68	70	70	72	73	74	74	76
9	10	11	12	13	14	15	16
8.5	10.5	10.5	12	13	14.5	14.5	16

77	77	78	79	79	80	82	83
17	18	19	20	21	22	23	24
17.5	17.5	19	20.5	20.5	22	23	24

84	85	85	87	88	88	89	92
25	26	27	28	29	30	31	32
25	26.5	26.5	28	29.5	29.5	31	32

Placing these ranks back into the original samples and computing the rank sum for each sample, we have:

Los Angeles

Data	Ranks
79	20.5
83	24
85	26.5
76	16
78	19
80	22
77	17.5
79	20.5
$R_1 =$	166.0

Dallas

Data	Ranks
68	8.5
65	5.5
65	5.5
70	10.5
67	7
58	4
55	2
$R_2 =$	43.0

Chicago

Data	Ranks
72	12
74	14.5
70	10.5
73	13
74	14.5
56	3
68	8.5
77	17.5
52	1
$R_3 =$	94.5

Boston

Data	Ranks
84	25
82	23
88	29.5
89	31
87	28
92	32
88	29.5
85	26.5
$R_4 =$	224.5

Putting these rank sums and the sample sizes

$$n_1 = 8, \quad n_2 = 7, \quad n_3 = 9, \quad n_4 = 8, \quad n = 8+7+9+8 = 32$$

into the formula for H, we have:

$$H = \frac{12}{n(n+1)}\left(\frac{R_1^2}{n_1}+\frac{R_2^2}{n_2}+\frac{R_3^2}{n_3}+\frac{R_4^2}{n_4}\right)-3(n+1)$$

$$= \frac{12}{32(32+1)}\left(\frac{166.0^2}{8}+\frac{43.0^2}{7}+\frac{94.5^2}{9}+\frac{224.5^2}{8}\right)-3(32+1)$$

$$= \frac{12}{32(33)}\left(\frac{27{,}556.00}{8}+\frac{1{,}849.00}{7}+\frac{8{,}930.25}{9}+\frac{50{,}400.25}{8}\right)-3(33)$$

$$\approx \frac{12}{1{,}056}(3{,}444.5+264.143+992.25+6{,}300.031)-99$$

$$= \frac{1}{88}(11{,}000.924)-99 \approx 125.010-99$$

$$= 26.010.$$

Checking this H-value against the critical chi-square value $\chi^2_{.01}$ (with d.f. $= k-1=4-1=3$), where the critical value from the table is 11.345, we see that

$$H = 26.010 > 11.345$$

so that at the level of significance $\alpha = 0.01$ we *reject* the null hypothesis that the population means are equal. The sample data are sufficiently strong evidence that the true mean prices are *not* uniform from city to city.

NEW TERMS IN THIS SECTION

Kruskal–Wallis *H*-test
the *H*-statistic
nonparametric analysis of variance

EXERCISES 8-C

1. Three workers (Brown, Jones, and Smith) are timed, in seconds, to see how long it takes them to perform a certain assembly-line task. Each worker is observed six times at random, and the times are recorded as follows:

Brown:	38	41	40	36	39	44
Jones:	43	42	35	45	37	39
Smith:	32	39	35	46	37	48

What are the sample means? The rank sums? Do a Kruskal–Wallis H-test at the level 0.05 to test the null hypothesis that, overall, the three workers perform at the same rate on the average.

2. Suppose in Exercise 8-C.1, the observed times had been:

Brown: 37 33 31 41 32 36

Jones: 48 39 47 43 38 43

Smith: 32 36 41 33 31 31

Then what would be the result of an H-test at the level 0.05? At the level 0.01?

3. The apples produced by three different growers are randomly sampled and measured (diameters in centimeters) to test the null hypothesis that the overall mean diameters are the same for all three growers. The observed diameters are as follows:

Grower 1: 8.6 9.2 8.3 9.6 9.8 9.4 8.8

Grower 2: 8.5 8.3 8.9 8.2 9.1 9.0 8.2 8.0 8.3

Grower 3: 9.5 10.1 9.6 9.8 9.2 8.9 9.7 10.0

Using a Kruskal–Wallis test at the level 0.01, test the hypothesis H_0: $\mu_1 = \mu_2 = \mu_3$.

4. Laboratory mice are tested to see what reaction times (measured in deciseconds) they have to each of four different stimuli. The reaction times are as follows:

Stimulus 1: 14 19 17 23 16 20 17

Stimulus 2: 18 18 21 24 19 22 16

Stimulus 3: 17 19 25 21 15 26 28

Stimulus 4: 14 18 19 24 16 29 22

What are the sample means? The rank sums? Compute H to test, at the level 0.05, the hypothesis H_0 ($\mu_1 = \mu_2 = \mu_3 = \mu_4$) that the overall mean reaction times are equal for all stimuli.

5. A statistics-minded walker, who customarily walks from his home to the Kingsport Town Library, times himself in minutes, walking each of four available routes a number of times; on each route, walking times vary because of traffic. The sample times are:

Route 1: 22 25 24 24 27 22

Route 2: 20 23 19 18 24 21 22

Route 3: 22 29 17 16 25 23 27

Route 4: 24 22 25 28 26 21

Compute the sample means. At the level of significance 0.05, test (by a Kruskal–Wallis H-test) the hypothesis that the overall mean walking times would be the same for all four routes.

6. Samples are taken at random of the heights, in centimeters, of first-grade school children after 2 years of four different dietary programs, and the observed heights are as follows:

Program A: 89 89 91 90 88 87

Program B: 90 95 96 90 98

Program C: 98 96 103 104 107 98

Program D: 97 89 91 90 92 89 89

What are the sample mean heights for the various programs? To determine if these are significantly different from each other at the level 0.01, compute the rank sums and do a Kruskal–Wallis H-test. The null hypothesis, of course, is $\mu_A = \mu_B = \mu_C = \mu_D$.

7. The weights, in kilograms, of employees at three different stores are randomly sampled, with results as follows:

Store 1: 72 77 70 80 65 68 74 69

Store 2: 66 70 75 72 81 63 71 71

Store 3: 62 68 66 80 76 74 81 77

To test (at the level 0.05) the null hypothesis that the overall mean weights are the same at the three stores, do a Kruskal–Wallis test.

8. In Exercise 8-C.7, what if the two people weighing 81 kilograms had weighed 181 kilograms (as unlikely as that is)? What would have been the effect on the value of H? What, then, can you say about the H-test and extreme or atypical values?

9. A motorist records her gas mileage figures at various times for three different gasoline brands, with results as follows:

Brand A: 16 17 19 20 16 17 15 16

Brand B: 14 17 16 19 15 21 14 22 15

Brand C: 18 20 22 19 18 21 22

What are the sample mean mileages? Compute the three rank sums involved and do an H-test at the level 0.05 for the hypothesis $\mu_A = \mu_B = \mu_C$. What if the level of significance were $\alpha = 0.01$?

10. In Exercise 8-C.9, what if each mileage reading for gasoline brand C had been one less? (Use the level of significance 0.05.)

PONDERABLES 8-C

1. In the computation of the Kruskal–Wallis statistic H, what would be the effect of adding a constant k to each item in all the samples? Explain.

2. Given that for any counting number n,

$$1+2+3+\ldots+n=\frac{n(n+1)}{2},$$

what midcourse *check* can be employed in the H-test procedure, at the point at which the rank sums have just been computed? (Compare Ponderables problem 2, Section 8-B, a similar observation on the Mann–Whitney test.)

D THE RUNS TEST

We look now at a nonparametric test that is designed to test *randomness* in a situation in which sample data are presented in order, and in which the data are of two different kinds. For example, if 20 people are hired one at a time by a company, some male (M) and some female (F), and if the order of hiring is

MFFMFMMFFFMMFFMMMFMF,

we may well want to know whether the sequence of Ms and Fs is random, and for the purpose of investigating this question we employ a statistical technique which is called the **runs test.** (This procedure is sometimes also called the *Wald–Wolfowitz test.*)

First, we will define what is meant by a **run.** A run is any instance in which either symbol (either M or F) occurs one time or more than one time in a row before being "interrupted" by the occurrence of the other symbol. For example, in the sample of 20 symbols just given, the runs are:

M, FF, M, F, MM, FFF, MM, FF, MMM, F, M, F

so that in this situation there happen to be 12 runs.

In the sequence

FFFFFFFFFFFFFFFFFFFF

there would be *one* run, since we have one "string" of Fs not broken by the occurrence of an M.

In the sequence

MFMFMFMFMFMFMFMFMFMF

there are 20 runs, because in such an alternation each M or F counts as a "string of one in a row," so to speak.

Sometimes sample data have to be reinterpreted or reexpressed to make them conform to the two-symbol form. For example, if a die is thrown 24 times in a row, yielding

3 5 1 6 1 3 2 2 4 5 2 6 5 3 2 1 4 5 3 6 2 3 5 1

we might be interested in observing which results are "low" numbers (1, 2, or 3) and which are "high" numbers (4, 5, or 6), so that the sample sequence could be written

L H L H L L L L H H L H

H L L L H H L H L L H L

and, as the reader may verify, there would be 15 runs.

Now, given such a sample sequence involving two symbols, say *A* and *B*, we will let n_1 and n_2 denote the numbers of occurrences of the symbols *A* and *B*, respectively. For instance, in the die sample (*H*s and *L*s) just given, we would have $n_1 = 10$ (10 *H*s) and $n_2 = 14$ (14 *L*s). It ultimately does not matter which designation, n_1 or n_2, is associated with which symbol.

If both n_1 and n_2 are at least 10, the number *R* of runs is approximately *normally distributed*, with mean

$$\mu_R = \frac{2n_1n_2}{n_1+n_2} + 1$$

and standard deviation

$$\sigma_R = \sqrt{\frac{2n_1n_2(2n_1n_2 - n_1 - n_2)}{(n_1+n_2)^2(n_1+n_2-1)}}.$$

(If either n_1 or n_2 falls below 10, the normality assumption is not safe to make, and special tables are needed to tell whether the number of runs argues for or against randomness.)

To proceed with the runs test, we change the number of runs to a *z*-score in the usual manner:

$$z = \frac{R - \mu_R}{\sigma_R}$$

and check the result against one of the usual cutoff scores (±1.96, ±2.58, 1.64 or −1.64, 2.33 or −2.33), depending on the level of significance, and depending on whether the test is to be two-sided or one-sided.

The null hypothesis H_0 to be tested may be that the sample sequence is random: There are neither too few nor too many runs. In this case the test is two-sided.

Or the null hypothesis may be that there are *not too few* runs, and a rejection by a one-sided test would mean that there *are* too few: This may mean that there is a nonrandom **trend,** for example, too many of one symbol in a row. Or we may test (one-sided) the hypothesis that there are *not too many* runs. For example, in the experiment of throwing a die, if the die were thrown by two sleight-of-hand experts in alternation, one expert might make it turn up "low" and the other might make it turn up "high" more often than what could randomly be expected, so that there would be an unusually strong tendency toward alternation and the number of runs might be unusually high. The one-sided test then might well show a rejection.

EXAMPLES

1. A coin is flipped 25 times in a row, with the results (heads *H* or tails *T*)

THHHTTHTHHTHTTTHHTHTTHHHT.

To test the null hypothesis that the sequentiality of heads and tails is random (in the sense that there are neither too few nor too many runs) we will employ a runs test at the level 0.05. An inspection of the sequence for runs gives

T HHH TT H T HH T H TTT HH T H TT HHH T

with 15 runs: $R=15$. Because there are $n_1=13$ heads and $n_2=12$ tails (neither number falling below 10), we proceed with the normal distribution approach, noting that

$$\mu_R=\frac{2n_1n_2}{n_1+n_2}+1=\frac{2(13)(12)}{13+12}+1$$

$$=\frac{312}{25}+1=13.48$$

and $$\sigma_R=\sqrt{\frac{2n_1n_2(2n_1n_2-n_1-n_2)}{(n_1+n_2)^2(n_1+n_2-1)}}$$

$$=\sqrt{\frac{2(13)(12)(2\cdot 13\cdot 12-13-12)}{(13+12)^2(13+12-1)}}$$

$$=\sqrt{\frac{2\cdot 13\cdot 12(312-25)}{25^2(24)}}=\sqrt{\frac{13(287)}{25^2}}$$

$$=\sqrt{\frac{3{,}731}{625}}\approx\sqrt{5.97}\approx 2.44.$$

The necessary z-score then is

$$z=\frac{R-\mu_R}{\sigma_R}=\frac{15-13.48}{2.44}$$

$$=\frac{1.52}{2.44}\approx 0.62$$

and because $0.62 < 1.96$ we do *not* reject, at the level 0.05, the hypothesis that the sequentiality of heads and tails is random.

2. A certain company has promoted, over the past few years, 26 people, and the order in which males (M) and females (F) have been promoted is:

FMMFFFFFMMMMFFFMMMMMFFFFMF

An interested party suspects that the promotion of males or females, in general, is not random in the company because the promotions tend to run in *trends*, so that there are *fewer runs* in the sequence than one would expect for randomness. The null hypothesis to be tested is that the sequence is random, in the specific sense that it does *not* contain too few runs, so that to test it a one-sided runs test is needed. A rejection would mean "suspicions confirmed" to the investigator. The level of significance $\alpha = 0.05$ will be used. The runs in the sequence are identified as

F MM FFFFF MMMM FFF MMMMM FFFF M F

and we see that there are nine of them: $R = 9$. Because there are $n_1 = 12$ Ms and $n_2 = 14$ Fs (both being at least 10) the normal curve approach may be pursued, with mean and standard deviation

$$\mu_R = \frac{2(12)(14)}{12+14} + 1$$

$$= \frac{336}{26} + 1 \approx 13.92$$

$$\sigma_R = \sqrt{\frac{2(12)(14)(2 \cdot 12 \cdot 14 - 12 - 14)}{(12+14)^2(12+14-1)}}$$

$$= \sqrt{\frac{2(12)(14)(310)}{26(26)(25)}} = \frac{\sqrt{336}\sqrt{310}}{26(5)}$$

$$= \frac{18.33(17.61)}{130} \approx 2.48$$

so that the required z-score is

$$z = \frac{9 - 13.92}{2.48} \approx -1.98$$

and because -1.98 is farther to the left than -1.64, the null hypothesis is rejected at the level 0.05. Thus the analyst may be 95% confident that the sequentiality is not random, there being trends causing too few runs to occur. The company seems to be hiring, at various times, "strings" of men or of women.

NEW TERMS IN THIS SECTION

run
runs test (Wald–Wolfowitz test)
trend

EXERCISES 8-D

1. The buyer of a certain kind of stock watches the progress of that stock on the stock market for a period of twenty-three days, recording *U* if the stock price goes up or remains the same, or *D* if it goes down. The sequence obtained is:

DUUDDUDUDDUUUDDUDUUDDDU

Using the runs test at the level 0.05, test the hypothesis that the sequence is random in the sense that there are neither too few nor too many runs.

2. Suppose, in Exercise 8-D.1, the observed sequence had been

DUDUUDDUDUDDUDUUDUDUDDU

Test again, at the level 0.05, the hypothesis that there are neither too few nor too many runs. What if the level of significance had been 0.01?

3. A line of people (male *M*, female *F*) standing at a box office is observed to be as follows:

MFMFFMFMFFFMMFMMFFMFFMFM

and the observer suspects that the sequence shows that in general there is a tendency, among such people waiting in line, for the *M*s and *F*s to alternate, giving too many runs for the sequence to suggest randomness. What is the result of testing (one-sided at the level 0.05) the hypothesis that there are *not* too many runs?

4. A statistics-minded commuter observes the arrival time of the commuter train on 30 consecutive days, writing *T* if the train is on time or early, or *L* if the train is late. The sequence turns out to be:

TTLLLTTTLLLLLTTTTLLTTLLTTTLLLT

The commuter suspects (perhaps because two different train engineers are involved, each driving for a few days at a time) that there are trends, making the number of runs too few for randomness. What is the result of a runs test (one-sided or two-sided?) at the level 0.05 on the hypothesis that there are *not* too few runs?

5. Suppose, in Exercise 8-D.4, the sequence had been

TLTLTLLLTTLTTLTTLLLTLTLTLLTTLT

and suppose the commuter suspects that the sequence is not random because there are too *many* runs (perhaps because there are two drivers involved, who drive in alternation for very short periods). What is the result of a runs test at the level 0.05 on the hypothesis that there are *not* too many runs?

6. It has been found that the overall median age of people using the Ulthar Public Library is 41.5. On a certain day a statistician observes the first 27 people entering the library after a certain time; she asks their ages, and records in each case whether the age is (*A*) above or (*B*) below the median age. The ages are:

39 43 46 27 42 14 30 23 51 44 16 25 45 60

47 17 44 26 46 63 29 47 35 54 36 40 61

The null hypothesis is that the sequence is random, with neither too few nor too many runs. Test this hypothesis at the level 0.05 with the runs test.

7. A golfer keeps records of the last 25 golf games he has played (*W* for win, *L* for lose), and suspects that the sequence is nonrandom in that it contains trends (thus too few runs) which would suggest that he has "off-streaks" and "on-streaks" at various times. The sequence is:

WWLLLWWLLLWWWWWLLWLLLLWWW

Test at the level 0.05 the hypothesis that there are no indicated trends—that is, that there are not too few runs.

8. In Exercise 8-D.7, suppose the game results had been

WWLLLWWLWWWLLLLWLWLLWWWWL

9. A driver in a certain large city makes a note, each time on coming to a traffic light, whether the light is red (*R*) or green (*G*), counting yellow lights as green if she gets through the intersection and as red if she does not. After 35 lights, the results are:

GRGRRRGRRGGRRRRGRGGGGGRRGRGRGGRRGRR.

Test by a runs test at the level 0.05 the hypothesis that the sequence is random, having neither too many nor too few runs.

10. The last 50 people hired by a certain company, with *M* for male and *F* for female, have been in the sequence

FMMFFMFMMFMMMFFMFFFMFMFMM

MFMFFMMFFFFMFMFFMFFMMMMFF.

Test, at the level of significance 0.05, the hypothesis that the sequence is random, in that there is significant indication neither of trends (too few runs) nor near-alternation (too many runs).

PONDERABLES 8-D

1. What runs test result (say at the level 0.01) would be obtained from the sequence

AAAAAAAAAAAABBBBBBBBBBBB

if the null hypothesis is that there are not too few runs? Should the result be surprising?

2. And what about the sequence

ABABABABABABABABABABABAB

if the null hypothesis is that there are not too many runs?

E SPEARMAN RANK CORRELATION

Now we will examine a final nonparametric technique (final for this book—there are many others which the interested reader may look into in more advanced or more specialized texts)—a test that has to do with correlation between two rankings called *Spearman* **rank correlation.**

Let us suppose that two pipe smokers try out a dozen different pipe tobaccos (which for convenience we will call Brands *A, B, C, D, E, F, G, H, I, J, K, L*), ranking them from 1 to 12. There are two extreme situations which could arise. One is that the two smokers might rank the 12 tobaccos in exactly the same way. For example, the rankings could be:

	A	*B*	*C*	*D*	*E*	*F*	*G*	*H*	*I*	*J*	*K*	*L*
Smoker 1's rankings:	9	12	6	2	10	1	7	3	11	4	8	5
Smoker 2's rankings:	9	12	6	2	10	1	7	3	11	4	8	5

The other extreme is that the two smokers could rank the 12 tobaccos in exactly the *reverse* or opposite fashion. For example, the rankings could be:

	A	*B*	*C*	*D*	*E*	*F*	*G*	*H*	*I*	*J*	*K*	*L*
Smoker 1's rankings:	9	1	4	7	5	11	12	2	6	3	10	8
Smoker 2's rankings:	4	12	9	6	8	2	1	11	7	10	3	5

so that in "best to worst" order Smoker 1 has ranked the tobaccos *B, H, J, C, E, I, D, L, A, K, F, G* while Smoker 2 has ranked them in precisely the opposite order: *G, F, K, A, L, D, I, E, C, J, H, B.*

Now, in the first case, with identical rankings, we would say that there is **perfect positive correlation** between the two rankings. And in the second case, with exactly opposite rankings, we would say that there is **perfect negative correlation** between the two rankings.

But, of course, it much more often happens that the two rankings do not copy each other exactly or reverse each other exactly. More often, they differ in *some* respects and present an intermediate comparison of some kind. If they are *close* to being identical, we may say that there is good indication of positive correlation, and if they are *close* to being opposite, we may say that there is good indication of negative correlation. That is, the suggestion of correlation of one kind or another may be statistically *significant*. Or it may not be; two rankings that neither "come close" to being identical or being opposite may show no significant indication of being correlated.

As usual in statistics, we must have a specific procedure for telling "where to draw the line"—we must decide: When are two rankings "nearly enough" identical or "nearly enough" opposite to be significantly correlated?

To answer this question, we compute a statistic called the **Spearman coefficient of rank correlation,** denoted r_s, also sometimes called **Spearman's rho.** The following illustration should show how it is computed and interpreted.

Suppose our two pipe smokers, trying out and ranking the 12 tobaccos, have ranked them as follows:

	A	*B*	*C*	*D*	*E*	*F*	*G*	*H*	*I*	*J*	*K*	*L*
Smoker 1's rankings:	10	1	7	3	2	12	6	4	11	9	5	8
Smoker 2's rankings:	9	3	6	4	1	12	8	2	10	11	5	7

(Actually, *tied* rankings are allowable, but do not occur here. That is, for instance, one smoker might not be able to decide which to rank higher, Brand *C* or Brand *L*, and might rank them both 7.5 instead of 7 for one and 8 for the other.)

Let us suppose the null hypothesis is that the two rankings are uncorrelated (i.e., have nothing to do with each other), neither nearly copying nor nearly reversing each other. We will test this hypothesis at the level 0.05.

We compute, for each tobacco, the *difference* (denoted *d*) of the two rankings, then compute d^2, the square of that difference, and add the quantities d^2 obtained:

	A	*B*	*C*	*D*	*E*	*F*	*G*	*H*	*I*	*J*	*K*	*L*
d:	1	−2	1	−1	1	0	−2	2	1	−2	0	1
d^2:	1	4	1	1	1	0	4	4	1	4	0	1

with $\sum(d^2)=1+4+1+1+1+0+4+4+1+4+0+1=22$.

Now we compute the Spearman rank correlation coefficient by the formula

$$r_s = 1 - \frac{6\sum(d^2)}{n(n^2-1)}$$

where n is the number of pairs of ranks, in this case 12. Thus:

$$r_s = 1 - \frac{6(22)}{12(12^2-1)} = 1 - \frac{132}{12(144-1)}$$

$$= 1 - \frac{132}{12(143)} = 1 - \frac{132}{1{,}716}$$

$$\approx 1 - 0.077 = 0.923.$$

Now, a value for r_s close to 1 suggests a good *positive* correlation (while a value close to -1 would suggest a good negative correlation). But we still must determine whether $r_s = 0.923$ is close enough to 1. To do this, we use the fact that as long as $n \geq 10$, the statistic r_s is approximately normally distributed, with mean $\mu_{r_s} = 0$ and standard deviation

$$\sigma_{r_s} = \frac{1}{\sqrt{n-1}}$$

so that the statistic r_s may be converted to a z-score by the formula

$$z = \frac{r_s - 0}{\frac{1}{\sqrt{n-1}}},$$

that is (multiplying both numerator and denominator by the square root expression):

$$z = r_s\sqrt{n-1}.$$

Because we are testing the hypothesis that there is no correlation one way or the other, and testing at the level of significance 0.05, we would reject the hypothesis only if z went beyond 1.96 or -1.96. In this example,

$$z = r_s\sqrt{n-1} = 0.923\sqrt{12-1}$$

$$= 0.923\sqrt{11} \approx 0.923(3.317) \approx 3.06 > 1.96$$

and we *reject* the null hypothesis: The two rankings are not uncorrelated. (In fact, the present result would be good enough even for a rejection at the level 0.01, because $3.06 > 2.58$.)

EXAMPLES

1. Suppose the two smokers discussed in the text had ranked the 12 tobaccos as follows:

	A	*B*	*C*	*D*	*E*	*F*	*G*	*H*	*I*	*J*	*K*	*L*
Smoker 1's rankings:	7	6	11	1	5	8	2	3	12	10	4	9
Smoker 2's rankings:	5	1	6	11	8	2	7	12	9	3	4	10

We will test (at the level 0.05) the hypothesis that there is no correlation between the two smokers' rankings. We compute the quantities d (differences in rankings) and d^2:

	A	B	C	D	E	F	G	H	I	J	K	L
d:	2	5	5	−10	−3	6	−5	−9	3	7	0	−1
d^2:	4	25	25	100	9	36	25	81	9	49	0	1

with $\sum(d^2) = 364$, so that we have

$$r_s = 1 - \frac{6\sum(d^2)}{n(n^2-1)}$$

$$= 1 - \frac{6(364)}{12(12^2-1)} = 1 - \frac{2{,}184}{12(143)}$$

$$= 1 - \frac{2{,}184}{1{,}716} \approx 1 - 1.273 = -0.273.$$

The z-score, then, is:

$$z = r_s\sqrt{n-1} = (-0.273)\sqrt{12-1}$$

$$= (-0.273)\sqrt{11} \approx (-0.273)(3.317) \approx -0.91,$$

which is not beyond −1.96 out to the left under the normal curve. We do not reject the null hypothesis; there is apparently no correlation between the two rankings.

2. It is claimed that a certain husband and wife have so consistently different preferences in television programs that if they were to rank 14 programs they would rank them in opposite, or very nearly opposite, order. They carry out the experiment of doing the rankings independently, with results as follows:

Program:	1	2	3	4	5	6	7	8	9	10	11	12	13	14
Husband:	12	14	13	2	6	10	3	11	7	1	9	4	8	5
Wife:	5	4	1	14	9	3	12	2	8	13	6	10	7	11

The hypothesis that there is *not a negative correlation* will be tested at the level 0.01; thus a one-sided test will be used, with rejection only if z goes beyond −2.33. The computations:

	1	2	3	4	5	6	7	8	9	10	11	12	13	14
d:	7	10	12	−12	−3	7	−9	9	−1	−12	3	−6	1	−6
d^2:	49	100	144	144	9	49	81	81	1	144	9	36	1	36

so that $\sum(d^2) = 884$, and:

$$r_s = 1 - \frac{6\sum(d^2)}{n(n^2-1)} = 1 - \frac{6(884)}{14(14^2-1)}$$

$$= 1 - \frac{5{,}304}{14(196-1)} = 1 - \frac{5{,}304}{14(195)} = 1 - \frac{5{,}304}{2{,}730}$$

$$\approx 1 - 1.943 = -0.943.$$

$$z = r_s\sqrt{n-1} = (-0.943)\sqrt{14-1} = (-0.943)\sqrt{13}$$

$$\approx (-0.943)(3.606) \approx -3.40.$$

This z-value is beyond -2.33, so that H_0, the hypothesis that there is not a negative correlation, is rejected. Evidently the husband and wife *do* strongly tend to rank television programs oppositely.

NEW TERMS IN THIS SECTION

rank correlation
positive, negative correlations of rankings
Spearman rank correlation coefficient (Spearman's rho)

EXERCISES 8-E

1. Two college students, Jane and Fred, discover in conversation that they have both taken the same 11 college courses and they decide to rank these courses. A mutual friend remarks that their tastes are so much alike that they will probably rank courses about the same, but they are skeptical of this remark. Independently they rank the courses as follows:

Course	Jane's ranking	Fred's ranking
World Literature	3	6
Elementary French	6	7
Western Civilization	2	1
Intro to Sociology	11	10
Intro to Psychology	10	8
Precalculus Math	4	5
General Biology	7	4
Speech	9	11
Intro to Philosophy	8	9
Statistics	5	3
Government	1	2

Compute the Spearman rank correlation coefficient r_s. At the level 0.01, test the hypothesis that the two students' rankings have *no positive* correlation. (Do a one-sided test.)

2. In Exercise 8-E.1, suppose Fred had ranked Speech first and Western Civilization last, instead of the other way around. Then what would have been the result of the test?

3. A town committee meeting to consider 12 different plans for reducing air pollution ranks the plans from 1 (best) to 12 (worst), and independently the mayor of the town also ranks the plans, as follows:

Plan:	*A*	*B*	*C*	*D*	*E*	*F*	*G*	*H*	*I*	*J*	*K*	*L*
Committee:	4	10	1	7	12	3	11	2	6	9	8	5
Mayor:	2	12	5	8	11	1	9	4	6	10	7	3

At the level 0.01, test the null hypothesis that there is no correlation (positive or negative) between the two rankings.

4. Two literary critics who have read 15 books currently on the market rank them from 1 to 15. These critics have a reputation for being so different in their outlooks that they might be expected to rank the books in opposite order. Their rankings are:

Book:	1	2	3	4	5	6	7	8	9	10	11	12	13	14	15
Critic 1:	13	6	15	7	1	14	8	5	2	12	9	11	3	10	4
Critic 2:	15	9	8	4	12	13	2	14	10	5	3	7	6	1	11

Test at the level 0.05 the claim that there is *no* negative correlation between the two rankings.

5. Suppose in Exercise 8-E.4 the two critics' rankings had been:

Book:	1	2	3	4	5	6	7	8	9	10	11	12	13	14	15
Critic 1:	9	13	8	1	14	7	11	3	6	2	12	15	4	10	5
Critic 2:	7	1	9	14	4	8	5	15	10	12	2	3	13	6	11

Then what would have been the result of a one-sided rank correlation test at the level 0.05?

6. A medical researcher ranks 13 patients in terms of their blood pressure and their weight, obtaining the following rankings:

Patient:	1	2	3	4	5	6	7	8	9	10	11	12	13
B.p. ranking:	2	7	11	1	9	10	5	13	12	3	6	8	4
Weight ranking:	2	9	10	5	8	11	4	13	12	1	6	7	3

Test the hypothesis, at the level 0.01, that there is no positive correlation between the rankings.

7. A college instructor ranks 16 students according to the order in which they finished an exam and the grade (rank 1 for highest, rank 16 for lowest) which they achieved. The rankings are:

Student:	1	2	3	4	5	6	7	8
Order of completion:	6	12	13	2	8	4	14	9
Grade ranking:	8	4	9	10	1	11	16	6

Student:	9	10	11	12	13	14	15	16
Order of completion:	15	16	5	10	3	7	11	1
Grade ranking:	14	12	15	2	13	3	7	5

Test, at the level of significance 0.05, the null hypothesis that there is no correlation between the order-of-completion ranking and the grade ranking.

8. Seventeen employees are ranked by their employer, according to how well they did on their aptitude testing when hired and how well they have performed on the job after 1 year. The rankings are:

Employee:	1	2	3	4	5	6	7	8	9
Test ranking:	12	11	2	15	17	10	9	1	8
Performance:	13	10	1	12	15	9	11	5	7

Employee:	10	11	12	13	14	15	16	17
Test ranking:	7	13	3	14	4	16	6	5
Performance:	6	14	4	16	2	17	8	3

At the level of significance $\alpha = 0.01$, test the hypothesis that there is no positive correlation between the two rankings.

9. Two people, Smith and Jones, are asked to rank the following languages in the order in which they would prefer learning them: French, German, Spanish, Italian, Portuguese, Russian, Chinese, Japanese, Arabic, Swedish, Polish, Greek, and Esperanto. Smith ranks them: Spanish, Esperanto, French, Portuguese, Italian, German, Swedish, Arabic, Polish, Russian, Greek, Japanese, and Chinese. Jones ranks them: Spanish, French, Esperanto, Italian, Portuguese, Russian, Polish, Swedish, Chinese, Japanese, German, Greek, and Arabic. Test the null hypothesis, at the level 0.05, that there is no correlation between the rankings. What about the level 0.01?

10. Two art students, Mary and John, are asked to rank, in order of their preference, the artists Van Gogh, Picasso, Rembrandt, Rubens, Monet, Goya, Degas, Pickman, Da Vinci, Van Dyck, El Greco, Doré, Cézanne, Renoir, Matisse, Whistler, and Raphael. Mary ranks them: Van Gogh, Monet, Rembrandt, Rubens, Pickman, Cézanne, Whistler, Degas, Raphael, El Greco, Doré, Renoir, Goya, Da Vinci, Van Dyck, Matisse, and Picasso. John ranks them: Cézanne, Da Vinci, Goya, Renoir, Van Gogh, Matisse, Whistler, Van Dyck, Degas, Doré, Monet, Rubens, Pickman, Rembrandt, El Greco, Picasso, and Raphael. Test at the level of significance 0.05 the hypothesis that there is no correlation between the two rankings.

PONDERABLES 8-E

1. In the situation described at the opening of Section 8-E, in which two pipe smokers independently rank 12 pipe tobaccos, suppose they rank them identically; compute the Spearman rank correlation coefficient r_s. Would the result be the same for *any* number of items being identically ranked by two people?

2. Again referring to the pipe smokers who rank pipe tobaccos, suppose they rank them in exact opposite order. For example, suppose Smoker 1 ranks them *G, A, J, K, B, I, D, H, C, L, F,* and *E*, and suppose Smoker 2 ranks them *E, F, L, C, H, D, I, B, K, J, A,* and *G*. Compute the rank correlation coefficient r_s. Do you think the result would be the same for *any* number of items oppositely ranked?

SUGGESTED READINGS

Statistics: Basic Techniques for Solving Applied Problems, by Stephen A. Book (McGraw-Hill Book Company, 1977). Chapter 11 provides good examples and exercises on nonparametric tests.

Applied Nonparametric Statistics, by Wayne W. Daniel (Houghton Mifflin Company, 1978). This book provides a very thorough treatment of nonparametric statistics, covering many different tests; this is a valuable reference volume for the student seeking a detailed study of this branch of statistics.

Statistics: An Applied Approach, by Neil R. Ullman (Xerox College Publishing, 1972). This is a useful reference work in which the student can look up a variety of nonparametric tests in various parts of the book, such as (in Chapter 15) the Kolmogorov–Smirnov test.

Tables

Digit Sequences for Square Roots

Normal Curve Areas between Mean and z

The t Distribution

Chi-square Distribution

F-ratios, $\alpha = 0.05$

F-ratios, $\alpha = 0.01$

Critical Values of r

Digit Sequences for Square Roots

100	1000	3162	150	1225	3873	200	1414	4472
101	1005	3178	151	1229	3886	201	1418	4483
102	1010	3194	152	1233	3899	202	1421	4494
103	1015	3209	153	1237	3912	203	1425	4506
104	1020	3225	154	1241	3924	204	1428	4517
105	1025	3240	155	1245	3937	205	1432	4528
106	1030	3256	156	1249	3950	206	1435	4539
107	1034	3271	157	1253	3962	207	1439	4550
108	1039	3286	158	1257	3975	208	1442	4561
109	1044	3302	159	1261	3987	209	1446	4572
110	1049	3317	160	1265	4000	210	1449	4583
111	1054	3332	161	1269	4012	211	1453	4593
112	1058	3347	162	1273	4025	212	1456	4604
113	1063	3362	163	1277	4037	213	1459	4615
114	1068	3376	164	1281	4050	214	1463	4626
115	1072	3391	165	1285	4062	215	1466	4637
116	1077	3406	166	1288	4074	216	1470	4648
117	1082	3421	167	1292	4087	217	1473	4658
118	1086	3435	168	1296	4099	218	1476	4669
119	1091	3450	169	1300	4111	219	1480	4680
120	1095	3464	170	1304	4123	220	1483	4690
121	1100	3479	171	1308	4135	221	1487	4701
122	1105	3493	172	1311	4147	222	1490	4712
123	1109	3507	173	1315	4159	223	1493	4722
124	1114	3521	174	1319	4171	224	1497	4733
125	1118	3536	175	1323	4183	225	1500	4743
126	1122	3550	176	1327	4195	226	1503	4754
127	1127	3564	177	1330	4207	227	1507	4764
128	1131	3578	178	1334	4219	228	1510	4775
129	1136	3592	179	1338	4231	229	1513	4785
130	1140	3606	180	1342	4243	230	1517	4796
131	1145	3619	181	1345	4254	231	1520	4806
132	1149	3633	182	1349	4266	232	1523	4817
133	1153	3647	183	1353	4278	233	1526	4827
134	1158	3661	184	1356	4290	234	1530	4837
135	1162	3674	185	1360	4301	235	1533	4848
136	1166	3688	186	1364	4313	236	1536	4858
137	1170	3701	187	1367	4324	237	1539	4868
138	1175	3715	188	1371	4336	238	1543	4879
139	1179	3728	189	1375	4347	239	1546	4889
140	1183	3742	190	1378	4359	240	1549	4899
141	1187	3755	191	1382	4370	241	1552	4909
142	1192	3768	192	1386	4382	242	1556	4919
143	1196	3782	193	1389	4393	243	1559	4930
144	1200	3795	194	1393	4405	244	1562	4940
145	1204	3808	195	1396	4416	245	1565	4950
146	1208	3821	196	1400	4427	246	1568	4960
147	1212	3834	197	1404	4438	247	1572	4970
148	1217	3847	198	1407	4450	248	1575	4980
149	1221	3860	199	1411	4461	249	1578	4990

Digit Sequences for Square Roots (*cont.*)

250	1581	5000	300	1732	5477	350	1871	5916
251	1584	5010	301	1735	5486	351	1873	5925
252	1587	5020	302	1738	5495	352	1876	5933
253	1591	5030	303	1741	5505	353	1879	5941
254	1594	5040	304	1744	5514	354	1881	5950
255	1597	5050	305	1746	5523	355	1884	5958
256	1600	5060	306	1749	5532	356	1887	5967
257	1603	5070	307	1752	5541	357	1889	5975
258	1606	5079	308	1755	5550	358	1892	5983
259	1609	5089	309	1758	5559	359	1895	5992
260	1612	5099	310	1761	5568	360	1897	6000
261	1616	5109	311	1764	5577	361	1900	6008
262	1619	5119	312	1766	5586	362	1903	6017
263	1622	5128	313	1769	5595	363	1905	6025
264	1625	5138	314	1772	5604	364	1908	6033
265	1628	5148	315	1775	5612	365	1910	6042
266	1631	5158	316	1778	5621	366	1913	6050
267	1634	5167	317	1780	5630	367	1916	6058
268	1637	5177	318	1783	5639	368	1918	6066
269	1640	5187	319	1786	5648	369	1921	6075
270	1643	5196	320	1789	5657	370	1924	6083
271	1646	5206	321	1792	5666	371	1926	6091
272	1649	5215	322	1794	5675	372	1929	6099
273	1652	5225	323	1797	5683	373	1931	6107
274	1655	5234	324	1800	5692	374	1934	6116
275	1658	5244	325	1803	5701	375	1936	6124
276	1661	5254	326	1806	5710	376	1939	6132
277	1664	5263	327	1808	5718	377	1942	6140
278	1667	5273	328	1811	5727	378	1944	6148
279	1670	5282	329	1814	5736	379	1947	6156
280	1673	5292	330	1817	5745	380	1949	6164
281	1676	5301	331	1819	5753	381	1952	6173
282	1679	5310	332	1822	5762	382	1954	6181
283	1682	5320	333	1825	5771	383	1957	6189
284	1685	5329	334	1828	5779	384	1960	6197
285	1688	5339	335	1830	5788	385	1962	6205
286	1691	5348	336	1833	5797	386	1965	6213
287	1694	5357	337	1836	5805	387	1967	6221
288	1697	5367	338	1838	5814	388	1970	6229
289	1700	5376	339	1841	5822	389	1972	6237
290	1703	5385	340	1844	5831	390	1975	6245
291	1706	5394	341	1847	5840	391	1977	6253
292	1709	5404	342	1849	5848	392	1980	6261
293	1712	5413	343	1852	5857	393	1982	6269
294	1715	5422	344	1855	5865	394	1985	6277
295	1718	5431	345	1857	5874	395	1987	6285
296	1720	5441	346	1860	5882	396	1990	6293
297	1723	5450	347	1863	5891	397	1992	6301
298	1726	5459	348	1865	5899	398	1995	6309
299	1729	5468	349	1868	5908	399	1997	6317

Digit Sequences for Square Roots (*cont.*)

400	2000	6325	450	2121	6708	500	2236	7071
401	2002	6332	451	2124	6716	501	2238	7078
402	2005	6340	452	2126	6723	502	2241	7085
403	2007	6348	453	2128	6731	503	2243	7092
404	2010	6356	454	2131	6738	504	2245	7099
405	2012	6364	455	2133	6745	505	2247	7106
406	2015	6372	456	2135	6753	506	2249	7113
407	2017	6380	457	2138	6760	507	2252	7120
408	2020	6387	458	2140	6768	508	2254	7127
409	2022	6395	459	2142	6775	509	2256	7134
410	2025	6403	460	2145	6782	510	2258	7141
411	2027	6411	461	2147	6790	511	2261	7148
412	2030	6419	462	2149	6797	512	2263	7155
413	2032	6427	463	2152	6804	513	2265	7162
414	2035	6434	464	2154	6812	514	2267	7169
415	2037	6442	465	2156	6819	515	2269	7176
416	2040	6450	466	2159	6826	516	2272	7183
417	2042	6458	467	2161	6834	517	2274	7190
418	2045	6465	468	2163	6841	518	2276	7197
419	2047	6473	469	2166	6848	519	2278	7204
420	2049	6481	470	2168	6856	520	2280	7211
421	2052	6488	471	2170	6863	521	2283	7218
422	2054	6496	472	2173	6870	522	2285	7225
423	2057	6504	473	2175	6877	523	2287	7232
424	2059	6512	474	2177	6885	524	2289	7239
425	2062	6519	475	2179	6892	525	2291	7246
426	2064	6527	476	2182	6899	526	2293	7253
427	2066	6535	477	2184	6907	527	2296	7259
428	2069	6542	478	2186	6914	528	2298	7266
429	2071	6550	479	2189	6921	529	2300	7273
430	2074	6557	480	2191	6928	530	2302	7280
431	2076	6565	481	2193	6935	531	2304	7287
432	2078	6573	482	2195	6943	532	2307	7294
433	2081	6580	483	2198	6950	533	2309	7301
434	2083	6588	484	2200	6957	534	2311	7308
435	2086	6595	485	2202	6964	535	2313	7314
436	2088	6603	486	2205	6971	536	2315	7321
437	2090	6611	487	2207	6979	537	2317	7328
438	2093	6618	488	2209	6986	538	2319	7335
439	2095	6626	489	2211	6993	539	2322	7342
440	2098	6633	490	2214	7000	540	2324	7348
441	2100	6641	491	2216	7007	541	2326	7355
442	2102	6648	492	2218	7014	542	2328	7362
443	2105	6656	493	2220	7021	543	2330	7369
444	2107	6663	494	2223	7029	544	2332	7376
445	2110	6671	495	2225	7036	545	2335	7382
446	2112	6678	496	2227	7043	546	2337	7389
447	2114	6686	497	2229	7050	547	2339	7396
448	2117	6693	498	2232	7057	548	2341	7403
449	2119	6701	499	2234	7064	549	2343	7409

Digit Sequences for Square Roots (*cont.*)

550	2345	7416	600	2449	7746	650	2550	8062
551	2347	7423	601	2452	7752	651	2551	8068
552	2349	7430	602	2454	7759	652	2553	8075
553	2352	7436	603	2456	7765	653	2555	8081
554	2354	7443	604	2458	7772	654	2557	8087
555	2356	7450	605	2460	7778	655	2559	8093
556	2358	7457	606	2462	7785	656	2561	8099
557	2360	7463	607	2464	7791	657	2563	8106
558	2362	7470	608	2466	7797	658	2565	8112
559	2364	7477	609	2468	7804	659	2567	8118
560	2366	7483	610	2470	7810	660	2569	8124
561	2369	7490	611	2472	7817	661	2571	8130
562	2371	7497	612	2474	7823	662	2573	8136
563	2373	7503	613	2476	7829	663	2575	8142
564	2375	7510	614	2478	7836	664	2577	8149
565	2377	7517	615	2480	7842	665	2579	8155
566	2379	7523	616	2482	7849	666	2581	8161
567	2381	7530	617	2484	7855	667	2583	8167
568	2383	7537	618	2486	7861	668	2585	8173
569	2385	7543	619	2488	7868	669	2587	8179
570	2387	7550	620	2490	7874	670	2588	8185
571	2390	7556	621	2492	7880	671	2590	8191
572	2392	7563	622	2494	7887	672	2592	8198
573	2394	7570	623	2496	7893	673	2594	8204
574	2396	7576	624	2498	7899	674	2596	8210
575	2398	7583	625	2500	7906	675	2598	8216
576	2400	7589	626	2502	7912	676	2600	8222
577	2402	7596	627	2504	7918	677	2602	8228
578	2404	7603	628	2506	7925	678	2604	8234
579	2406	7609	629	2508	7931	679	2606	8240
580	2408	7616	630	2510	7937	680	2608	8246
581	2410	7622	631	2512	7944	681	2610	8252
582	2412	7629	632	2514	7950	682	2612	8258
583	2415	7635	633	2516	7956	683	2613	8264
584	2417	7642	634	2518	7962	684	2615	8270
585	2419	7649	635	2520	7969	685	2617	8276
586	2421	7655	636	2522	7975	686	2619	8283
587	2423	7662	637	2524	7981	687	2621	8289
588	2425	7668	638	2526	7987	688	2623	8295
589	2427	7675	639	2528	7994	689	2625	8301
590	2429	7681	640	2530	8000	690	2627	8307
591	2431	7688	641	2532	8006	691	2629	8313
592	2433	7694	642	2534	8012	692	2631	8319
593	2435	7701	643	2536	8019	693	2632	8325
594	2437	7707	644	2538	8025	694	2634	8331
595	2439	7714	645	2540	8031	695	2636	8337
596	2441	7720	646	2542	8037	696	2638	8343
597	2443	7727	647	2544	8044	697	2640	8349
598	2445	7733	648	2546	8050	698	2642	8355
599	2447	7740	649	2548	8056	699	2644	8361

Digit Sequences for Square Roots (*cont.*)

700	2646	8367	750	2739	8660	800	2828	8944
701	2648	8373	751	2740	8666	801	2830	8950
702	2650	8379	752	2742	8672	802	2832	8955
703	2651	8385	753	2744	8678	803	2834	8961
704	2653	8390	754	2746	8683	804	2835	8967
705	2655	8396	755	2748	8689	805	2837	8972
706	2657	8402	756	2750	8695	806	2839	8978
707	2659	8408	757	2751	8701	807	2841	8983
708	2661	8414	758	2753	8706	808	2843	8989
709	2663	8420	759	2755	8712	809	2844	8994
710	2665	8426	760	2757	8718	810	2846	9000
711	2666	8432	761	2759	8724	811	2848	9006
712	2668	8438	762	2760	8729	812	2850	9011
713	2670	8444	763	2762	8735	813	2851	9017
714	2672	8450	764	2764	8741	814	2853	9022
715	2674	8456	765	2766	8746	815	2855	9028
716	2676	8462	766	2768	8752	816	2857	9033
717	2678	8468	767	2769	8758	817	2858	9039
718	2680	8473	768	2771	8764	818	2860	9044
719	2681	8479	769	2773	8769	819	2862	9050
720	2683	8485	770	2775	8775	820	2864	9055
721	2685	8491	771	2777	8781	821	2865	9061
722	2687	8497	772	2778	8786	822	2867	9066
723	2689	8503	773	2780	8792	823	2869	9072
724	2691	8509	774	2782	8798	824	2871	9077
725	2693	8515	775	2784	8803	825	2872	9083
726	2694	8521	776	2786	8809	826	2874	9088
727	2696	8526	777	2787	8815	827	2876	9094
728	2698	8532	778	2789	8820	828	2877	9099
729	2700	8538	779	2791	8826	829	2879	9105
730	2702	8544	780	2793	8832	830	2881	9110
731	2704	8550	781	2795	8837	831	2883	9116
732	2706	8556	782	2796	8843	832	2884	9121
733	2707	8562	783	2798	8849	833	2886	9127
734	2709	8567	784	2800	8854	834	2888	9132
735	2711	8573	785	2802	8860	835	2890	9138
736	2713	8579	786	2804	8866	836	2891	9143
737	2715	8585	787	2805	8871	837	2893	9149
738	2717	8591	788	2807	8877	838	2895	9154
739	2718	8597	789	2809	8883	839	2897	9160
740	2720	8602	790	2811	8888	840	2898	9165
741	2722	8608	791	2812	8894	841	2900	9171
742	2724	8614	792	2814	8899	842	2902	9176
743	2726	8620	793	2816	8905	843	2903	9182
744	2728	8626	794	2818	8911	844	2905	9187
745	2729	8631	795	2820	8916	845	2907	9192
746	2731	8637	796	2821	8922	846	2909	9198
747	2733	8643	797	2823	8927	847	2910	9203
748	2735	8649	798	2825	8933	848	2912	9209
749	2737	8654	799	2827	8939	849	2914	9214

Digit Sequences for Square Roots (*cont.*)

850	2915	9220	900	3000	9487	950	3082	9747
851	2917	9225	901	3002	9492	951	3084	9752
852	2919	9230	902	3003	9497	952	3085	9757
853	2921	9236	903	3005	9503	953	3087	9762
854	2922	9241	904	3007	9508	954	3089	9767
855	2924	9247	905	3008	9513	955	3090	9772
856	2926	9252	906	3010	9518	956	3092	9778
857	2927	9257	907	3012	9524	957	3094	9783
858	2929	9263	908	3013	9529	958	3095	9788
859	2931	9268	909	3015	9534	959	3097	9793
860	2933	9274	910	3017	9539	960	3098	9798
861	2934	9279	911	3018	9545	961	3100	9803
862	2936	9284	912	3020	9550	962	3102	9808
863	2938	9290	913	3022	9555	963	3103	9813
864	2939	9295	914	3023	9560	964	3105	9818
865	2941	9301	915	3025	9566	965	3106	9823
866	2943	9306	916	3027	9571	966	3108	9829
867	2944	9311	917	3028	9576	967	3110	9834
868	2946	9317	918	3030	9581	968	3111	9839
869	2948	9322	919	3031	9586	969	3113	9844
870	2950	9327	920	3033	9592	970	3114	9849
871	2951	9333	921	3035	9597	971	3116	9854
872	2953	9338	922	3036	9602	972	3118	9859
873	2955	9343	923	3038	9607	973	3119	9864
874	2956	9349	924	3040	9612	974	3121	9869
875	2958	9354	925	3041	9618	975	3122	9874
876	2960	9359	926	3043	9623	976	3124	9879
877	2961	9365	927	3045	9628	977	3126	9884
878	2963	9370	928	3046	9633	978	3127	9889
879	2965	9375	929	3048	9638	979	3129	9894
880	2966	9381	930	3050	9644	980	3130	9899
881	2968	9386	931	3051	9649	981	3132	9905
882	2970	9391	932	3053	9654	982	3134	9910
883	2972	9397	933	3055	9659	983	3135	9915
884	2973	9402	934	3056	9664	984	3137	9920
885	2975	9407	935	3058	9670	985	3138	9925
886	2977	9413	936	3059	9675	986	3140	9930
887	2978	9418	937	3061	9680	987	3142	9935
888	2980	9423	938	3063	9685	988	3143	9940
889	2982	9429	939	3064	9690	989	3145	9945
890	2983	9434	940	3066	9695	990	3146	9950
891	2985	9439	941	3068	9701	991	3148	9955
892	2987	9445	942	3069	9706	992	3150	9960
893	2988	9450	943	3071	9711	993	3151	9965
894	2990	9455	944	3072	9716	994	3153	9970
895	2992	9460	945	3074	9721	995	3154	9975
896	2993	9466	946	3076	9726	996	3156	9980
897	2995	9471	947	3077	9731	997	3158	9985
898	2997	9476	948	3079	9737	998	3159	9990
899	2998	9482	949	3081	9742	999	3161	9995

Normal Curve Areas between Mean and z

z	.00	.01	.02	.03	.04	.05	.06	.07	.08	.09
0.0	.0000	.0040	.0080	.0120	.0160	.0199	.0239	.0279	.0319	.0359
0.1	.0398	.0438	.0478	.0517	.0557	.0596	.0636	.0675	.0714	.0753
0.2	.0793	.0832	.0871	.0910	.0948	.0987	.1026	.1064	.1103	.1141
0.3	.1179	.1217	.1255	.1293	.1331	.1368	.1406	.1443	.1480	.1517
0.4	.1554	.1591	.1628	.1664	.1700	.1736	.1772	.1808	.1844	.1879
0.5	.1915	.1950	.1985	.2019	.2054	.2088	.2123	.2157	.2190	.2224
0.6	.2257	.2291	.2324	.2357	.2389	.2422	.2454	.2486	.2517	.2549
0.7	.2580	.2611	.2642	.2673	.2704	.2734	.2764	.2794	.2823	.2852
0.8	.2881	.2910	.2939	.2967	.2995	.3023	.3051	.3078	.3106	.3133
0.9	.3159	.3186	.3212	.3238	.3264	.3289	.3315	.3340	.3365	.3389
1.0	.3413	.3438	.3461	.3485	.3508	.3531	.3554	.3577	.3599	.3621
1.1	.3643	.3665	.3686	.3708	.3729	.3749	.3770	.3790	.3810	.3830
1.2	.3849	.3869	.3888	.3907	.3925	.3944	.3962	.3980	.3997	.4015
1.3	.4032	.4049	.4066	.4082	.4099	.4115	.4131	.4147	.4162	.4177
1.4	.4192	.4207	.4222	.4236	.4251	.4265	.4279	.4292	.4306	.4319
1.5	.4332	.4345	.4357	.4370	.4382	.4394	.4406	.4418	.4429	.4441
1.6	.4452	.4463	.4474	.4484	.4495	.4505	.4515	.4525	.4535	.4545
1.7	.4554	.4564	.4573	.4582	.4591	.4599	.4608	.4616	.4625	.4633
1.8	.4641	.4649	.4656	.4664	.4671	.4678	.4686	.4693	.4699	.4706
1.9	.4713	.4719	.4726	.4732	.4738	.4744	.4750	.4756	.4761	.4767
2.0	.4772	.4778	.4783	.4788	.4793	.4798	.4803	.4808	.4812	.4817
2.1	.4821	.4826	.4830	.4834	.4838	.4842	.4846	.4850	.4854	.4857
2.2	.4861	.4864	.4868	.4871	.4875	.4878	.4881	.4884	.4887	.4890
2.3	.4893	.4896	.4898	.4901	.4904	.4906	.4909	.4911	.4913	.4916
2.4	.4918	.4920	.4922	.4925	.4927	.4929	.4931	.4932	.4934	.4936
2.5	.4938	.4940	.4941	.4943	.4945	.4946	.4948	.4949	.4951	.4952
2.6	.4953	.4955	.4956	.4957	.4959	.4960	.4961	.4962	.4963	.4964
2.7	.4965	.4966	.4967	.4968	.4969	.4970	.4971	.4972	.4973	.4974
2.8	.4974	.4975	.4976	.4977	.4977	.4978	.4979	.4979	.4980	.4981
2.9	.4981	.4982	.4982	.4983	.4984	.4984	.4985	.4985	.4986	.4986
3.0	.4987	.4987	.4987	.4988	.4988	.4989	.4989	.4989	.4990	.4990

The *t* Distribution

d.f.	$t_{.100}$	$t_{.050}$	$t_{.025}$	$t_{.010}$	$t_{.005}$	d.f.
1	3.078	6.314	12.706	31.821	63.657	1
2	1.886	2.920	4.303	6.965	9.925	2
3	1.638	2.353	3.182	4.541	5.841	3
4	1.533	2.132	2.776	3.747	4.604	4
5	1.476	2.015	2.571	3.365	4.032	5
6	1.440	1.943	2.447	3.143	3.707	6
7	1.415	1.895	2.365	2.998	3.499	7
8	1.397	1.860	2.306	2.896	3.355	8
9	1.383	1.833	2.262	2.821	3.250	9
10	1.372	1.812	2.228	2.764	3.169	10
11	1.363	1.796	2.201	2.718	3.106	11
12	1.356	1.782	2.179	2.681	3.055	12
13	1.350	1.771	2.160	2.650	3.012	13
14	1.345	1.761	2.145	2.624	2.977	14
15	1.341	1.753	2.131	2.602	2.947	15
16	1.337	1.746	2.120	2.583	2.921	16
17	1.333	1.740	2.110	2.567	2.898	17
18	1.330	1.734	2.101	2.552	2.878	18
19	1.328	1.729	2.093	2.539	2.861	19
20	1.325	1.725	2.086	2.528	2.845	20
21	1.323	1.721	2.080	2.518	2.831	21
22	1.321	1.717	2.074	2.508	2.819	22
23	1.319	1.714	2.069	2.500	2.807	23
24	1.318	1.711	2.064	2.492	2.797	24
25	1.316	1.708	2.060	2.485	2.787	25
26	1.315	1.706	2.056	2.479	2.779	26
27	1.314	1.703	2.052	2.473	2.771	27
28	1.313	1.701	2.048	2.467	2.763	28
29	1.311	1.699	2.045	2.462	2.756	29
inf.	1.282	1.645	1.960	2.326	2.576	inf.

Source: Adapted with permission of Macmillan Publishing Co., Inc. from Table IV of *Statistical Methods for Research Workers*, 14th ed., by R. A. Fisher.

Chi-Square Distribution

d.f.	$\chi^2_{.05}$	$\chi^2_{.01}$	d.f.*
1	3.841	6.635	1
2	5.991	9.210	2
3	7.815	11.345	3
4	9.488	13.277	4
5	11.070	15.086	5
6	12.592	16.812	6
7	14.067	18.475	7
8	15.507	20.090	8
9	16.919	21.666	9
10	18.307	23.209	10
11	19.675	24.725	11
12	21.026	26.217	12
13	22.362	27.688	13
14	23.685	29.141	14
15	24.996	30.578	15
16	26.296	32.000	16
17	27.587	33.409	17
18	28.869	34.805	18
19	30.144	36.191	19
20	31.410	37.566	20
21	32.671	38.932	21
22	33.924	40.289	22
23	35.172	41.638	23
24	36.415	42.980	24
25	37.652	44.314	25
26	38.885	45.642	26
27	40.113	46.963	27
28	41.337	48.278	28
29	42.557	49.588	29
30	43.773	50.892	30

Source: This table is based on Table 8 of *Biometrika Tables for Statisticians*, Vol. I (New York: Cambridge University Press, 1966) by permission of the *Biometrika* trustees.

* d.f. $= (r-1)(c-1)$

F Ratios, $\alpha = 0.05$

d.f. for Denominator	Degrees of Freedom for Numerator 1	2	3	4	5	6	7	8	9	10	12	15	20	24	30	40	60	120	M
1	161	200	216	225	230	234	237	239	241	242	244	246	248	249	250	251	252	253	254
2	18.5	19.0	19.2	19.2	19.3	19.3	19.4	19.4	19.4	19.4	19.4	19.4	19.4	19.5	19.5	19.5	19.5	19.5	19.5
3	10.1	9.55	9.28	9.12	9.01	8.94	8.89	8.85	8.81	8.79	8.74	8.70	8.66	8.64	8.62	8.59	8.57	8.55	8.53
4	7.71	6.94	6.59	6.39	6.26	6.16	6.09	6.04	6.00	5.96	5.91	5.86	5.80	5.77	5.75	5.72	5.69	5.66	5.63
5	6.61	5.79	5.41	5.19	5.05	4.95	4.88	4.82	4.77	4.74	4.68	4.62	4.56	4.53	4.50	4.46	4.43	4.40	4.37
6	5.99	5.14	4.76	4.53	4.39	4.28	4.21	4.15	4.10	4.06	4.00	3.94	3.87	3.84	3.81	3.77	3.74	3.70	3.67
7	5.59	4.74	4.35	4.12	3.97	3.87	3.79	3.73	3.68	3.64	3.57	3.51	3.44	3.41	3.38	3.34	3.30	3.27	3.23
8	5.32	4.46	4.07	3.84	3.69	3.58	3.50	3.44	3.39	3.35	3.28	3.22	3.15	3.12	3.08	3.04	3.01	2.97	2.93
9	5.12	4.26	3.86	3.63	3.48	3.37	3.29	3.23	3.18	3.14	3.07	3.01	2.94	2.90	2.86	2.83	2.79	2.75	2.71
10	4.96	4.10	3.71	3.48	3.33	3.22	3.14	3.07	3.02	2.98	2.91	2.85	2.77	2.74	2.70	2.66	2.62	2.58	2.54
11	4.84	3.98	3.59	3.36	3.20	3.09	3.01	2.95	2.90	2.85	2.79	2.72	2.65	2.61	2.57	2.53	2.49	2.45	2.40
12	4.75	3.89	3.49	3.26	3.11	3.00	2.91	2.85	2.80	2.75	2.69	2.62	2.54	2.51	2.47	2.43	2.38	2.34	2.30
13	4.67	3.81	3.41	3.18	3.03	2.92	2.83	2.77	2.71	2.67	2.60	2.53	2.46	2.42	2.38	2.34	2.30	2.25	2.21
14	4.60	3.74	3.34	3.11	2.96	2.85	2.76	2.70	2.65	2.60	2.53	2.46	2.39	2.35	2.31	2.27	2.22	2.18	2.13
15	4.54	3.68	3.29	3.06	2.90	2.79	2.71	2.64	2.59	2.54	2.48	2.40	2.33	2.29	2.25	2.20	2.16	2.11	2.07
16	4.49	3.63	3.24	3.01	2.85	2.74	2.66	2.59	2.54	2.49	2.42	2.35	2.28	2.24	2.19	2.15	2.11	2.06	2.01
17	4.45	3.59	3.20	2.96	2.81	2.70	2.61	2.55	2.49	2.45	2.38	2.31	2.23	2.19	2.15	2.10	2.06	2.01	1.96
18	4.41	3.55	3.16	2.93	2.77	2.66	2.58	2.51	2.46	2.41	2.34	2.27	2.19	2.15	2.11	2.06	2.02	1.97	1.92
19	4.38	3.52	3.13	2.90	2.74	2.63	2.54	2.48	2.42	2.38	2.31	2.23	2.16	2.11	2.07	2.03	1.98	1.93	1.88
20	4.35	3.49	3.10	2.87	2.71	2.60	2.51	2.45	2.39	2.35	2.28	2.20	2.12	2.08	2.04	1.99	1.95	1.90	1.84
21	4.32	3.47	3.07	2.84	2.68	2.57	2.49	2.42	2.37	2.32	2.25	2.18	2.10	2.05	2.01	1.96	1.92	1.87	1.81
22	4.30	3.44	3.05	2.82	2.66	2.55	2.46	2.40	2.34	2.30	2.23	2.15	2.07	2.03	1.98	1.94	1.89	1.84	1.78
23	4.28	3.42	3.03	2.80	2.64	2.53	2.44	2.37	2.32	2.27	2.20	2.13	2.05	2.01	1.96	1.91	1.86	1.81	1.76
24	4.26	3.40	3.01	2.78	2.62	2.51	2.42	2.36	2.30	2.25	2.18	2.11	2.03	1.98	1.94	1.89	1.84	1.79	1.73
25	4.24	3.39	2.99	2.76	2.60	2.49	2.40	2.34	2.28	2.24	2.16	2.09	2.01	1.96	1.92	1.87	1.82	1.77	1.71
30	4.17	3.32	2.92	2.69	2.53	2.42	2.33	2.27	2.21	2.16	2.09	2.01	1.93	1.89	1.84	1.79	1.74	1.68	1.62
40	4.08	3.23	2.84	2.61	2.45	2.34	2.25	2.18	2.12	2.08	2.00	1.92	1.84	1.79	1.74	1.69	1.64	1.58	1.51
60	4.00	3.15	2.76	2.53	2.37	2.25	2.17	2.10	2.04	1.99	1.92	1.84	1.75	1.70	1.65	1.59	1.53	1.47	1.39
120	3.92	3.07	2.68	2.45	2.29	2.18	2.09	2.02	1.96	1.91	1.83	1.75	1.66	1.61	1.55	1.50	1.43	1.35	1.25
∞	3.84	3.00	2.60	2.37	2.21	2.10	2.01	1.94	1.88	1.83	1.75	1.67	1.57	1.52	1.46	1.39	1.32	1.22	1.00

Source: This table is reproduced from M. Merrington and C. M. Thompson, "Tables of percentage points of the inverted beta (*F*) distribution," *Biometrika*, Vol. 33, 1943 (also in *Biometrika Tables for Statisticians*, Vol. I, Table 18), by permission of the *Biometrika* trustees.

F Ratios, $\alpha = 0.01$

d.f. for Denominator	Degrees of Freedom for Numerator 1	2	3	4	5	6	7	8	9	10	12	15	20	24	30	40	60	120	∞
1	4,052	5,000	5,403	5,625	5,764	5,859	5,928	5,982	6,023	6,056	6,106	6,157	6,209	6,235	6,261	6,287	6,313	6,339	6,366
2	98.5	99.0	99.2	99.2	99.3	99.3	99.4	99.4	99.4	99.4	99.4	99.4	99.4	99.5	99.5	99.5	99.5	99.5	99.5
3	34.1	30.8	29.5	28.7	28.2	27.9	27.7	27.5	27.3	27.2	27.1	26.9	26.7	26.6	26.5	26.4	26.3	26.2	26.1
4	21.2	18.0	16.7	16.0	15.5	15.2	15.0	14.8	14.7	14.5	14.4	14.2	14.0	13.9	13.8	13.7	13.7	13.6	13.5
5	16.3	13.3	12.1	11.4	11.0	10.7	10.5	10.3	10.2	10.1	9.89	9.72	9.55	9.47	9.38	9.29	9.20	9.11	9.02
6	13.7	10.9	9.78	9.15	8.75	8.47	8.26	8.10	7.98	7.87	7.72	7.56	7.40	7.31	7.23	7.14	7.06	6.97	6.88
7	12.2	9.55	8.45	7.85	7.46	7.19	6.99	6.84	6.72	6.62	6.47	6.31	6.16	6.07	5.99	5.91	5.82	5.74	5.65
8	11.3	8.65	7.59	7.01	6.63	6.37	6.18	6.03	5.91	5.81	5.67	5.52	5.36	5.28	5.20	5.12	5.03	4.95	4.86
9	10.6	8.02	6.99	6.42	6.06	5.80	5.61	5.47	5.35	5.26	5.11	4.96	4.81	4.73	4.65	4.57	4.48	4.40	4.31
10	10.0	7.56	6.55	5.99	5.64	5.39	5.20	5.06	4.94	4.85	4.71	4.56	4.41	4.33	4.25	4.17	4.08	4.00	3.91
11	9.65	7.21	6.22	5.67	5.32	5.07	4.89	4.74	4.63	4.54	4.40	4.25	4.10	4.02	3.94	3.86	3.78	3.69	3.60
12	9.33	6.93	5.95	5.41	5.06	4.82	4.64	4.50	4.39	4.30	4.16	4.01	3.86	3.78	3.70	3.62	3.54	3.45	3.36
13	9.07	6.70	5.74	5.21	4.86	4.62	4.44	4.30	4.19	4.10	3.96	3.82	3.66	3.59	3.51	3.43	3.34	3.25	3.17
14	8.86	6.51	5.56	5.04	4.70	4.46	4.28	4.14	4.03	3.94	3.80	3.66	3.51	3.43	3.35	3.27	3.18	3.09	3.00
15	8.68	6.36	5.42	4.89	4.56	4.32	4.14	4.00	3.89	3.80	3.67	3.52	3.37	3.29	3.21	3.13	3.05	2.96	2.87
16	8.53	6.23	5.29	4.77	4.44	4.20	4.03	3.89	3.78	3.69	3.55	3.41	3.26	3.18	3.10	3.02	2.93	2.84	2.75
17	8.40	6.11	5.19	4.67	4.34	4.10	3.93	3.79	3.68	3.59	3.46	3.31	3.16	3.08	3.00	2.92	2.83	2.75	2.65
18	8.29	6.01	5.09	4.58	4.25	4.01	3.84	3.71	3.60	3.51	3.37	3.23	3.08	3.00	2.92	2.84	2.75	2.66	2.57
19	8.19	5.93	5.01	4.50	4.17	3.94	3.77	3.63	3.52	3.43	3.30	3.15	3.00	2.92	2.84	2.76	2.67	2.58	2.49
20	8.10	5.85	4.94	4.43	4.10	3.87	3.70	3.56	3.46	3.37	3.23	3.09	2.94	2.86	2.78	2.69	2.61	2.52	2.42
21	8.02	5.78	4.87	4.37	4.04	3.81	3.64	3.51	3.40	3.31	3.17	3.03	2.88	2.80	2.72	2.64	2.55	2.46	2.36
22	7.95	5.72	4.82	4.31	3.99	3.76	3.59	3.45	3.35	3.26	3.12	2.98	2.83	2.75	2.67	2.58	2.50	2.40	2.31
23	7.88	5.66	4.76	4.26	3.94	3.71	3.54	3.41	3.30	3.21	3.07	2.93	2.78	2.70	2.62	2.54	2.45	2.35	2.26
24	7.82	5.61	4.72	4.22	3.90	3.67	3.50	3.36	3.26	3.17	3.03	2.89	2.74	2.66	2.58	2.49	2.40	2.31	2.21
25	7.77	5.57	4.68	4.18	3.86	3.63	3.46	3.32	3.22	3.13	2.99	2.85	2.70	2.62	2.53	2.45	2.36	2.27	2.17
30	7.56	5.39	4.51	4.02	3.70	3.47	3.30	3.17	3.07	2.98	2.84	2.70	2.55	2.47	2.39	2.30	2.21	2.11	2.01
40	7.31	5.18	4.31	3.83	3.51	3.29	3.12	2.99	2.89	2.80	2.66	2.52	2.37	2.29	2.20	2.11	2.02	1.92	1.80
60	7.08	4.98	4.13	3.65	3.34	3.12	2.95	2.82	2.72	2.63	2.50	2.35	2.20	2.12	2.03	1.94	1.84	1.73	1.60
120	6.85	4.79	3.95	3.48	3.17	2.96	2.79	2.66	2.56	2.47	2.34	2.19	2.03	1.95	1.86	1.76	1.66	1.53	1.38
∞	6.63	4.61	3.78	3.32	3.02	2.80	2.64	2.51	2.41	2.32	2.18	2.04	1.88	1.79	1.70	1.59	1.47	1.32	1.00

Source: This table is reproduced from M. Merrington and C. M. Thompson, "Tables of percentage points of the inverted beta (F) distribution," *Biometrika*, Vol. 33, 1943 (also in *Biometrika Tables for Statisticians*, Vol. I, Table 18), by permission of the *Biometrika* trustees.

Critical Values of *r*

d.f.	$r_{.05}$	$r_{.01}$
1	0.997	1.000
2	0.950	0.990
3	0.878	0.959
4	0.811	0.917
5	0.754	0.874
6	0.707	0.834
7	0.666	0.798
8	0.632	0.765
9	0.602	0.735
10	0.576	0.708
11	0.553	0.684
12	0.532	0.661
13	0.514	0.641
14	0.497	0.623
15	0.482	0.606
16	0.468	0.590
17	0.456	0.575
18	0.444	0.561
19	0.433	0.549
20	0.423	0.537
21	0.413	0.526
22	0.404	0.515
23	0.396	0.505
24	0.388	0.496
25	0.381	0.487
26	0.374	0.478
27	0.367	0.470
28	0.361	0.463
29	0.355	0.456
30	0.349	0.449
35	0.325	0.418
40	0.304	0.393
45	0.288	0.372
50	0.273	0.354
60	0.250	0.325
70	0.232	0.302
80	0.217	0.283
90	0.205	0.267
100	0.195	0.254
125	0.174	0.228
150	0.159	0.208
200	0.138	0.181
300	0.113	0.148
400	0.098	0.128
500	0.088	0.115
1000	0.062	0.081

Source: Reprinted by permission from *Statistical Methods* by G. W. Snedecor and W. G. Cochran, sixth edition.

Answers to Exercises

EXERCISES 1-A

1. (a) $\dfrac{238}{12} \approx 19.8$ (b) $\dfrac{597}{11} \approx 54.6$ (c) $\dfrac{220.1}{7} \approx 31.44$ (d) $\dfrac{934}{9} \approx 103.8$

   (e) $\dfrac{769.4}{8} = 96.175 \approx 96.18$ (because 8 is even)

   (f) $\dfrac{5{,}634}{13} \approx 433.4$

   (g) $\dfrac{222.50}{8} = 27.8125 \approx 27.812$ (because 2 is even)

   (h) $\dfrac{0.136}{9} \approx 0.0151$

2. (a) 19.0 (b) 56.0 (c) 31.80 (d) 103.0 (e) 94.8
   (f) 433.0 (g) 28.060 (h) 0.0130

3. The presence of an atypical datum each time "distorts" the mean but has little effect on the median.
   (a) $\bar{x} = 76.8$, $\tilde{x} = 50.5$ (b) $\bar{x} = 634.7$, $\tilde{x} = 735.0$ (c) $\bar{x} = 0.1440$, $\tilde{x} = 0.0825$

4. (a) $\bar{x} = 50.0$, $\tilde{x} = 50.0$ (b) $\bar{x} = 738.0$, $\tilde{x} = 737.0$ (c) $\bar{x} = 0.0809$, $\tilde{x} = 0.0810$

5. $\tilde{x}$

6. (a) $\bar{x} = \dfrac{106}{47} \approx 2.3$ (b) $\tilde{x} = 2.0$ (c) $\hat{x} = 2.0$

7. $\bar{x} = \dfrac{1{,}610}{49} \approx 32.9$, $\tilde{x} = 32.0$, $\hat{x} = 29.0$

8. Reliability: comparatively little chance fluctuation occurs in the means of successive random samples.

9. For Jones, $\bar{x} = 78.6$, $\tilde{x} = 79.5$, $\hat{x} = 80.0$.
   For Smith, $\bar{x} = 84.8$, $\tilde{x} = 79.0$, $\hat{x} = 77.0$.
   For Brown, $\bar{x} = 83.0$, $\tilde{x} = 84.5$, $\hat{x} = 75.0$.
   For *means*, the ranking is Smith, Brown, Jones; for *medians*, the ranking is Brown, Jones, Smith; for *modes*, the ranking is Jones, Smith, Brown.

EXERCISES 1-B

1. (a) 19.52 (b) 6.173 (c) 1.952 (d) 0.6173
 (e) 0.1952 (f) 7.563 (g) 75.63 (h) 23.92
 (i) 2.392 (j) 0.007563 (k) 2.718 (l) 8.591
 (m) 27.18 (n) 271.8 (o) 0.2718 (p) 9.524
 (q) 27.48 (r) 3.038 (s) 0.02802 (t) 423.1

2. (a) $\bar{x}=40.0$, $s^2=\frac{322}{7}=46.0$, $s\approx 6.8$

 (b) $\bar{x}=79.0$, $s^2=\frac{392}{9}\approx 43.6$, $s\approx 6.6$

 (c) $\bar{x}=20.0$, $s^2=\frac{524}{10}=52.4$, $s\approx 7.2$

 (d) $\bar{x}=101.0$, $s^2=\frac{648}{8}=81$, $s=9.0$

 (e) $\bar{x}=60.5$, $s^2=\frac{437}{11}\approx 39.7$, $s\approx 6.3$

 (f) $\bar{x}\approx 63.1$, $s^2=\frac{1{,}548.94}{13}\approx 119.1$, $s\approx 10.9$

 (g) $\bar{x}\approx 68.3$, $s^2=\frac{759.43}{6}\approx 126.6$, $s\approx 11.3$

 (h) $\bar{x}\approx 20.2$, $s^2=\frac{93.72}{12}=7.81$, $s\approx 2.8$

 (i) $\bar{x}=0.0205$, $s^2=\frac{0.001638}{15}=0.0001092$, $s\approx 0.0104$

 (j) $\bar{x}=205.28$, $s^2=\frac{124.956}{9}=13.884$, $s\approx 3.73$

3. First student: $\bar{x}=34.9$ volts, $s\approx 1.5$ volts.
 Second student: $\bar{x}=35.1$ volts, $s\approx 1.3$ volts.

 The second student's results are slightly more consistent because their standard deviation is slightly smaller.

4. Drug *A*: $\bar{x}=11.80$ millimeters, $s\approx 1.32$ millimeters.
 Drug *B*: $\bar{x}=11.80$ millimeters, $s\approx 1.55$ millimeters.

 The drugs have lowered blood pressure by the same mean amount, and Drug *A* appears a little more consistent in its effect than Drug *B*. (However, the samples are small.)

5. (a) Sets *A* and *B* each have range 26 and standard deviation $s=\sqrt{62}\approx 7.9$, but different means ($\bar{x}_A=36$, $\bar{x}_B=91$).
 (b) Same ranges (20) but different standard deviations:

$$s_A=\sqrt{\frac{374}{9}}\approx 5.5,\qquad s_B=\sqrt{\frac{336}{9}}\approx 6.1.$$

 (Means are also different: $\bar{x}_A=20.0$, $\bar{x}_B=29.0$.)
 (c) Ranges are different: 8 for *A*, 4 for *B*. But the standard deviations are exactly equal: in each case,

$$s=\sqrt{\frac{16}{3}}\approx\sqrt{5.33}\approx 2.3.$$

 The means are equal: 24.

(d) Range of A is 25, larger than the range of B, which is 20. But with sample standard deviations the comparison is the other way around:

$$s_A \approx \sqrt{24.2} \approx 4.9$$
$$s_B = \sqrt{50} \approx 7.1.$$

(Means are equal: 49.0.)

6. Both 0.

EXERCISES 1-C

1. (a)

class	f
82 to 85	3
86 to 89	9
90 to 93	18
94 to 97	18
98 to 101	8
102 to 105	11
106 to 109	5

(b)

class	f
82 to 86	6
87 to 91	16
92 to 96	24
97 to 101	10
102 to 106	11
107 to 111	5

(c)

class	f
83 to 85	3
86 to 88	5
89 to 91	14
92 to 94	16
95 to 97	10
98 to 100	5
101 to 103	8
104 to 106	6
107 to 109	5

2. (a) $\bar{X} = \dfrac{6{,}876}{72} = 95.5 \qquad \tilde{X} = 93.5 + \dfrac{6}{18}(4) \approx 94.8$

(b) $\bar{X} = \dfrac{6{,}863}{72} \approx 95.3 \qquad \tilde{X} = 91.5 + \dfrac{14}{24}(5) \approx 94.4$

(c) $\bar{X} = \dfrac{6{,}870}{72} \approx 95.4 \qquad \tilde{X} = 91.5 + \dfrac{8}{16}(3) = 93.0$

3. $\bar{x} = \dfrac{6{,}873}{72} \approx 95.46 \approx 95.5 \qquad \tilde{x} = 94.0$

4. (a)

class	f
20 to 24	12
25 to 29	36
30 to 34	18
35 to 39	10
40 to 44	3
45 to 49	2

(b) $\bar{X} = \dfrac{2{,}402}{81} \approx 29.7$

$$\tilde{X} = 24.5 + \frac{28.5}{36}(5)$$
$$= 24.5 + 3.96 = 28.46 \approx 28.5$$

(c)

class	f
21 to 24	12
25 to 28	31
29 to 32	18
33 to 36	10
37 to 40	6
41 to 44	2
45 to 48	2

(d) $\bar{X} = \dfrac{2{,}394.5}{81} \approx 29.6$

$$\tilde{X} = 24.5 + \frac{28.5}{31}(4) \approx 24.5 + 3.68$$
$$= 28.18 \approx 28.2$$

(Remember, the medial class really runs from 24.5 all the way to 28.5, thus includes 28.2.)

(e)

class	f
22 to 24	12
25 to 27	21
28 to 30	21
31 to 33	10
34 to 36	7
37 to 39	5
40 to 42	2
43 to 45	1
46 to 48	2

(f) $\bar{X} = \dfrac{2{,}406}{81} \approx 29.7$

$$\tilde{X} = 27.5 + \frac{7.5}{21}(3) \approx 27.5 + 1.07$$

$$= 28.57 \approx 28.6$$

5. $\bar{x} = \dfrac{2{,}403}{81} \approx 29.7 \qquad \tilde{x} = 28.0$

6. (a) The numbers shown: 70 and 76, 77 and 83, etc.
   (b) 69.5 and 76.5; 76.5 and 83.5; 83.5 and 90.5; 90.5 and 97.5; 97.5 and 104.5; 104.5 and 111.5.
   (c) 73, 80, 87, 94, 101, and 108
   (d) 7
   (e) 85
   (f) $\bar{X} = \dfrac{7{,}623}{85} \approx 89.7$. It cannot be checked against $\bar{x}$ because the raw data are unavailable.
   (g) $\tilde{X} = 90.5 + \dfrac{1.5}{20}(7) = 90.5 + 0.525 \approx 91.0$. It cannot be checked against $\tilde{x}$ because the raw data are unavailable.

7. (a) $\bar{X} = \dfrac{11{,}485}{200} \approx 57.4 \qquad \tilde{X} = 55.5 + \dfrac{12}{43}(7) \approx 55.5 + 1.954 \approx 57.5$
   (b) $\bar{X} = \dfrac{3{,}450}{100} = 34.5 \qquad \tilde{X} = 29.5 + \dfrac{20}{24}(5) \approx 29.5 + 4.17 \approx 33.7$
   (c) $\bar{X} = \dfrac{10{,}659}{150} = 71.06 \approx 71.1 \qquad \tilde{X} = 65.5 + \dfrac{33}{38}(6) \approx 65.5 + 5.21 \approx 70.7$

8. $d = 0$, so that $\tilde{X} = B + \dfrac{d}{f}(w) = B + 0 = B = 34.5$.

9. (a)

class	f
<40	21
<45	54
<50	100
<55	182
<60	279
<65	354
<70	405
<75	435

(b)

class	f
≤ 39	21
≤ 44	54
≤ 49	100
≤ 54	182
≤ 59	279
≤ 64	354
≤ 69	405
≤ 74	435

10. (a) 105 (b) Cannot tell. (c) $243 - 105 = 138$
    (d) Cannot tell. (e) $201 - 177 = 24$

11. (a) Cannot tell. (b) $135 - 62 = 73$ (c) Cannot tell.
    (d) $135 - 126 = 9$ (e) Cannot tell.

EXERCISES 2-A

1. $3 \cdot 2 \cdot 2 \cdot 2 \cdot 2 \cdot 3 = 144$

2. $15 \cdot 4 \cdot 3 \cdot 2 = 360$

3. $4 \cdot 6 \cdot 4 \cdot 7 \cdot 3 \cdot 5 = 10{,}080$

4. $26 \cdot 26 \cdot 9 \cdot 10 \cdot 10 \cdot 10 = 6{,}084{,}000$

5. $(16+4)(16+4) = 20 \cdot 20 = 400$

6. $2 \cdot 2 \cdot 2 = 8$ outcomes: *HHH, HHT, HTH, THH, HTT, THT, TTH, TTT.*

7. $2 \cdot 2 \cdot 2 \cdot 2 = 16$ outcomes: HHHH, HHHT, HHTH, HTHH, THHH, HHTT, HTTH, TTHH, HTHT, THTH, THHT, HTTT, THTT, TTHT, TTTH, TTTT.

8. (a) $5! = 120$ (b) $7! = 5{,}040$ (c) $9! = 362{,}880$
   (d) $11! = 39{,}916{,}800$ (e) $12! = 479{,}001{,}600$
   (f) $15! = 1{,}307{,}674{,}368{,}000$

9. $6! = 720$; $5! = 120$.

10. $\dfrac{10!}{(10-6)!} = \dfrac{10!}{4!} = 151{,}200.$

11. $\dfrac{14!}{(14-4)!} = \dfrac{14!}{10!} = 14 \cdot 13 \cdot 12 \cdot 11 = 24{,}024$

12. $\dfrac{20!}{(20-3)!} = \dfrac{20!}{17!} = 20 \cdot 19 \cdot 18 = 6{,}840$. And $\dfrac{6{,}840}{365.25} \approx 18.7$ years.

13. $\dbinom{20}{3} = \dfrac{20!}{3!\,(20-3)!} = \dfrac{20!}{3!\,17!} = 1{,}140$. And $\dfrac{1{,}140}{365.25} \approx 3.1$ years.

14. (a) 21 (b) 21 (c) 56 (d) 56 (e) 126 (f) 126
    (g) 120 (h) 120 (i) 1 (j) 1 (k) 11 (l) 11

15. $\dbinom{14}{4} = 1{,}001$

16. $\dbinom{16}{5} = 4{,}368$

17. Without regard to order: $\dbinom{13}{4} = 715$. With order important: $\dfrac{13!}{9!} = 17{,}160$.

18. Order unimportant: $\dbinom{18}{6} = 18{,}564$. Order important: $\dfrac{18!}{12!} = 13{,}366{,}080$.

19. $2^{10} = 1{,}024$; $2^{11} = 2{,}048$; $2^{12} = 4{,}096$. Each additional flip of the coin doubles the size of the sample space.

20. $\dbinom{11}{7} = 330$. $7! = 5{,}040$.

EXERCISES 2-B

1. Sample space: {*HHHH, HHHT, HHTH, HTHH, THHH, HHTT, HTTH, TTHH, HTHT, THTH, THHT, TTTH, TTHT, THTT, HTTT, TTTT*}.
   (a) {*HHTT, HTTH, TTHH, HTHT, THTH, THHT*}

(b) {$HHTT$, $TTHH$, $THHT$}
(c) {$HHHT$, $HHTH$, $HTHH$, $THHH$}
(d) {$HHHT$, $THHH$}
(e) {$HHHT$, $HHTH$, $HTHH$, $THHH$, $HHHH$}
(f) {$HHHT$, $THHH$, $HHHH$}
(g) {$HTHT$, $THTH$}
(h) {$THHH$, $TTHH$, $THTH$, $THHT$, $TTTH$, $TTHT$, $THTT$, $TTTT$}
(i) {$TTHH$, $TTTH$, $TTHT$}
(j) {$HHHH$, $HHHT$, $HHTH$, $HTHH$, $THHH$, $HTHT$, $THTH$, $THHT$}

2. (a) $P(A)=\frac{6}{16}=\frac{3}{8}$ (b) $P(B)=\frac{3}{16}$ (c) $P(C)=\frac{4}{16}=\frac{1}{4}$ (d) $P(D)=\frac{2}{16}=\frac{1}{8}$
(e) $P(E)=\frac{5}{16}$ (f) $P(F)=\frac{3}{16}$ (g) $P(G)=\frac{2}{16}=\frac{1}{8}$ (h) $P(H)=\frac{8}{16}=\frac{1}{2}$
(i) $P(I)=\frac{3}{16}$ (j) $P(J)=\frac{8}{16}=\frac{1}{2}$

3. They are mutually exclusive.

4. (a) $\frac{12}{84}=\frac{1}{7}$ (b) $\frac{19}{84}$ (c) $\frac{21}{84}=\frac{1}{4}$ (d) $\frac{17}{84}$ (e) $\frac{15}{84}=\frac{5}{28}$
(f) $\frac{40}{84}=\frac{10}{21}$ (g) 0 (h) $\frac{27}{84}=\frac{9}{28}$ (i) $\frac{51}{84}=\frac{17}{28}$ (j) $\frac{72}{84}=\frac{6}{7}$

5. They are complementary.

6. (a) 0.11 (b) 0.23 (c) 0.34 (d) 0.89 (e) 0.77 (f) 0.66

7. Not mutually exclusive (unless English and chemistry are the only majors existing). $P(D \text{ or } E)=P(D)+P(E)-P(D \text{ and } E)=0.89+0.77-0.66=1$; the meaning is that everyone on campus either fails to be a chemistry major *or* fails to be an English major (because no one is *both*).

8. (a) {1 and 1, 1 and 2, 2 and 1, 3 and 1, 1 and 3, 4 and 1, 1 and 4, 5 and 1, 1 and 5, 6 and 1, 1 and 6, 2 and 2, 2 and 3, 3 and 2, 2 and 4, 4 and 2, 2 and 5, 5 and 2, 2 and 6, 6 and 2, 3 and 3, 3 and 4, 4 and 3, 3 and 5, 5 and 3, 3 and 6, 6 and 3, 4 and 4, 4 and 5, 5 and 4, 4 and 6, 6 and 4, 5 and 5, 5 and 6, 6 and 5, 6 and 6}
(b) For 12: $\frac{1}{36}$. For 11: $\frac{2}{36}=\frac{1}{18}$. For 10: $\frac{3}{36}=\frac{1}{12}$. For 9: $\frac{4}{36}=\frac{1}{9}$. For 8: $\frac{5}{36}$. For 7: $\frac{6}{36}=\frac{1}{6}$.
For 6: $\frac{5}{36}$. For 5: $\frac{4}{36}=\frac{1}{9}$. For 4: $\frac{3}{36}=\frac{1}{12}$. For 3: $\frac{2}{36}=\frac{1}{18}$. For 2: $\frac{1}{36}$. For 1: 0.
(c) $\frac{18}{36}=\frac{1}{2}$ (d) $\frac{1}{2}$ (e) 1; 0. (f) $\frac{30}{36}=\frac{5}{6}$; $\frac{33}{36}=\frac{11}{12}$.

9. (a) $\binom{13}{5}=1{,}287$ (b) $\binom{12}{4}=495$ (c) $\frac{495}{1{,}287}=\frac{5}{13}$ (d) $\frac{1}{1{,}287}$
(e) $\frac{1{,}286}{1{,}287}$ (f) $\binom{10}{2}\Big/\binom{13}{5}=\frac{45}{1{,}287}=\frac{5}{143}$
(g) $\binom{9}{1}\Big/\binom{13}{5}=\frac{9}{1{,}287}=\frac{1}{143}$ (h) $\binom{11}{5}\Big/\binom{13}{5}=\frac{462}{1{,}287}=\frac{154}{429}=\frac{14}{39}$

10. (a) $5!=120$ (b) $\frac{1}{120}$ (c) $\frac{119}{120}$ (d) $4!=24$

(e) $\frac{24}{120}=\frac{1}{5}$ (f) $\frac{4}{5}$ (g) $3!=6$ (h) $\frac{6}{120}=\frac{1}{20}$

(i) $\frac{1}{20}$ (j) $\frac{1}{20}+\frac{1}{20}=\frac{1}{10}$ (k) $\frac{9}{10}$

EXERCISES 2-C

1. (a) $\binom{6}{0}\left(\frac{3}{5}\right)^0\left(\frac{2}{5}\right)^6=\frac{64}{15{,}625}\approx 0.004$ (b) $\binom{6}{1}\left(\frac{3}{5}\right)^1\left(\frac{2}{5}\right)^5=\frac{576}{15{,}625}\approx 0.037$

(c) $\binom{6}{2}\left(\frac{3}{5}\right)^2\left(\frac{2}{5}\right)^4=\frac{2{,}160}{15{,}625}\approx 0.138$ (d) $\binom{6}{3}\left(\frac{3}{5}\right)^3\left(\frac{2}{5}\right)^3=\frac{4{,}320}{15{,}625}\approx 0.276$

(e) $\binom{6}{4}\left(\frac{3}{5}\right)^4\left(\frac{2}{5}\right)^2=\frac{4{,}860}{15{,}625}\approx 0.311$ (f) $\binom{6}{5}\left(\frac{3}{5}\right)^5\left(\frac{2}{5}\right)^1=\frac{2{,}916}{15{,}625}\approx 0.187$

(g) $\binom{6}{6}\left(\frac{3}{5}\right)^6\left(\frac{2}{5}\right)^0=\frac{729}{15{,}625}\approx 0.047$ (h) $\frac{11{,}980}{15{,}625}\approx 0.767$

(i) $\frac{7{,}120}{15{,}625}\approx 0.456$ (j) $\frac{12{,}825}{15{,}625}\approx 0.821$ (k) $\frac{8{,}505}{15{,}625}\approx 0.544$

(l) $\frac{3{,}645}{15{,}625}\approx 0.233$ (m) $\frac{2{,}800}{15{,}625}\approx 0.179$ (n) $\frac{11{,}340}{15{,}625}\approx 0.726$

(o) $\frac{15{,}561}{15{,}625}\approx 0.996$

3. Probability of three or four successes:

$$\frac{256}{625}+\frac{256}{625}=\frac{512}{625}\approx 0.819=81.9\%<85\%.$$

The condition for use is not met.

4. (a) $\binom{7}{0}\left(\frac{1}{2}\right)^0\left(\frac{1}{2}\right)^7=\frac{1}{128}\approx 0.008$ (b) $\binom{7}{1}\left(\frac{1}{2}\right)^1\left(\frac{1}{2}\right)^6=\frac{7}{128}\approx 0.055$

(c) $\binom{7}{2}\left(\frac{1}{2}\right)^2\left(\frac{1}{2}\right)^5=\frac{21}{128}\approx 0.164$ (d) $\binom{7}{3}\left(\frac{1}{2}\right)^3\left(\frac{1}{2}\right)^4=\frac{35}{128}\approx 0.273$

(e) $\binom{7}{4}\left(\frac{1}{2}\right)^4\left(\frac{1}{2}\right)^3=\frac{35}{128}\approx 0.273$ (f) $\binom{7}{5}\left(\frac{1}{2}\right)^5\left(\frac{1}{2}\right)^2=\frac{21}{128}\approx 0.164$

(g) $\binom{7}{6}\left(\frac{1}{2}\right)^6\left(\frac{1}{2}\right)^1=\frac{7}{128}\approx 0.055$ (h) $\binom{7}{7}\left(\frac{1}{2}\right)^7\left(\frac{1}{2}\right)^0=\frac{1}{128}\approx 0.008$

5. With $\pi=\frac{1}{2}$ there is exact symmetry, so that $x=7$ and $x=0$ have the same probability; similarly, $x=1$ and $x=6$ have the same probability; and so on. The histogram should have perfect left–right symmetry.

6. (a) $\frac{64}{128}=\frac{1}{2}=0.500$ (b) $\frac{29}{128}\approx 0.227$ (c) $\frac{99}{128}\approx 0.773$

(d) $\frac{64}{128}=\frac{1}{2}=0.500$ (e) $\frac{112}{128}=\frac{7}{8}=0.875$ (f) $\frac{127}{128}\approx 0.992$

7. (a) $\binom{8}{0}\left(\frac{1}{2}\right)^0\left(\frac{1}{2}\right)^8=\frac{1}{256}\approx 0.004$ (b) $\binom{8}{1}\left(\frac{1}{2}\right)^1\left(\frac{1}{2}\right)^7=\frac{8}{256}\approx 0.031$

(c) $\binom{8}{2}\left(\frac{1}{2}\right)^2\left(\frac{1}{2}\right)^6=\frac{28}{256}\approx 0.109$ (d) $\binom{8}{3}\left(\frac{1}{2}\right)^3\left(\frac{1}{2}\right)^5=\frac{56}{256}\approx 0.218$

(e) $\binom{8}{4}\left(\frac{1}{2}\right)^4\left(\frac{1}{2}\right)^4=\frac{70}{256}\approx 0.273$ (f) $\binom{8}{5}\left(\frac{1}{2}\right)^5\left(\frac{1}{2}\right)^3=\frac{56}{256}\approx 0.218$

(g) $\binom{8}{6}\left(\frac{1}{2}\right)^6\left(\frac{1}{2}\right)^2=\frac{28}{256}\approx 0.109$ (h) $\binom{8}{7}\left(\frac{1}{2}\right)^7\left(\frac{1}{2}\right)^1=\frac{8}{256}\approx 0.031$

(i) $\binom{8}{8}\left(\frac{1}{2}\right)^8\left(\frac{1}{2}\right)^0=\frac{1}{256}\approx 0.004$

9. (a) $\binom{5}{0}\left(\frac{1}{3}\right)^0\left(\frac{2}{3}\right)^5=\frac{32}{243}\approx 0.132$ (b) $\binom{5}{1}\left(\frac{1}{3}\right)^1\left(\frac{2}{3}\right)^4=\frac{80}{243}\approx 0.329$

(c) $\binom{5}{2}\left(\frac{1}{3}\right)^2\left(\frac{2}{3}\right)^3=\frac{80}{243}\approx 0.329$ (d) $\binom{5}{3}\left(\frac{1}{3}\right)^3\left(\frac{2}{3}\right)^2=\frac{40}{243}\approx 0.165$

(e) $\binom{5}{4}\left(\frac{1}{3}\right)^4\left(\frac{2}{3}\right)^1=\frac{10}{243}\approx 0.041$ (f) $\binom{5}{5}\left(\frac{1}{3}\right)^5\left(\frac{2}{3}\right)^0=\frac{1}{243}\approx 0.004$

11. (a) $\frac{211}{243}\approx 0.868$ (b) $\frac{131}{243}\approx 0.539$ (c) $\frac{232}{243}\approx 0.955$ (d) $\frac{192}{243}\approx 0.790$

(e) $\frac{11}{243}\approx 0.045$ (f) $\frac{242}{243}\approx 0.996$ (g) $\frac{232}{243}\approx 0.955$ (h) $\frac{200}{243}\approx 0.823$

12. Anderson has somewhat the better bet with one car, because the probability of one success out of eight trials is

$$\binom{8}{1}\left(\frac{1}{4}\right)^1\left(\frac{3}{4}\right)^7\approx 0.267$$

while the probability of three successes is

$$\binom{8}{3}\left(\frac{1}{4}\right)^3\left(\frac{3}{4}\right)^5\approx 0.208.$$

EXERCISES 3-A

1. (a) 1.75 (b) −1.75 (c) 2.50 (d) −2.50 (e) 0.00
(f) 1.46 (g) −1.46 (h) 1.29 (i) −1.29 (j) 0.67
(k) −0.67 (l) 1.65 (m) −1.65 (n) 1.25 (o) −1.25
(p) 0.08 (q) −0.08 (r) 2.40 (s) −2.40 (t) 0.00

2. (a) 0.4279 (b) 0.2454 (c) 0.0871 (d) 0.4904 (e) 0.4279
(f) 0.4988 (g) 0.3085 (h) 0.1003 (i) 0.0041 (j) 0.0301
(k) 0.0035 (l) 0.3446

3. (a) 0.2518 (b) 0.4897 (c) 0.0048 (d) 0.3465 (e) 0.4108
(f) 0.5080 (g) 0.9342 (h) 0.2179 (i) 0.0079 (j) 0.4259
(k) 0.9974 (l) 0.0160

4. (a) $(z=1.11)$ $0.3665+0.5000=0.8665$
(b) $(z=0.35)$ $0.1368+0.5000=0.6368$

(c) ($z=0.67$) $0.2486+0.5000=0.7486$
(d) ($z=1.47$) $0.4292+0.5000=0.9292$
(e) ($z=0.98$) $0.3365+0.5000=0.8365$
(f) ($z=-1.03$) $0.3485+0.5000=0.8485$
(g) ($z=-0.66$) $0.2454+0.5000=0.7454$
(h) ($z=-1.87$) $0.4693+0.5000=0.9693$
(i) ($z=-0.14$) $0.0557+0.5000=0.5557$
(j) ($z=-3.00$) $0.4987+0.5000=0.9987$

5. (a) ($z=1.62$, because 1.625 is rounded off *even*) 0.4474
(b) ($z=-0.66$) 0.2454
(c) ($z=-0.44$ and $z=0.19$) $0.1700+0.0753=0.2453$
(d) ($z=-0.94$ and $z=0.94$) $0.3264+0.3264=0.6528$
(e) ($z=-2.00$ and $z=2.00$) $0.4772+0.4772=0.9544$
(f) ($z=1.34$ and $z=1.97$) $0.4756-0.4099=0.0657$
(g) ($z=-1.22$ and $z=-0.47$) $0.3888-0.1808=0.2080$
(h) ($z=0.03$ and $z=3.06$) $0.4989-0.0120=0.4869$
(i) ($z=0.34$) $0.5000-0.1331=0.3669$
(j) ($z=-1.75$) $0.5000-0.4599=0.0401$
(k) ($z=-1.34$) $0.4099+0.5000=0.9099$
(l) ($z=2.81$) $0.5000+0.4975=0.9975$

6. For example, in 3-A.5(a), 44.74% of the trees could be expected to be between 16.4 and 21.6 meters tall.

7. (a) $0.2684=26.84\%$ (b) $0.1747=17.47\%$ (c) $0.7888=78.88\%$
(d) $0.9544=95.44\%$ (e) $0.9974=99.74\%$ (f) $0.0122=1.22\%$
(g) $0.9878=98.78\%$ (h) $0.0668=6.68\%$ (i) $0.9332=93.32\%$
(j) $0.9970=99.70\%$

8. Area beyond $x=69$ ($z=2.14$): 0.0162; thus 1.6% may find they are too long-armed. Similarly, area beyond $x=39$ ($z=-2.14$) is 0.0162, so that 1.62% may find they are too short-armed. The two groups together make 3.24%.

9. For $x=2{,}400$, $z=1.58$; area beyond is 0.0571; 5.71% of the time.

10. Between 61 and 79 kilograms; between 52 and 88 kilograms; between 43 and 97 kilograms.

11. (a) ($z=0.45$ and $z=1.05$) $0.3531-0.1736=0.1795$
(b) ($z=-0.55$ and $z=0.55$) $0.2088+0.2088=0.4176$
(c) ($z=0.05$ and $z=1.55$) $0.4394-0.0199=0.4195$
(d) ($z=-2.05$ and $z=2.05$) $0.4798+0.4798=0.9596$
(e) ($z=-0.05$ and $z=0.05$) $0.0199+0.0199=0.0298$
(f) ($z=0.45$) $0.5000-0.1736=0.3264$
(g) ($z=0.35$) $0.5000-0.1368=0.3632$
(h) ($z=0.95$) $0.5000+0.3289=0.8289$
(i) ($z=1.05$) $0.5000+0.3531=0.8531$
(j) ($z=0.15$ and $z=0.25$) $0.0987-0.0596=0.0391$
(k) ($z=-0.25$ and $z=-0.15$) $0.0987-0.0596=0.0391$
(l) ($z=-0.85$) $0.5000-0.3023=0.1977$
(m) ($z=-0.75$) $0.5000-0.2734=0.2266$
(n) ($z=-0.75$) $0.5000+0.2734=0.7734$
(o) ($z=-0.85$) $0.5000+0.3023=0.8023$

12. (a) ($z=-0.45$ and $z=0.45$) $0.1736+0.1736=0.3472$
(b) ($z=-0.92$ and $z=0.92$) $0.3212+0.3212=0.6424$

(c) ($z=-1.50$ and $z=1.50$) $0.4332+0.4332=0.8664$
(d) ($z=-2.66$ and $z=2.66$) $0.4961+0.4961=0.9922$
(e) ($z=-0.03$ and $z=0.03$) $0.0120+0.0120=0.0240$
(f) ($z=-0.08$ and $z=-0.03$) $0.0319-0.0120=0.0199$
(g) ($z=0.03$ and $z=0.08$) $0.0319-0.0120=0.0199$
(h) ($z=0.08$ and $z=1.18$) $0.3810-0.0319=0.3491$
(i) ($z=0.03$ and $z=2.97$) $0.4985-0.0120=0.4865$
(j) ($z=-2.13$ and $z=-0.13$) $0.4834-0.0517=0.4317$
(k) ($z=2.73$) $0.5000-0.4968=0.0032$
(l) ($z=2.71$) $0.5000-0.4966=0.0034$
(m) ($z=2.87$) $0.5000+0.4979=0.9979$
(n) ($z=-2.87$) $0.5000+0.4979=0.9979$
(o) ($z=-2.55$) $0.5000-0.4946=0.0054$

EXERCISES 3-B

1. $\mu=60\left(\frac{5}{8}\right)=\frac{75}{2}=37.5.$ $\sigma=\sqrt{60\left(\frac{5}{8}\right)\left(\frac{3}{8}\right)}=\sqrt{\frac{225}{16}}=\frac{15}{4}=3.75.$

(a) 0.4082	(b) 0.8164	(c) 0.4962	(d) 0.9924	(e) 0.2119
(f) 0.2981	(g) 0.7019	(h) 0.7881	(i) 0.3936	(j) 0.5000
(k) 0.9452	(l) 0.9693	(m) 0.1064	(n) 0.1064	(o) 0.0696

2. $\mu=48\left(\frac{1}{4}\right)=12.0.$ $\sigma=\sqrt{48\left(\frac{1}{4}\right)\left(\frac{3}{4}\right)}=\sqrt{9}=3.0.$

(a) 0.1350	(b) 0.1052	(c) 0.1052	(d) 0.8454	(e) 0.4302
(f) 0.9876	(g) 0.0668	(h) 0.1210	(i) 0.8790	(j) 0.9332
(k) 0.9664	(l) 0.9850	(m) 0.4325	(n) 0.3085	(o) 0.8454

3. $\mu=2{,}100(0.3)=630.0.$ $\sigma=\sqrt{2{,}100(0.3)(0.7)}=\sqrt{441}=21.0.$

(a) 0.0268	(b) 0.0301	(c) 0.0082	(d) 0.0091	(e) 0.6730
(f) 0.9464	(g) 0.9960	(h) 0.9495	(i) 0.9545	(j) 0.0160
(k) 0.8869	(l) 0.8790	(m) 0.6573	(n) 0.6180	(o) 0.1872

5. $\pi=\frac{1{,}440}{4{,}000}=0.36.$ $\mu=100(0.36)=36.0.$

$\sigma=\sqrt{100(0.36)(0.64)}=\sqrt{100}\sqrt{0.36}\sqrt{0.64}=10(0.6)(0.8)=4.8.$

(a) 0.7498	(b) 0.9714	(c) 0.9974	(d) 0.0594	(e) 0.0885
(f) 0.3015	(g) 0.3783	(h) 0.0796	(i) 0.9616	(j) 0.9761
(k) 0.9857	(l) 0.9761	(m) 0.0768	(n) 0.8230	(o) 0.9987

EXERCISES 4-A

1. (a) $27.4<\mu<30.6$ (b) $26.9<\mu<31.1$

2. (a) $28.2<\mu<29.8$ (b) $27.9<\mu<30.1$

3. (a) $14{,}342.6<\mu<14{,}793.4$ (b) $14{,}271.3<\mu<14{,}864.7$

4. From $\sigma_{\bar{x}}\approx 479.2$: (a) $13{,}628.7<\mu<15{,}507.3$ (b) $13{,}331.6<\mu<15{,}804.4$

5. From $\sigma_{\bar{x}}\approx 254.5$: (a) $80{,}500<\mu<81{,}500$ (b) $80{,}300<\mu<81{,}700$

6. (a) From $\sigma_{\bar{x}}\approx 569.2$: 95%, $29{,}800<\mu<32{,}200$. 99%, $29{,}500<\mu<32{,}500$
   (b) From $\sigma_{\bar{x}}\approx 161.0$: 95%, $30{,}600<\mu<31{,}400$. 99%, $30{,}500<\mu<31{,}500$

7. (a) $205.5<\mu<254.5$ (b) $210.9<\mu<249.1$ (c) $212.8<\mu<247.2$
   (d) $220.0<\mu<240.0$ ($\sigma_{\bar{x}}\approx 5.051\approx 5.1$) (e) $221.3<\mu<238.7$
   (f) $223.1<\mu<236.9$ (g) $224.3<\mu<235.7$ (h) $225.6<\mu<234.4$
   (i) $227.8<\mu<232.2$ (j) $228.6<\mu<231.4$

8. (a) $21.07<\mu<23.83$ (b) $21.17<\mu<23.73$ (c) $21.27<\mu<23.63$
   (d) $21.37<\mu<23.53$ (e) $21.47<\mu<23.43$ (f) $21.56<\mu<23.34$
   (g) $21.66<\mu<23.24$ (h) $21.76<\mu<23.14$ (i) $21.86<\mu<23.04$
   (j) $21.96<\mu<23.94$ (k) $22.05<\mu<22.85$

9. (a) $\$18.21<\mu<\$19.19$ (b) $\$18.06<\mu<\$19.34$

10. (a) 95%: $\$18.19<\mu<\$19.21$. 99%: $\$18.02<\mu<\$19.38$.
    (b) 95%: $\$18.15<\mu<\$19.25$. 99%: $\$17.98<\mu<\$19.42$
    (c) 95%: $\$18.12<\mu<\$19.28$. 99%: $\$17.93<\mu<\$19.47$.
    (d) 95%: $\$18.07<\mu<\$19.33$. 99%: $\$17.88<\mu<\$19.52$
    (e) 95%: $\$18.01<\mu<\$19.39$. 99%: $\$17.80<\mu<\$19.60$.
    (f) 95%: $\$17.93<\mu<\$19.47$. 99%: $\$17.69<\mu<\$19.71$.
    (g) 95%: $\$17.81<\mu<\$19.59$. 99%: $\$17.53<\mu<\$19.87$.
    (h) 95%: $\$17.62<\mu<\$19.78$. 99%: $\$17.27<\mu<\$20.13$.

11. (a) $4{,}749<\mu<4{,}851$ (b) $4{,}732<\mu<4{,}868$

12. (a) 95%: $4{,}751<\mu<4{,}849$. 99%: $4{,}735<\mu<4{,}865$.
    (b) 95%: $4{,}752<\mu<4{,}848$. 99%: $4{,}737<\mu<4{,}863$.
    (c) 95%: $4{,}754<\mu<4{,}846$. 99%: $4{,}740<\mu<4{,}860$.
    (d) 95%: $4{,}755<\mu<4{,}845$. 99%: $4{,}741<\mu<4{,}859$.
    (e) 95%: $4{,}757<\mu<4{,}843$. 99%: $4{,}743<\mu<4{,}857$.
    (f) 95%: $4{,}758<\mu<4{,}842$. 99%: $4{,}745<\mu<4{,}855$.
    (g) 95%: $4{,}759<\mu<4{,}841$. 99%: $4{,}746<\mu<4{,}854$.
    (h) 95%: $4{,}760<\mu<4{,}840$. 99%: $4{,}748<\mu<4{,}852$.
    (i) 95%: $4{,}772<\mu<4{,}828$. 99%: $4{,}763<\mu<4{,}837$.
    (j) 95%: $4{,}791<\mu<4{,}809$. 99%: $4{,}788<\mu<4{,}812$.

13. (a) 0.3413 (b) 0.3830 (c) 0.3345 (d) 0.9624 (e) 0.9964
    (f) 0.3310 (g) 0.0475 (h) 0.0188 (i) 0.9525 (j) 0.9429
    (k) 0.0188 (l) 0.3256

14. (a) 0.3264 (b) 0.3015 (c) 0.2743 (d) 0.2483 (e) 0.2266
    (f) 0.1314 (g) 0.0668 (h) 0.0301 (i) 0.0122 (j) 0.0013

EXERCISES 4-B

1. (a) $27.2<\mu<29.6$ (b) $26.8<\mu<30.0$

2. (a) $15.9<\mu<18.7$ (b) $15.4<\mu<19.2$

3. (a) $10.6<\mu<14.4$ (b) $9.8<\mu<15.2$

4. (a) $\$35.04<\mu<\$36.36$ (b) $\$34.80<\mu<\$36.60$

5. (a) $212.3<\mu<220.3$ (b) $210.8<\mu<221.8$

6. (a) $443.4<\mu<453.4$ (b) $441.6<\mu<455.2$

7. (a) $7.20<\mu<10.00$ (b) $6.62<\mu<10.58$

8. (a) $20.00<\mu<21.80$ (b) $19.98<\mu<21.82$ (c) $19.96<\mu<21.84$

(d) $19.95<\mu<21.85$ (e) $19.93<\mu<21.87$ (f) $19.90<\mu<21.90$
(g) $19.88<\mu<21.92$ (h) $19.86<\mu<21.94$ (i) $19.83<\mu<21.97$
(j) $19.80<\mu<22.00$ (k) $19.77<\mu<22.03$ (l) $19.74<\mu<22.06$
(m) $19.70<\mu<22.10$ (n) $19.66<\mu<22.14$ (o) $19.62<\mu<22.18$
(p) $19.57<\mu<22.23$ (q) $19.51<\mu<22.29$ (r) $19.44<\mu<22.36$
(s) $19.37<\mu<22.43$ (t) $19.28<\mu<22.52$

9. (a) $19.69<\mu<22.11$ (b) $19.66<\mu<22.14$ (c) $19.64<\mu<22.16$
(d) $19.61<\mu<22.19$ (e) $19.58<\mu<22.22$ (f) $19.55<\mu<22.25$
(g) $19.52<\mu<22.28$ (h) $19.48<\mu<22.32$ (i) $19.45<\mu<22.35$
(j) $19.41<\mu<22.39$ (k) $19.36<\mu<22.44$ (l) $19.31<\mu<22.49$
(m) $19.26<\mu<22.54$ (n) $19.19<\mu<22.61$ (o) $19.13<\mu<22.67$
(p) $19.05<\mu<22.75$ (q) $18.96<\mu<22.84$ (r) $18.86<\mu<22.94$
(s) $18.74<\mu<23.06$ (t) $18.60<\mu<23.20$

EXERCISES 4-C

1. $p=0.19$, $\hat{\sigma}_p=\sqrt{0.000081}=0.009$.
(a) $0.172<\pi<0.208$ (b) $0.166<\pi<0.214$

2. $p=0.19$ each time.
(a) $\hat{\sigma}_p=\sqrt{0.0000905}\approx 0.010$. $0.170<\pi<0.210$.
(b) $\hat{\sigma}_p\approx\sqrt{0.000118}\approx 0.011$. $0.168<\pi<0.212$.
(c) $\hat{\sigma}_p\approx\sqrt{0.000140}\approx 0.012$. $0.166<\pi<0.214$
(d) $\hat{\sigma}_p\approx\sqrt{0.000192}\approx 0.014$. $0.162<\pi<0.218$.
(e) $\hat{\sigma}_p\approx\sqrt{0.000256}=0.016$. $0.158<\pi<0.222$.
(f) $\hat{\sigma}_p\approx\sqrt{0.000385}\approx 0.020$. $0.150<\pi<0.230$.
(g) $\hat{\sigma}_p\approx\sqrt{0.000770}\approx 0.028$. $0.135<\pi<0.245$.
(h) $\hat{\sigma}_p\approx\sqrt{0.00154}\approx 0.039$. $0.113<\pi<0.267$.

3. (a) $0.164<\pi<0.216$ (b) $0.161<\pi<0.219$ (c) $0.159<\pi<0.221$
(d) $0.153<\pi<0.227$ (e) $0.148<\pi<0.332$ (f) $0.138<\pi<0.242$
(g) $0.117<\pi<0.263$ (h) $0.089<\pi<0.291$

4. As n decreases the confidence interval widens, because $\hat{\sigma}_p$ is *increased* by dividing (under the radical sign) by smaller and smaller n's, so that $1.96\hat{\sigma}_p$ or $2.58\hat{\sigma}_p$ each successive time is a larger radius for the interval.

5. $p=0.76$, $\hat{\sigma}_p\approx\sqrt{0.000811}\approx 0.028$.
(a) $0.705<\pi<0.815$ (b) $0.687<\pi<0.833$

6. $p=0.76$ each time.
(a) $\hat{\sigma}_p\approx\sqrt{0.000730}\approx 0.027$. $0.707<\pi<0.813$.
(b) $\hat{\sigma}_p\approx\sqrt{0.000663}\approx 0.026$. $0.709<\pi<0.811$.
(c) $\hat{\sigma}_p=\sqrt{0.000608}\approx 0.025$. $0.711<\pi<0.809$.
(d) $\hat{\sigma}_p\approx\sqrt{0.000561}\approx 0.024$. $0.712<\pi<0.808$.
(e) $\hat{\sigma}_p\approx\sqrt{0.000521}\approx 0.023$. $0.714<\pi<0.806$.
(f) $\hat{\sigma}_p\approx\sqrt{0.000486}\approx 0.022$. $0.716<\pi<0.804$.
(g) $\hat{\sigma}_p=\sqrt{0.000456}\approx 0.021$. $0.718<\pi<0.802$.
(h) $\hat{\sigma}_p\approx\sqrt{0.000429}\approx 0.021$. $0.718<\pi<0.802$.

(Difference between results of parts g and h with appropriate round-offs does not show up.)

(i) $\hat{\sigma}_p \approx \sqrt{0.000405} \approx 0.020$. $0.720 < \pi < 0.800$.
(j) $\hat{\sigma}_p = \sqrt{0.000384} \approx 0.020$. $0.720 < \pi < 0.800$.

(Difference between parts *i* and *j* are too small to show up.)

(k) $\hat{\sigma}_p \approx \sqrt{0.000365} \approx 0.019$. $0.722 < \pi < 0.798$.
(l) $\hat{\sigma}_p \approx \sqrt{0.000182} \approx 0.013$. $0.734 < \pi < 0.786$.
(m) $\hat{\sigma}_p \approx \sqrt{0.0000182} \approx 0.004$. $0.752 < \pi < 0.768$.

7. As *n* increases, the confidence interval narrows, because larger *n*'s make for smaller values of $\hat{\sigma}_p$ (because of division by *n*), so that in adding and subtracting $1.96\hat{\sigma}_p$ or $2.58\hat{\sigma}_p$ one is adding and subtracting successively less and less.

8. $p = 0.26$, $\hat{\sigma}_p = \sqrt{0.000481} \approx 0.022$.
(a) $0.216 < \pi < 0.304$ (b) $0.203 < \pi < 0.317$

9. $p = 0.26$ in each case.
(a) $\hat{\sigma}_p \approx \sqrt{0.000385} \approx 0.020$. $0.208 < \pi < 0.312$.
(b) $\hat{\sigma}_p \approx \sqrt{0.000321} \approx 0.018$. $0.213 < \pi < 0.307$.
(c) $\hat{\sigma}_p \approx \sqrt{0.000275} \approx 0.017$. $0.216 < \pi < 0.304$.
(d) $\hat{\sigma}_p \approx \sqrt{0.000240} \approx 0.015$. $0.221 < \pi < 0.299$.
(e) $\hat{\sigma}_p \approx \sqrt{0.000192} \approx 0.014$. $0.223 < \pi < 0.297$.
(f) $\hat{\sigma}_p \approx \sqrt{0.000128} \approx 0.011$. $0.231 < \pi < 0.289$.

10. $p = 0.13$ each time.
(a) $\hat{\sigma}_p \approx \sqrt{0.000113} \approx 0.011$. $0.108 < \pi < 0.152$.
(b) $\hat{\sigma}_p \approx \sqrt{0.000141} \approx 0.012$. $0.106 < \pi < 0.154$.
(c) $\hat{\sigma}_p \approx \sqrt{0.000188} \approx 0.014$. $0.102 < \pi < 0.158$.
(d) $\hat{\sigma}_p \approx \sqrt{0.000283} \approx 0.017$. $0.096 < \pi < 0.164$.
(e) $\hat{\sigma}_p \approx \sqrt{0.000566} \approx 0.024$. $0.082 < \pi < 0.178$.
(f) $\hat{\sigma}_p \approx \sqrt{0.00113} \approx 0.034$. $0.063 < \pi < 0.197$.

11. $\sigma_p = 0.048$, $z \approx 0.83$, probability 0.2033.

12. (a) $\sigma_p = 0.06$, $z \approx 0.67$, probability 0.2514.
(b) $\sigma_p = 0.024$, $z \approx 1.67$, probability 0.0475.

13. (a) $\sigma_p \approx 0.043$, $z \approx \pm 0.47$, probability 0.3616.
(b) $\sigma_p \approx 0.031$, $z \approx \pm 0.65$, probability 0.4844.
(c) $\sigma_p = 0.025$, $z = \pm 0.80$, probability 0.5762.
(d) $\sigma_p \approx 0.022$, $z \approx \pm 0.91$, probability 0.6372.
(e) $\sigma_p \approx 0.019$, $z \approx \pm 1.05$, probability 0.7062.
(f) $\sigma_p \approx 0.018$, $z \approx \pm 1.11$, probability 0.7330.
(g) $\sigma_p \approx 0.016$, $z = \pm 1.25$, probability 0.7888.
(h) $\sigma_p \approx 0.015$, $z \approx \pm 1.33$, probability 0.8164.

EXERCISES 5-A

1. Two-sided. $z = 1.24 < 1.96$, no rejection.

2. $z = 2.07 > 1.96$, reject H_0. Intuitively, the sample mean $\bar{x}$ should not be as far away from \$5,100 as it is, if H_0 is true and \$5,100 is the true mean. If H_0 were true, it would be less likely to get an $\bar{x}$ \$80 away from μ at a larger sample size 225 than at a smaller sample size 81. Mathematically, *n* being larger makes $\sigma_{\bar{x}}$ smaller and thus makes $\dfrac{\bar{x}-\mu}{\sigma_{\bar{x}}}$ a larger *z*-score.

3. (a) $z = 0.40 < 2.58$; no rejection. (b) $z = 0.80 < 2.58$; no rejection.
   (c) $z = 1.20 < 2.58$; no rejection. (d) $z = 1.60 < 2.58$; no rejection.
   (e) $z = 2.00 < 2.58$; no rejection. (f) $z = 2.40 < 2.58$; no rejection.
   (g) $z = 2.80 > 2.58$; reject H_0.

4. At (e) $z = 2.00 > 1.96$.

5. $z = -2.14 < -1.96$; reject H_0.

6. No rejection at $\alpha = 0.01$, because -2.14 is not out beyond -2.58.

7. $z = -2.36$
   (a) Not beyond -2.58; no rejection at the level 0.01.
   (b) z *is* to the left of -1.96; reject H_0 at the level 0.05.

8. With $n = 60$, $z = 2.582 > 2.58$; reject H_0 (barely: accepting the full 0.01 risk of being wrong).

9. One-sided. $z = -0.83$, not beyond -1.64; no rejection.

10. (a) $z = -0.93$ compared with -1.64; no rejection.
    (b) $z = -1.02$ compared with -1.64; no rejection.
    (c) $z = -1.10$ compared with -1.64; no rejection.
    (d) $z = -1.18$ compared with -1.64; no rejection.
    (e) $z = -1.25$ compared with -1.64; no rejection.
    (f) $z = -1.32$ compared with -1.64; no rejection.
    (g) $z = -1.38$ compared with -1.64; no rejection.
    (h) $z = -1.44$ compared with -1.64; no rejection.
    (i) $z = -1.50$ compared with -1.64; no rejection.
    (j) $z = -1.56$ compared with -1.64; no rejection.
    (k) $z = -1.61$ compared with -1.64; no rejection.
    (l) $z = -1.67$ compared with -1.64; reject H_0.

11. $z = 1.67 < 2.33$; no rejection at the level 0.01.

12. $z = 1.67 > 1.64$; rejection at the level 0.05.

13. (a) $z = 1.86 < 2.33$; no rejection. (b) $z = 2.04 < 2.33$; no rejection.
    (c) $z = 2.20 < 2.33$; no rejection. (d) $z = 2.24 < 2.33$; no rejection.
    (e) $z = 2.30 < 2.33$; no rejection. (f) $z = 2.36 > 2.33$; rejection of H_0.

14. One-sided test: $z = 4.00 > 1.64$, reject H_0.

15. (a) $z = 3.74 > 1.64$; reject H_0. (b) $z = 3.46 > 1.64$; reject H_0.
    (c) $z = 3.16 > 1.64$; reject H_0. (d) $z = 2.83 > 1.64$; reject H_0.
    (e) $z = 2.45 > 1.64$; reject H_0. (f) $z = 2.00 > 1.64$; reject H_0.
    (g) $z = 1.73 > 1.64$; reject H_0. (h) $z = 1.60 < 1.64$; no rejection.

16. $z = -2.11$, not out beyond -2.33, no rejection. Difference is not significant at this level.

17. (a) $z = -2.15$, not beyond -2.33; no rejection.
    (b) $z = -2.19$, not beyond -2.33; no rejection.
    (c) $z = -2.23$, not beyond -2.33; no rejection.
    (d) $z = -2.27$, not beyond -2.33; no rejection.
    (e) $z = -2.31$, not beyond -2.33; no rejection (but close).
    (f) $z = -2.35$, out beyond -2.33; reject H_0.

18. Comparison would have been with $z = -1.64$; all rejections (even at the original smallest sample size 250).

19. $z = 2.50 > 2.33$; reject H_0.

20. (a) $z = 2.40 > 2.33$; still reject H_0. (b) $z = 2.37 > 2.33$; still reject H_0.
    (c) $z = 2.34 > 2.33$; still reject H_0.
    (d) $z = 2.31 < 2.33$; no rejection at the level 0.01.

21. $z = 1.88 < 2.58$; no rejection.

22. (a) $z = 1.97 < 2.58$; no rejection. (b) $z = 2.11 < 2.58$; no rejection.
    (c) $z = 2.24 < 2.58$; no rejection. (d) $z = 2.36 < 2.58$; no rejection.
    (e) $z = 2.47 < 2.58$; no rejection.
    (f) $z = 2.582 > 2.58$; reject H_0 at the level 0.01.

EXERCISES 5-B

1. $z = -1.20$, not beyond -1.96; no rejection.

2. (a) $z = -1.26$. not beyond -1.96; no rejection.
   (b) $z = -1.41$, not beyond -1.96; no rejection.
   (c) $z = -1.55$, not beyond -1.96; no rejection.
   (d) $z = -1.68$, not beyond -1.96; no rejection.
   (e) $z = -1.73$, not beyond -1.96; no rejection.
   (f) $z = -1.79$, not beyond -1.96; no rejection.
   (g) $z = -1.85$, not beyond -1.96; no rejection.
   (h) $z = -1.90$, not beyond -1.96; no rejection.
   (i) $z = -1.95$, not quite beyond -1.96; strictly speaking, no rejection at level 0.05, but very close.
   (j) $z = -2.00$, beyond -1.96; reject H_0.

3. $z = 2.99 > 2.58$; reject H_0; the mean speeds are different.

4. (a) $z = 2.82 > 2.58$; reject H_0. (b) $z = 2.72 > 2.58$; reject H_0.
   (c) $z = 2.47 < 2.58$; do not reject H_0 at level 0.01.

5. $z = 0.94 < 1.96$; no rejection; the mean times have not been shown to be different.

6. (a) $z = 1.19 < 1.96$; no rejection. (b) $z = 1.37 < 1.96$; no rejection.
   (c) $z = 1.53 < 1.96$; no rejection. (d) $z = 1.68 < 1.96$; no rejection.
   (e) $z = 1.81 < 1.96$; no rejection.
   (f) $z = 1.94 < 1.96$; no rejection (but close).
   (g) $z = 1.97 > 1.96$; reject H_0; the times are different.

7. $z = 0.83 < 1.64$; no rejection.

8. (a) $z = 0.93 < 1.64$; no rejection. (b) $z = 1.14 < 1.64$; no rejection.
   (c) $z = 1.32 < 1.64$; no rejection. (d) $z = 1.47 < 1.64$; no rejection.
   (e) $z = 1.61 < 1.64$; no rejection, but close.
   (f) $z = 1.66 > 1.64$; reject H_0 at level 0.05.

9. $z = -1.93$, beyond -1.64, reject H_0; sample means are significantly different at the level 0.05.

10. $z = -1.41$, not beyond -1.64; no rejection.

11. $z = -2.38$, beyond -2.33; reject H_0.

12. $z = -1.89$, not beyond -2.33; no rejection at level 0.01.

13. $z = 2.97 > 2.33$; reject H_0.

14. (a) $z = 2.97 > 2.33$; reject H_0. (b) $z = 2.48 > 2.33$; reject H_0.
    (c) $z = 1.98 < 2.33$; no rejection.

15. $z = 3.68 > 2.33$; reject H_0.

16. (a) $z = 3.48 > 2.33$; reject H_0.
(b) $z = 3.29 > 2.33$; reject H_0.
(c) $z = 3.09 > 2.33$; reject H_0.
(d) $z = 2.90 > 2.33$; reject H_0.
(e) $z = 2.71 > 2.33$; reject H_0.
(f) $z = 2.51 > 2.33$; reject H_0.
(g) $z = 2.32 < 2.33$; strictly speaking no rejection at the level 0.01, but very close.
(h) $z = 2.13 < 2.33$; no rejection at level 0.01.

17. $z = -0.49$, not beyond -1.64; do not reject H_0.

18. (a) $z = -0.98$, not beyond -1.64; no rejection.
(b) $z = -1.48$, not beyond -1.64; no rejection.
(c) $z = -1.97$, beyond -1.64; reject H_0.

19. $z = 1.63 < 2.58$; no rejection.

20. (a) $z = 1.73 < 2.58$; no rejection.
(b) $z = 1.84 < 2.58$; no rejection.
(c) $z = 1.94 < 2.58$; no rejection.
(d) $z = 2.04 < 2.58$; no rejection.
(e) $z = 2.13 < 2.58$; no rejection.
(f) $z = 2.21 < 2.58$; no rejection.
(g) $z = 2.30 < 2.58$; no rejection.
(h) $z = 2.38 < 2.58$; no rejection.
(i) $z = 2.45 < 2.58$; no rejection.
(j) $z = 2.53 < 2.58$; no rejection.
(k) $z = 2.61 > 2.58$; reject H_0 at level 0.01.

EXERCISES 5-C

1. $p = \dfrac{126}{300} = 0.42$. Two-sided. $z = 0.71 < 1.96$; no rejection.

2. $p = 0.42$ each time.
(a) $z = 0.82 < 1.96$; no rejection.
(b) $z = 1.15 < 1.96$; no rejection.
(c) $z = 1.41 < 1.96$; no rejection.
(d) $z = 1.63 < 1.96$; no rejection.
(e) $z = 1.83 < 1.96$; no rejection.
(f) $z = 2.00 > 1.96$; reject H_0.

3. One-sided. $p = 0.66$, $z = -2.00$, beyond -1.64; reject H_0.

4. $p = 0.66$ each time.
(a) $z = -1.82$, beyond -1.64; reject H_0.
(b) $z = -1.74$, beyond -1.64; reject H_0.
(c) $z = -1.51$, not beyond -1.64; do not reject H_0.

5. One-sided. $p = 0.21$, $z = 1.00 < 1.64$; no rejection of the claim.

6. $p = 0.21$ each time.
(a) $z = 1.12 < 1.64$; no rejection.
(b) $z = 1.22 < 1.64$; no rejection.
(c) $z = 1.32 < 1.64$; no rejection.
(d) $z = 1.41 < 1.64$; no rejection.
(e) $z = 1.50 < 1.64$; no rejection.
(f) $z = 1.58 < 1.64$; no rejection.
(g) $z = 1.66 > 1.64$; reject H_0 at the level 0.05.

7. Two-sided. $p = 0.60$, $z = -2.19$, not beyond -2.58; no rejection; *the sample is too small* for the difference 15% to be significant at the level 0.01.

8. Yes, because $z = -2.19$ *is* out beyond -1.96 to the left, H_0 would be rejected at the level 0.05.

9. $p = 0.60$ each time.
(a) $z = -2.32$, not beyond -2.58; no rejection at level 0.01.
(b) $z = -2.45$, not beyond -2.58; no rejection at level 0.01.
(c) $z = -2.68$, beyond -2.58; reject H_0 at level 0.01.

10. $p=0.49$, $z=-0.40$, not beyond -2.33; no rejection of the claim.

11. $p=0.49$ each time, and no rejections at the level 0.01; with $p=0.49$ the sample size would have to be over 13,500 for such a rejection. Specifically:
(a) $z=-0.45$ (b) $z=-0.63$ (c) $z=-0.89$
(d) $z=-1.27$ (e) $z=-1.79$

12. $p=0.36$, $z=2.93>2.33$, reject H_0. The person is (one may feel 99% confident) more sensitive in general than the directors have predicted.

13. $p=0.36$ each time.
(a) $z=2.62>2.33$; reject H_0. (b) $z=2.45>2.33$; reject H_0.
(c) $z=2.27<2.33$; no rejection at level 0.01.

14. $p=0.70$, $z=-2.82$, beyond -2.33; reject H_0.

15. (a) $z=-2.38$; reject H_0. (b) $z=-2.38$; reject H_0.
(c) $z=-2.26$; do not reject H_0.

16. $p=0.79$, $z=-0.35$, not beyond -2.33; no rejection of the claim.

17. (a) $p=0.78$, $z=-0.71$; no rejection. (b) $p=0.77$, $z=-1.06$; no rejection.
(c) $p=0.76$, $z=-1.41$; no rejection. (d) $p=0.75$, $z=-1.77$; reject the claim.

18. $p=0.128$, $z=2.95>2.33$; reject the claim.

19. (a) $p=0.127$, $z=2.84$; reject H_0. (b) $p=0.126$, $z=2.74$; reject H_0.
(c) $p=0.125$, $z=2.63$; reject H_0. (d) $p=0.124$, $z=2.53$; reject H_0.
(e) $p=0.123$, $z=2.42$; reject H_0.
(f) $p=0.122$, $z=2.32$; strictly speaking not a rejection at level 0.01, but very close.
(g) $p=0.121$, $z=2.21$; do not reject H_0 at level 0.01.

20. $p=0.89$, $z=2.50>1.96$; reject H_0.

21. (a) $p=0.888$, $z=2.38$; reject H_0. (b) $p=0.886$, $z=2.25$; reject H_0.
(c) $p=0.884$, $z=2.12$; reject H_0. (d) $p=0.882$, $z=2.00$; reject H_0.
(e) $p=0.880$, $z=1.88$; do not reject H_0.

EXERCISES 5-D

1. $\bar{x}_1=299.0$, $s_1^2=707.0$; $\bar{x}_2=292.0$, $s_2^2=870.8$.
$F=870.8/707.0\approx 1.23<2.98=F_{.05}$; no rejection.

2. $\bar{x}_1=12.70$, $s_1^2=0.3028/12\approx 0.02523$;
$\bar{x}_2=12.73$, $s_2^2=0.1676/12\approx 0.01397$.
$F=0.02523/0.01397\approx 1.81<2.69$, no rejection.

3. $\bar{x}_A=48.0$, $s_A^2=19.78$; $\bar{x}_B=48.0$, $s_B^2=347.33$.
$F=17.56>5.35$; H_0 is rejected.

4. $\bar{x}_1=21.0$, $s_1^2=5.90$; $\bar{x}_2=21.0$, $s_2^2=5.87$.
$F=1.01<2.33$; no rejection.

5. $\bar{x}_1=170.0$, $s_1^2=67.73$; $\bar{x}_2=165.0$, $s_2^2=12.17$.
$F=5.57>2.89$; H_0 is rejected. With 99% confidence one may conclude that $\sigma_1^2\neq\sigma_2^2$.

6. (a) 2.26 (b) 2.21 (c) 2.05 (d) 2.51 (e) 1.96
(f) 1.98 (g) 1.95 (h) 2.01 (i) 2.02 (j) 2.23
(k) 2.04 (l) 1.96 (m) 2.00 (n) 1.82 (o) 1.92
(p) 1.90 (q) 1.82 (r) 2.00 (s) 1.70 (t) 1.93

(u) 1.87	(v) 1.78	(w) 1.56	(x) 1.86	(y) 1.88
(z) 1.56				

7.

(a) 3.22	(b) 3.11	(c) 2.79	(d) 3.74	(e) 2.62
(f) 2.65	(g) 2.60	(h) 2.72	(i) 2.72	(j) 3.12
(k) 2.76	(l) 2.63	(m) 2.70	(n) 2.34	(o) 2.52
(p) 2.51	(q) 2.36	(r) 2.70	(s) 2.14	(t) 2.57
(u) 2.45	(v) 2.27	(w) 1.89	(x) 2.43	(y) 2.48
(z) 1.88				

8. $\bar{x}_1 = 151.0$, $s_1^2 = 201.56$; $\bar{x}_2 = 147.0$, $s_2^2 = 229.20$.
$F = 1.14 < 2.36$; no rejection.

9. $\bar{x}_1 = 18.00$, $s_1^2 = 1.069$; $\bar{x}_2 = 18.00$, $s_2^2 = 1.846$.
$F = 1.73 < 2.51$; no rejection. The conclusion $\sigma_1^2 \neq \sigma_2^2$ cannot be drawn with 95% confidence, because the null hypothesis is not rejected at the level 0.05.

10. $\bar{x}_A = 8.60$, $s_A^2 = 0.64$; $\bar{x}_B = 8.70$, $s_B^2 = 1.49$.
$F = 2.33 > 2.29$; reject H_0 at the level 0.05. But because $2.33 < 3.28$, no rejection at the level 0.01.

EXERCISES 6-A

1. $\bar{x}_A = 9.0$, $\bar{x}_B = 11.0$, $\bar{x}_C = 8.0$; $F = 1.62 < 3.68$; do not reject H_0.
2. $\bar{x}_A = 6.5$, $\bar{x}_B = 11.0$, $\bar{x}_C = 7.5$; $F = 10.66 > 5.78$; reject H_0.
3. $\bar{x}_A = 7.5$, $\bar{x}_B = 7.0$, $\bar{x}_C = 12.0$, $\bar{x}_D = 8.0$; $F = 2.89 < 3.10$; do not reject H_0.
4. $\bar{x}_A = 6.5$, $\bar{x}_B = 6.0$, $\bar{x}_C = 11.0$, $\bar{x}_D = 7.0$. Same SS_{total} as before: 311.625. Same SS_{factor} as before: 94.125. Then $SS_{\text{error}} = 217.5$, $F = 2.89$ as before. No rejection; the differences involved are all unchanged by the uniform subtraction of 1 from each datum.
5. $\bar{x}_A = 4.9$, $\bar{x}_B = 4.9$, $\bar{x}_C = 6.9$; $F = 13.44 > 5.39$; reject H_0. Source C can claim to have larger p'ums than the others.
6. $\bar{x}_A = 5.2$, $\bar{x}_B = 5.8$, $\bar{x}_C = 5.4$, $\bar{x}_D = 5.4$; $F = 0.08 < 3.24$; no rejection of the hypothesis $\mu_A = \mu_B = \mu_C = \mu_D$.
7. $\bar{x}_A = 7.7$, $\bar{x}_B = 10.6$, $\bar{x}_C = 5.6$, $\bar{x}_D = 5.7$; $F = 19.76 > 4.72$; reject the null hypothesis; the population means are decidedly not all the same.
8. $\bar{x}_A = 68.0$, $\bar{x}_B = 73.0$, $\bar{x}_C = 72.0$, $\bar{x}_D = 71.0$; $F = 0.84 < 3.10$; do not reject H_0. The sample means are not significantly different at the level 0.05.
9. $\bar{x}_A = 15.0$, $\bar{x}_B = 10.0$, $\bar{x}_C = 17.0$; $F = 19.23 > 5.61$; reject H_0: The sample means are significantly different at the level 0.01.
10. $\bar{x}_A = 2.9$, $\bar{x}_B = 2.8$, $\bar{x}_C = 3.2$, $\bar{x}_D = 3.0$; $F = 3.28$; reject the hypothesis $\mu_A = \mu_B = \mu_C = \mu_D$. For $\text{d.f.}_{\text{num}} = 28$, $F_{.05} \approx 2.95$: three-fifths of the way between 2.99 (from d.f. = 25) and 2.92 (d.f. = 30), because 28 is three-fifths of the way between 25 and 30.
11. $\bar{x}_A = 12$, $\bar{x}_B = 11$, $\bar{x}_C = 12$, $\bar{x}_D = 10$; $F = 6.42 > 4.72$; reject the hypothesis that the population means are equal; the sample means are significantly different at the level 0.01.
12. $\bar{x}_A = 25.0$, $\bar{x}_B = 22.0$, $\bar{x}_C = 28.0$; $F = 5.02 > 3.55$; reject the hypothesis $\mu_A = \mu_B = \mu_C$, because the sample means are significantly different at the level 0.05.
13. $F = 5.02 < 6.01$; no rejection at the level 0.01. The sample means, though different

enough to warrant 95% confidence in rejecting H_0, are not different enough to warrant 99% confidence.

14. $\bar{x}_A = 13.0$, $\bar{x}_B = 12.0$, $\bar{x}_C = 12.0$, $\bar{x}_D = 13.0$; $F = 1.36 < 2.84$; do not reject the hypothesis that the population means are equal.

15. $\bar{x}_A = 3.0$, $\bar{x}_B = 3.7$, $\bar{x}_C = 3.3$; $F = 1.30 < 3.15$; do not reject H_0; the sample means are not significantly different at level 0.05.

16. $\bar{x}_A = 7.0$, $\bar{x}_B = 5.0$, $\bar{x}_C = 6.0$, $\bar{x}_D = 5.0$, $\bar{x}_E = 7.0$; $F = 0.78 < 2.69$; do not reject the hypothesis that the five population means are equal. It has not been shown that it makes a difference, in general, where one buys a car, as far as length of service without repair is concerned.

17. (a) 2.37 (b) 2.25 (c) 2.76 (d) 2.12 (e) 2.57

18. (a) 3.34 (b) 3.12 (c) 4.13 (d) 2.89 (e) 3.81

EXERCISES 6-B

1. Each expected value is $\frac{300}{6} = 50$. $\chi^2 = 1.240 < 11.070$, no rejection of H_0. The die has not been shown to be unfair.

2. (a) $\chi^2 = 1.800 < 11.070$; no rejection. (b) $\chi^2 = 2.80 < 11.070$; no rejection.
(c) $\chi^2 = 3.52 < 11.070$; no rejection. (d) $\chi^2 = 4.92 < 11.070$; no rejection.
(e) $\chi^2 = 7.16 < 11.070$; no rejection. (f) $\chi^2 = 9.12 < 11.070$; no rejection.
(g) $\chi^2 = 10.56 < 11.070$; no rejection.
(h) $\chi^2 = 11.400 > 11.070$; reject H_0. At the 0.05 level the die is deemed unfair.

3. Each expected value is $\frac{140}{7} = 20$. $\chi^2 = 15.600 > 12.592$; reject H_0 at the level 0.05, but not at level 0.01, which would require $\chi^2 > 16.812$.

4. Expected values $\frac{500}{4} = 125$. $\chi^2 = \frac{494}{125} = 3.952 < 7.815$; no rejection. The game has not been shown to be unfair.

5. $\chi^2 = \frac{1{,}976}{250} = 7.904 > 7.815$; reject H_0: With 95% confidence the game may be declared unfair. The difference this time is the larger sample size, at which the differences from the expected uniform $\frac{1}{4}$ proportions are more significant than at the smaller sample size.

6. Expected values $\frac{400}{5} = 80$. $\chi^2 = \frac{980}{80} = 12.250 > 9.488$; reject H_0. The proportions overall are not regarded as the same.

7. Expected values:

20	30	31	19
40	60	62	38
20	30	31	19

$\chi^2 = 17.619 > 16.812$; reject H_0 at level 0.01; brand choice is regional.

8. Expected values:

88	68	84	96	64
66	51	63	72	48
66	51	63	72	48

$\chi^2 = 21.002 > 15.507$; reject H_0 at level 0.05; channel preference is not unrelated to age bracket.

9. Because $\chi^2 = 21.002 > 20.090$, we would still reject H_0, even at the level 0.01.

10. Expected values:

32	34	24	30
64	68	48	60
48	51	36	45
16	17	12	15

$\chi^2 = 12.534 < 16.919$; do not reject H_0. No connection between IQ and hair color is indicated by the sample.

11. Expected values:

75	90	105	30
50	60	70	20
125	150	175	50

$\chi^2 = 3.996 < 12.592$; no rejection. There is no indication that subject inclination is related to astrological sign.

12. Expected values:

130	55	65
52	22	26
78	33	39

$\chi^2 = 22.145 > 13.277$; reject H_0. Fur color is apparently related to hair color, even for unicorns.

13. Expected values:

212	164	184	152	88
265	205	230	190	110
53	41	46	38	22

$\chi^2 = 2.031 < 15.507$; do not reject H_0.

14. Expected values:

129	48	63
215	80	105
86	32	42

$\chi^2 = 23.665 > 13.277$; reject H_0. Mode of transportation is related to amount of sick days.

15. Expected values:

88	68	72	92	80
22	17	18	23	20
66	51	54	69	60
44	34	36	46	40

$\chi^2 = 8.885 < 21.026$; do not reject H_0; no relation between sign color and first purchase has been evidenced.

16. Expected values:

27	31	41	28	23
81	93	123	84	69
54	62	82	56	46

$\chi^2 = 29.508 > 20.090$; reject H_0. (H_0 would have stated that network preference is independent of education level.)

17. Expected values:

126	168	150	156
105	140	125	130
42	56	50	52
42	56	50	52

$\chi^2 = 10.886 < 16.919$; no rejection. H_0 states that mode of habitation is independent of choice of insurance company.

18. Expected values:

90	80	120	140	70
144	128	192	224	112
126	112	168	196	98

$\chi^2 = 32.235 > 20.090$; reject H_0 (which would have stated that the age classification and the other classification are independent).

EXERCISES 7-A

1. $n = 11$, $\sum x = 37$, $\sum y = 102$, $(\sum x)^2 = 1{,}369$, $(\sum y)^2 = 10{,}404$, $\sum (xy) = 354$, $\sum (x^2) = 137$, $\sum (y^2) = 966$.

$$r = \frac{120}{\sqrt{138}\sqrt{222}} \approx 0.685 > 0.602, \text{ positive correlation}$$

significant at the level 0.05 (but not at 0.01, because $0.685 < 0.735$).

2. $n = 13$, $\sum x = 100$, $\sum y = 51$, $(\sum x)^2 = 10{,}000$, $(\sum y)^2 = 2{,}601$, $\sum (xy) = 396$, $\sum (x^2) = 798$, $\sum (y^2) = 239$.

$$r = \frac{48}{\sqrt{374}\sqrt{506}} \approx 0.110 < 0.553; \text{ no correlation shown.}$$

3. $n=10$, $\sum x=43$, $\sum y=31$, $(\sum x)^2=1{,}849$, $(\sum y)^2=961$, $\sum (xy)=120$, $\sum (x^2)=197$, $\sum (y^2)=121$.

$$r=-\frac{133}{\sqrt{121}\sqrt{249}}\approx -0.766$$

which is farther to the left than −0.765 (barely) so that there is a negative correlation shown, at level 0.01.

4. $n=12$, $\sum x=72$, $\sum y=118$, $(\sum x)^2=5{,}184$, $(\sum y)^2=13{,}924$, $\sum (xy)=724$, $\sum (x^2)=458$, $\sum (y^2)=1{,}256$.

$$r=\frac{192}{\sqrt{312}\sqrt{1{,}148}}\approx\frac{192}{\sqrt{312}\sqrt{1{,}150}}\approx 0.321<0.576,$$

no correlation shown.

5. $n=14$, $\sum x=26$, $\sum y=32.5$, $(\sum x)^2=676$, $(\sum y)^2=1{,}056.25$, $\sum (xy)=73.25$, $\sum (x^2)=65.5$, $\sum (y^2)=88.75$.

$$r=\frac{180.5}{\sqrt{241}\sqrt{186.25}}\approx\frac{180.5}{\sqrt{241}\sqrt{186}}\approx 0.853>0.661,$$

correlation good at the level 0.01.

6. $n=15$, $\sum x=85$, $\sum y=91$, $(\sum x)^2=7{,}225$, $(\sum y)^2=8{,}281$, $\sum (xy)=500$, $\sum (x^2)=515$, $\sum (y^2)=589$.

$$r=-\frac{235}{\sqrt{500}\sqrt{554}}\approx -0.446$$

which is not between −0.514 and −1; no correlation shown.

7. $n=10$, $\sum x=45$, $\sum y=52$, $(\sum x)^2=2{,}025$, $(\sum y)^2=2{,}704$, $\sum (xy)=181$, $\sum (x^2)=285$, $\sum (y^2)=314$.

$$r=-\frac{530}{\sqrt{825}\sqrt{436}}\approx -0.884$$

which is between −0.765 and −1; a good negative correlation (!) is indicated.

8. $n=13$, $\sum x=89$, $\sum y=97$, $(\sum x)^2=7{,}921$, $(\sum y)^2=9{,}409$, $\sum (xy)=680$, $\sum (x^2)=657$, $\sum (y^2)=783$.

$$r=\frac{207}{\sqrt{620}\sqrt{770}}\approx 0.300<0.553,$$

no correlation shown.

9. $n=16$, $\sum x=530$, $\sum y=2{,}670$, $(\sum x)^2=280{,}900$, $(\sum y)^2=7{,}128{,}900$, $\sum (xy)=87{,}429$, $\sum (x^2)=20{,}298$, $\sum (y^2)=446{,}832$.

$$r=-\frac{16{,}236}{\sqrt{43{,}868}\sqrt{20{,}412}}\approx -\frac{16{,}236}{\sqrt{43{,}900}\sqrt{20{,}400}}$$

$$\approx -0.543,$$

beyond -0.497 to the left; good negative correlation at the level 0.05 (but not at the level 0.01, which requires going leftward beyond -0.623).

10. $n = 25$, $\sum x = 990$, $\sum y = 80$, $(\sum x)^2 = 980{,}100$, $(\sum y)^2 = 6{,}400$, $\sum (xy) = 3{,}110$, $\sum (x^2) = 40{,}926$, $\sum (y^2) = 348$.

$$r = -\frac{1{,}450}{\sqrt{43{,}050}\,\sqrt{2{,}300}} \approx -\frac{1{,}450}{\sqrt{43{,}000}\,\sqrt{2{,}300}} \approx -0.146,$$

not beyond -0.396 to the left; no correlation indicated.

11. $n = 15$, $\sum x = 140$, $\sum y = 120$, $(\sum x)^2 = 19{,}600$, $(\sum y)^2 = 14{,}400$, $\sum (xy) = 1{,}216$, $\sum (x^2) = 1{,}532$, $\sum (y^2) = 1{,}014$.

$$r = \frac{1{,}440}{\sqrt{3{,}380}\,\sqrt{810}} \approx 0.870 > 0.641,$$

good 0.01-level correlation.

EXERCISES 7-B

1. (a) 5.47 seconds (b) 5.95 seconds (c) 6.91 seconds (d) 7.23 seconds (e) 7.71 seconds (f) 8.03 seconds (g) 8.35 seconds (h) 8.51 seconds (i) 9.31 seconds (j) 9.47 seconds

2. $b = \frac{120}{138} \approx 0.87.$ $a = \frac{102(137) - 37(354)}{138} = \frac{876}{138} \approx 6.35.$

   Line: $y = 0.87x + 6.35$.

3. $b = \frac{180.5}{241} \approx 0.75.$ $a = \frac{32.5(65.5) - 26(73.25)}{241}$

   $= \frac{224.25}{241} \approx 0.93.$

   Line: $y = 0.75x + 0.93$.

4. (a) 1.46 millimeters (b) 1.83 millimeters (c) 2.28 millimeters (d) 2.66 millimeters (e) 2.73 millimeters (f) 2.88 millimeters (g) 3.10 millimeters (h) 3.40 millimeters (i) 3.63 millimeters (j) 3.78 millimeters

5. $b = \frac{-16{,}236}{43{,}868} \approx -0.37.$ $a = \frac{2{,}670(20{,}298) - 530(87{,}429)}{43{,}868}$

   $= \frac{54{,}195{,}660 - 46{,}337{,}370}{43{,}868}$

   $= \frac{7{,}858{,}290}{43{,}868} \approx 179.13.$

   Line: $y = -0.37x + 179.13$.

6. (a) 174 centimeters (b) 172 centimeters (c) 170 centimeters (d) 168 centimeters (e) 167 centimeters (f) 165 centimeters (g) 163 centimeters (h) 162 centimeters (i) 160 centimeters (j) 159 centimeters

7. 0.05 centimeters, approximately the height of a flea. 484 cigarettes per day would be (assuming 8 hours off to sleep) about one every 2 minutes, even during meals.

8. $b = \frac{1,440}{3,380} \approx 0.43.$ $a = \frac{120(1,532) - 140(1,216)}{3,380}$

$$= \frac{13,600}{3,380} \approx 4.02.$$

Line: $y = 0.43x + 4.02$.

9. (a) 5 (b) 7 (c) 7 (d) 9 (e) 10

10. $n = 14$, $\sum x = 185$, $\sum y = 34.0$, $(\sum x)^2 = 34,225$, $(\sum y)^2 = 1,156$, $\sum (xy) = 512.2$, $\sum (x^2) = 2,921$, $\sum (y^2) = 91.76$.

$$r = \frac{880.8}{\sqrt{6,669}\sqrt{128.64}} \approx 0.949. \qquad b = \frac{880.8}{6,669} \approx 0.13.$$

$$a = \frac{34(2,921) - 185(512.2)}{6,669} = \frac{4,557}{6,669} \approx 0.68.$$

Line: $y = 0.13x + 0.68$.

11. (a) 1.46 (b) 1.52 (c) 1.66 (d) 1.72 (e) 1.92
(f) 1.98 (g) 2.24 (h) 2.63 (i) 2.89 (j) 3.15
(k) 3.41 (l) 3.54 (m) 3.67 (n) 3.80

EXERCISES 8-A

1. $n = 18$, $\mu = \frac{n}{2} = 9$. Two-sided test. Area beyond 11.5 and 6.5 is not less than 0.05, because for $x = 11.5$, $z = 1.18 < 1.96$. No rejection of H_0.

2. $n = 18$, $\mu = 9$. One-sided test. For $x = 13.5$, $z = 2.03 > 1.64$; reject H_0.

3. $n = 17$, $\mu = 8.5$. For 12 minus signs, $x = 11.5$ gives $z = 1.46 < 1.64$. H_0 is not rejected at the level 0.05.

4. For 13 minus signs, $x = 12.5$ gives $z = 1.94 > 1.64$; reject H_0 at the level 0.05.

5. $n = 25$, $\mu = 12.5$. For 18 plus signs, the total area beyond 17.5 and 7.5 is less than 0.05, because for $x = 17.5$, $z = 2.00 > 1.96$. Reject H_0.

6. $n = 18$, $\mu = 9$. For 11 plus signs, $x = 10.5$ becomes $z = 0.71 < 1.64$; no rejection of H_0.

7. For 13 plus signs, $x = 12.5$ becomes $z = 1.65 > 1.64$, reject H_0 at level 0.05.

8. $n = 30$, $\mu = 15$. For 19 plus signs (Maze *A* minus Maze *B*), the area beyond $x = 18.5$ and $x = 11.5$ combined does not exceed 0.05, because for $x = 18.5$, $z = 1.28 < 1.96$. No rejection of H_0.

9. For 21 plus signs, $x = 20.5$ gives $z = 2.01 > 1.96$; reject H_0. It is 95% probable that the overall average maze times are different. But because $2.01 < 2.58$, the result is not good enough for a rejection at the level 0.01.

10. (a) For $x = 26.5$, $z = 0.42 < 1.96$; no rejection.
(b) For $x = 28.5$, $z = 0.99 < 1.96$; no rejection.
(c) For $x = 30.5$, $z = 1.55 < 1.96$; no rejection.
(d) For $x = 32.5$, $z = 2.12 > 1.96$; reject H_0.

EXERCISES 8-B

1. $\bar{x}_1 = 49.5$, $\bar{x}_2 = 46.7$. $R_1 = 158.5$, $R_2 = 117.5$. $\mu_U = 66$, $\sigma_U = \sqrt{264} \approx 16.25$. $U = 39.5$ or $U' = 92.5$, $z = \pm 1.63$; not beyond ± 1.96; no rejection.

2. $\bar{x}_1 = 53.7$, $\bar{x}_2 = 42.3$. $R_1 = 194.0$, $R_2 = 82.0$. $\mu_U = 66$, $\sigma_U = 16.25$ as before. $U = 4$ or $U' = 128$. $z = \pm 3.82$, beyond ± 1.96; reject H_0 at level 0.05, and even at level 0.01, because $3.82 > 2.58$.

3. $\bar{x}_A = 16.9$, $\bar{x}_B = 16.6$. $R_1 = 206.0$, $R_2 = 200.0$. $U = 95$ or $U' = 101$. $\mu_U = 98$, $\sigma_U = 21.78$. $z = \pm 0.14$, not beyond ± 1.96; no rejection.

4. No each time, because these values would still have rank 28, giving the same rank sum, U-value, and z-score. The Mann-Whitney test thus is not sensitive to (or vulnerable to) such an extreme value as $x = 300$ minutes.

5. $\bar{x}_1 = 81.0$, $\bar{x}_2 = 70.0$. $R_1 = 391.0$, $R_2 = 204.0$. $\mu_U = 144.5$, $\sigma_U = 29.03$. $U = 51$ or $U' = 238$. $z = \pm 3.22$, beyond ± 2.58. Reject H_0.

6. $\bar{x}_A = 5.0$, $\bar{x}_B = 4.0$. $R_1 = 257.0$, $R_2 = 178.0$. $\mu_U = 104$, $\sigma_U = 22.80$. $U = 42$ or $U' = 166$. $z = \pm 2.72$, beyond ± 2.58. Reject H_0.

7. $\bar{x}_1 = 400.0$, $\bar{x}_2 = 420.0$. $R_1 = 385.0$, $R_2 = 435.0$. $\mu_U = 200$, $\sigma_U = 37.01$. $U = 225$ or $U' = 175$. $z = \pm 0.68$, not beyond ± 1.96; no rejection.

8. $R_1 = 210$, $R_2 = 610$. $U = 400$, $U' = 0$. $z = \pm 5.40$, far beyond ± 1.96; reject H_0. In this extreme example, the sample standard deviations s_1 and s_2 would be zero, making the difference in sample means highly significant.

9. $\bar{x}_1 = 74.7$, $\bar{x}_2 = 77.1$. $R_1 = 241.5$, $R_2 = 319.5$. $\mu_U = 135$, $\sigma_U = 27.66$. $U = 199.5$ or $U' = 70.5$. $z = \pm 2.33$, beyond ± 1.96; reject H_0 at level 0.05 but not at level 0.01.

10. $\bar{x}_1 = 17.5$, $\bar{x}_2 = 16.5$. $R_1 = 184.0$, $R_2 = 116.0$. $\mu_U = 72$, $\sigma_U = 17.32$. $U = 38$ or $U' = 106$. $z = \pm 1.963$, beyond ± 1.96 (barely); reject H_0.

EXERCISES 8-C

1. $\bar{x}_1 = 39.7$, $\bar{x}_2 = 40.2$, $\bar{x}_3 = 39.5$. $R_1 = 58.0$, $R_2 = 60.0$, $R_3 = 53.0$. $H = 0.152 < 5.991$; no rejection of H_0.

2. $\bar{x}_1 = 35.0$, $\bar{x}_2 = 43.0$, $\bar{x}_3 = 34.0$. $R_1 = 45.0$, $R_2 = 89.0$, $R_3 = 37.0$. $H = 9.170 > 5.991$; reject H_0 at level 0.05. But because $9.170 < 9.210$, no rejection at level 0.01.

3. $\bar{x}_1 = 9.1$, $\bar{x}_2 = 8.5$, $\bar{x}_3 = 9.6$. $R_1 = 92.5$, $R_2 = 58.5$, $R_3 = 149.0$. $H = 12.554 > 9.210$; reject H_0.

4. $\bar{x}_1 = 18.0$, $\bar{x}_2 = 19.7$, $\bar{x}_3 = 21.6$, $\bar{x}_4 = 20.3$. $R_1 = 76.0$, $R_2 = 104.0$, $R_3 = 122.0$, $R_4 = 104.0$. $H = 2.286 < 7.815$; no rejection.

5. $\bar{x}_1 = 24.0$, $\bar{x}_2 = 21.0$, $\bar{x}_3 = 22.7$, $\bar{x}_4 = 24.3$. $R_1 = 96.5$, $R_2 = 58.5$, $R_3 = 96.0$, $R_4 = 100.0$. $H = 4.883 < 7.815$; no rejection.

6. $\bar{x}_A = 89.0$, $\bar{x}_B = 92.0$, $\bar{x}_C = 101.0$, $\bar{x}_D = 91.0$. $R_A = 38.0$, $R_B = 60.0$, $R_C = 127.5$, $R_D = 74.5$. $H = 14.259 > 11.345$; reject H_0.

7. $\bar{x}_1 = 71.9$, $\bar{x}_2 = 71.1$, $\bar{x}_3 = 73.0$. $R_1 = 97.0$, $R_2 = 93.0$, $R_3 = 110.0$. $H = 0.395 < 5.991$; no rejection.

8. H would be unchanged, since the two 181s would have rank 23.5 again. The Kruskal-Wallis test is thus shielded against the "distorting" effects of bizarre values at the extremes.

9. $\bar{x}_A = 17.0$, $\bar{x}_B = 17.0$, $\bar{x}_C = 20.0$. $R_A = 83.0$, $R_B = 89.0$, $R_C = 128.0$. $H = 6.636 > 5.991$; reject H_0 at level 0.05, but not at level 0.01, because $6.636 < 9.210$.

10. $\bar{x}_A = 17.0$ and $\bar{x}_B = 17.0$ as before, but now $\bar{x}_C = 19.0$ and $R_A = 87.0$, $R_B = 93.5$, $R_C = 119.5$, and $H = 4.150 < 5.991$; no rejection even at the 0.05 level.

EXERCISES 8-D

1. $R = 14$. $n_1 = 12$ *D*s, $n_2 = 11$ *U*s. $\mu_R = 12.48$, $\sigma_R = \sqrt{5.47} \approx 2.34$. $z = 0.65 < 1.96$; no rejection.

2. $R = 18$. n_1, n_2, μ_R, and σ_R as in Exercise 8-D.1. $z = 2.36 > 1.96$; reject H_0 at level 0.05; the sequence is not random. But the case is not strong enough for an 0.01-level rejection, because $2.36 < 2.58$.

3. $R = 17$. $n_1 = 11$ *M*s, $n_2 = 13$ *F*s. $\mu_R = 12.92$, $\sigma_R = \sqrt{5.66} \approx 2.38$. $z = 1.71 > 1.64$; reject H_0. There does appear to be a tendency toward alternation.

4. $R = 11$. $n_1 = 15$ *T*s, $n_2 = 15$ *L*s. $\mu_R = 16.0$, $\sigma_R = \sqrt{7.24} \approx 2.69$. $z = -1.86$, beyond -1.64 to the left; reject H_0. The suspicion of trends is probably correct.

5. $R = 21$. n_1, n_2, μ_R, and σ_R as in Exercise 8-D.4. $z = 1.86 > 1.64$; reject H_0. The suspicion of near-alternation is probably correct.

6. $R = 18$. $n_1 = 14$ *A*s, $n_2 = 13$ *B*s. $\mu_R = 14.48$, $\sigma_R = \sqrt{6.47} \approx 2.54$. $z = 1.39 < 1.96$; no rejection; no strong argument against the claim that the sequence is random.

7. $R = 9$. $n_1 = 13$ *W*s, $n_2 = 12$ *L*s. $\mu_R = 13.48$, $\sigma_R = \sqrt{5.97} \approx 2.44$. $z = -1.84$, beyond -1.64; reject H_0; there is significant indication of trends.

8. $R = 12$, all else as in Exercise 8-D.7, but $z = -0.61$, not beyond -1.64; no rejection; no significant indication of trends.

9. $R = 20$. $n_1 = 16$ *G*s, $n_2 = 19$ *R*s. $\mu_R = 18.37$, $\sigma_R = \sqrt{8.36} \approx 2.89$. $z = 0.56 < 1.96$; no rejection.

10. $R = 30$. $n_1 = 26$ *F*s, $n_2 = 24$ *M*s. $\mu_R = 25.96$, $\sigma_R = \sqrt{12.2} \approx 3.49$. $z = 1.18 < 1.96$; no rejection.

EXERCISES 8-E

1. $\sum(d^2) = 36$, $r_s = 0.836$, $z = 2.64 > 2.33$; reject H_0. The rankings are significantly correlated positively.

2. $\sum(d^2) = 176$, $r_s = 0.2$, $z = 0.63 < 2.33$; no rejection of the claim that the rankings fail to show a positive correlation.

3. $\sum(d^2) = 44$, $r_s = 0.846$, $z = 2.81 > 2.58$; reject H_0.

4. $\sum(d^2) = 604$, $r_s = -0.079$, $z = -0.030$, not beyond -1.64; no rejection of the claim that no negative correlation is indicated.

5. $\sum(d^2) = 1{,}092$, $r_s = -0.95$, $z = -3.55$, beyond -1.64 for a rejection at the 0.05 level, and even beyond -2.33 for a rejection at the 0.01 level.

6. $\sum(d^2) = 30$, $r_s = 0.918$, $z = 3.18 > 2.33$; reject H_0. There is significant indication of positive correlation.

7. $\sum(d^2) = 588$, $r_s = 0.135$, $z = 0.52 < 1.96$; do not reject H_0. No correlation is indicated at the level 0.05.

8. $\sum(d^2) = 58$, $r_s = 0.929$, $z = 3.72 > 2.33$; reject H_0.

9. $\sum(d^2) = 96$, $r_s = 0.736$, $z = 2.55 > 1.96$, reject H_0 at level 0.05. Because $2.55 < 2.58$, the result is not *quite* a rejection at the level 0.01, but very close.

10. $\sum(d^2) = 920$, $r_s = -0.127$, $z = -0.51$, not beyond -1.64; no rejection; no correlation is indicated at the level 0.05.

Index

Sample mean: $\bar{x} = \frac{\sum x}{n}$ **Population mean:** $\mu = \frac{\sum x}{N}$

Sample standard deviation: $s = \sqrt{\frac{\sum (x - \bar{x})^2}{n-1}}$

Population standard deviation: $\sigma = \sqrt{\frac{\sum (x - \mu)^2}{N}}$

Frequency distribution mean: $\bar{X} = \frac{\sum fx'}{\sum f}$

Frequency distribution median: $\tilde{X} = B + \frac{d}{f}(w)$

Frequency distribution standard deviation: $S = \sqrt{\frac{\sum (x' - \bar{X})^2 f}{(\sum f) - 1}}$

Permutations of *n* objects *r* at a time: $\frac{n!}{(n-r)!}$ $(0! = 1)$

Combinations of *n* objects *r* at a time: $\frac{n!}{r!(n-r)!}$, denoted $\binom{n}{r}$

Binomial probability formula: $P(x) = \binom{n}{x} \pi^x (1-\pi)^{n-x}$

Standard units for a normally distributed variable *x*: $z = \frac{x - \mu}{\sigma}$

Critical *z*-values:

	Two-sided	One-sided
$\alpha = 0.05$	± 1.96	1.64 or -1.64
$\alpha = 0.01$	± 2.58	2.33 or -2.33

Normal approximation to binomial: use $\mu = n\pi$, $\sigma = \sqrt{n\pi(1-\pi)}$, and continuity correction

Confidence intervals for μ:

Large sample

$$95\%: \quad \bar{x} - 1.96\frac{\sigma}{\sqrt{n}} < \mu < \bar{x} + 1.96\frac{\sigma}{\sqrt{n}}$$

$$99\%: \quad \bar{x} - 2.58\frac{\sigma}{\sqrt{n}} < \mu < \bar{x} + 2.58\frac{\sigma}{\sqrt{n}}$$

Small sample (d.f. $= n - 1$)

$$95\%: \quad \bar{x} - t_{.025}\left(\frac{s}{\sqrt{n}}\right) < \mu < \bar{x} + t_{.025}\left(\frac{s}{\sqrt{n}}\right)$$

$$99\%: \quad \bar{x} - t_{.005}\left(\frac{s}{\sqrt{n}}\right) < \mu < \bar{x} + t_{.005}\left(\frac{s}{\sqrt{n}}\right)$$

Confidence intervals for population proportion π, from sample proportion *p*:

$$95\%: \quad p - 1.96\sqrt{\frac{p(1-p)}{n}} < \pi < p + 1.96\sqrt{\frac{p(1-p)}{n}}$$

$$99\%: \quad p - 2.58\sqrt{\frac{p(1-p)}{n}} < \pi < p + 2.58\sqrt{\frac{p(1-p)}{n}}$$

Large- and small-sample hypothesis tests concerning μ:

$$z = \frac{\bar{x} - \mu}{\frac{\sigma}{\sqrt{n}}}, \quad t = \frac{\bar{x} - \mu}{\frac{s}{\sqrt{n}}} \text{ (and use d.f.} = n - 1)$$